T0309318

Linear and Nonlinear Optimization using Spreadsheets

Examples for Prescriptive, Predictive and Descriptive Analytics

Linear and Nonlinear Optimization using Spreadsheets

Examples for Prescriptive, Predictive and Descriptive Analytics

Michael J Brusco
Florida State University, USA

Stephanie Stahl

World Scientific

NEW JERSEY · LONDON · SINGAPORE · BEIJING · SHANGHAI · HONG KONG · TAIPEI · CHENNAI · TOKYO

Published by

World Scientific Publishing Co. Pte. Ltd.
5 Toh Tuck Link, Singapore 596224
USA office: 27 Warren Street, Suite 401-402, Hackensack, NJ 07601
UK office: 57 Shelton Street, Covent Garden, London WC2H 9HE

Library of Congress Control Number: 2024021625

British Library Cataloguing-in-Publication Data
A catalogue record for this book is available from the British Library.

LINEAR AND NONLINEAR OPTIMIZATION USING SPREADSHEETS
Examples for Prescriptive, Predictive and Descriptive Analytics

ISBN 978-981-12-9404-4 (hardcover)
ISBN 978-981-12-9405-1 (ebook for institutions)
ISBN 978-981-12-9406-8 (ebook for individuals)

For any available supplementary material, please visit
https://www.worldscientific.com/worldscibooks/10.1142/13863#t=suppl

Desk Editors: Nambirajan Karuppiah/Pui Yee Lum

Typeset by Stallion Press
Email: enquiries@stallionpress.com

Preface

Linear and nonlinear optimization models and methods are central to classic applications of mathematical programming in the fields of management science and operations management, as well as other business disciplines such as marketing and finance. Although it is sometimes less emphasized, they also support several multivariate statistical methods that are used in predictive and descriptive business analytics. For both mathematical programming and statistical applications, spreadsheets are a powerful tool for gaining a detailed understanding of the models and methods.

Originally, our plan was to write two separate books. The first book was to be a collection of linear, integer, and nonlinear programming examples in business. Unlike traditional management science books, the plan for this book was to focus solely on mathematical programming. Moreover, we envisioned an organizational structure based on the area of application rather than the modeling topic. In other words, instead of having marketing examples spread across chapters on linear programming, integer programming, and nonlinear programming, all of the marketing examples (regardless of the type of model) would be in one chapter. Other chapters would focus on examples dedicated to finance, sports management, and various areas of operational decision-making.

The second book was intended to be an optimization perspective of multivariate statistical methods with an emphasis on those methods that are popular in business analytics. Although there are many useful software packages for implementing these methods, the mechanics of the methods are typically opaque to business analysts.

We planned a spreadsheet-based book of examples that would enable business analysts to have a better grasp of the statistical models and methods they implement for improved prediction and description using multivariate data.

Ultimately, we opted to combine the two planned books into a single book that focuses on linear and nonlinear optimization using spreadsheets. One of the benefits of this integration is that it emphasizes the point that optimization is a unifying framework for mathematical programming and multivariate statistics, which are often presented and studied independently of one another.

After an introductory chapter that establishes the scope of the book and some fundamental principles regarding optimization and the Excel Solver, the book is organized into two parts. Part I presents a diverse array of mathematical programming examples spanning 10 different topic areas organized by chapter: (Chapter 2) production planning, (Chapter 3) workforce planning, (Chapter 4) continuous facility location, (Chapter 5) discrete facility location, (Chapter 6) routing, (Chapter 7) facility layout, (Chapter 8) project scheduling, (Chapter 9) marketing, (Chapter 10) finance, and (Chapter 11) sports. The Excel Solver program is used to obtain solutions for the examples in Part I. Although the Simplex LP engine is, by far, the most frequently used Solver engine, the GRG Nonlinear and Evolutionary engines are also used.

Part II of the book provides an optimization-centered coverage of popular multivariate statistical models and methods used in business analytics. The specific topics covered are as follows: (Chapter 12) regression, (Chapter 13) logistic regression, (Chapter 14) linear discriminant analysis, (Chapter 15) factor analysis, and (Chapter 16) cluster analysis. The GRG Nonlinear engine is used for some examples, such as those requiring maximum likelihood estimation (e.g., logistic regression or confirmatory factor analysis). However, we also demonstrate optimization tools outside of the Excel Solver: (i) eigenvalue/eigenvector estimation using the power rule (for discriminant analysis and principal component analysis), (ii) the E–M algorithm for latent class (cluster) analysis, and (iii) a VBA macro implementation of K-means clustering.

We are indebted to the authors of several management science, business analytics, and multivariate statistics textbooks, from which we have drawn both insight and ideas for some of our examples. These books include (but are not necessarily limited to) *An Introduction to Management Science* (by Anderson, Sweeney, Williams, Camm, Cochran, Fry, and Ohlmann, 2019, 15th Edition), *Spreadsheet Modeling and Decision Analysis: A Practical Introduction to Business Analytics* (by Ragsdale, 2018, 8th Edition), and *Applied Multivariate Statistical Analysis* (by Johnson and Wichern, 2007, 6th Edition). Another important source of inspiration was Baker's (2015) book titled *Optimization Modeling with Spreadsheets* (3rd Edition). We also express our sincerest gratitude to the publishing team at World Scientific for their guidance with the development of this book: David Sharp (Senior Commissioning Editor), Pui Yee Lum (In-house Desk Editor), and Sanjay Varadharajan (Book Editor).

All of the Excel workbooks described in this book, as well as a few others that are not explicitly covered, are available from a public repository. We hope that readers find the assortment of examples in this book interesting and informative.

About the Authors

Michael J. Brusco is a Haywood & Betty Taylor Eminent Scholar in Business Administration and a professor in the Department of Business Analytics, Information Systems, and Supply Chain at Florida State University's College of Business. His academic specialty is operations management, and he teaches courses in the MBA and Master of Science in Business Analytics (MS-BA) programs. His research has focused primarily on models and methods for scheduling, clustering, sequencing, and variable selection in multivariate statistics. He has published papers on these and other topics in *Annals of Operations Research, Computational Statistics and Data Analysis, Computers and Operations Research, Decision Sciences, Decision Support Systems, European Journal of Operational Research, IIE Transactions, Journal of Abnormal Psychology, Journal of Marketing, Journal of Marketing Research, Journal of Mathematical Psychology, Journal of Operations Management, Journal of Supply Chain Management, Management Science, Marketing Science, Naval Research Logistics, Operations Research, Psychological Methods, Psychometrika, Science, Statistical Analysis and Data Mining, Technometrics*, and numerous other journals. More recently, he has worked on the development of Excel spreadsheets for business analytics and operations management courses, and this work has appeared in pedagogical journals such as *Decision Sciences Journal*

of Innovative Education, INFORMS Transactions on Education, and *Spreadsheets in Education.*

Stephanie Stahl is an author and researcher with years of experience in computer programming, writing, editing, and quantitative psychology research. Her experience spans both academic and non-academic domains and her written work includes both fiction and non-fiction. Her academic research includes the 2005 Springer monograph titled *Branch-and-Bound Applications in Combinatorial Data Analysis.*

She has also published a number of academic articles on topics such as clustering, seriation and scaling, and permutation tests. Her work has been published in a variety of journals including *British Journal of Mathematical and Statistical Psychology, European Journal of Parapsychology, Journal of Classification, Journal of Marketing Research, Journal of Parapsychology,* and *Psychometrika.*

Contents

Chapter 1

Introduction

1.1 What is Optimization?

Whether we are aware of it or not, optimization is something we often seek to accomplish in our daily lives. When we finish work for the day, we might seek to *minimize* the amount of time it takes to get home or the travel time to our favorite restaurant. When we plan a vacation, we might seek to *maximize* the number of days that we can stay at our favorite resort. Alternatively, we might seek to maximize the number of historic sites that we can visit in a particular city or country or maximize the number of attractions we visit at a particular theme park.

In most (if not all) instances, our ability to optimize is *constrained* (or restricted) by one or more factors. For example, when driving home after work, our travel time is constrained by speed limits on the various roads we must traverse, by various traffic signals (stop lights, stop signs, and yield signs), as well as other aspects of road travel (roundabouts, bridges, railroad crossings, etc.). Likewise, our length of stay at our favorite resort might be constrained by a number of factors, such as the number of days we have available for vacation, our finances, flight schedules to the resort location, and availability of rooms. In general, constraints can take many forms and may involve considerations such as time, money, infrastructure, and the needs of other people.

Simply put, there are many instances in our lives where we must make *decisions* to minimize or maximize some *objective*. Most of the time, our ability to optimize the objective is constrained by one or

1

more conditions. As a general rule, we approach these optimization problems informally. That is, we do not meticulously lay out all the decisions under our control (e.g., the route to take home and how fast to drive) and all of the explicit constraints that must be considered. Instead, we adopt a more ad hoc or heuristic approach to the problem and often this works very well.

Just as there are optimization problems in our daily lives, there are also many practical problems encountered in business and industry that require *decisions* to be made so as to *minimize* or *maximize* some objective *function* subject to one or more *constraints*. Obtaining optimal (or near-optimal) solutions to these problems can have tremendous benefits for the business. These benefits are often directly financial in nature but they might also accrue from things like better customer service or greater employee satisfaction.

Our book is replete with examples of different types of optimization problems that arise in business and industry. The principal focus is establishing coherent formulations of the optimization problems and showing how Microsoft Excel (Microsoft Corporation, 2018) spreadsheets can be used to efficiently obtain solutions.

1.2   What this Book is Not

There are many different ways to cover the topic of optimization. For example, in a mathematics or engineering program of study, there might be formal coverage of the methods and algorithms used to solve optimization problems. This might include the theory behind the methods for linear and nonlinear optimization (complete with theorems and proofs), as well as the verification of the convergence of the algorithms used to solve the optimization problems. Although our book will, at times, offer presentation of some algorithms for specific examples (e.g., Newton's method for single-facility location in Chapter 4, eigenvalue/eigenvector estimation using the power method in Chapters 14 and 15, and the K-means and E–M algorithms in Chapter 16), the book is certainly not a replacement for a text on the formal theory of optimization.

Our book is also not a replacement for textbooks on the topics of management science, operations research, or business analytics. It is true that many of the examples in our book are similar to those that might be found in these textbooks and, in fact, some of our examples are based on those we have found in such textbooks (e.g., Anderson *et al.*, 2019). However, the organization of the first two-thirds of our book differs markedly from the layout of most textbooks. Specifically, most management science and operations textbooks are organized by methodological topics (e.g., linear programming, integer programming, nonlinear programming, simulation, decision analysis, and queuing). By contrast, Chapters 2–11 in our book focus only on the optimization methods and are organized based on the type of application rather than by methods. Other features that differentiate our book from traditional management science and operations research textbooks are (i) the absence of end-of-chapter problems in our book and (ii) our rigorous coverage of optimization problems in the area of multivariate statistics in Chapters 12–16. The topics covered in Chapters 12–16 are fundamental methods in the areas of predictive and descriptive analytics and might be found in business analytics textbooks. However, our treatment of these topics is more directed at demonstrating what is *under the hood* of these methods rather than their implementation as analytics tools.

Although we provide a formal treatment of several important multivariate statistical methods, our book should also not be viewed as a replacement for multivariate statistics textbooks because it lacks both the breadth of coverage and the 'statistical' emphasis of such textbooks. Our motivation for the inclusion of multivariate statistical methods is to drive home the point that they too are grounded by optimization models. More specifically, although management science and multivariate statistics are often considered to be markedly different topics, both have their foundation in optimization. During the past 10–15 years, there have been a number of efforts to bring to bear state-of-the-art optimization tools for important problems in multivariate statistics (Aloise *et al.*, 2012; Bertsimas & King, 2016, 2017; Bertsimas *et al.*, 2020a, 2020b; Bertsimas & Van Parys, 2020; du Merle *et al.*, 2000).

Finally, although there is an abundance of examples demonstrating the use of Excel to tackle practical problems, our book should not be viewed as an Excel primer. Throughout the book, we make extensive use of the Excel Solver. In the latter chapters of the book, we also incorporate a variety of Excel functions for matrix multiplication, matrix inverses, matrix determinants, and matrix transposes. We also use a small number of VBA macros. Nevertheless, a basic understanding of Excel spreadsheets and commands is assumed.

1.3   What this Book is

Succinctly, this is a book of examples. A major goal of the book is to sell readers on why it is so important to understand optimization, and the large collection of examples for a diverse range of business decision-making areas (e.g., production planning and scheduling, workforce planning and scheduling, location and supply chain distribution, location of emergency services, assembly line balancing, vehicle routing, project scheduling, revenue management, advertising, product design, payout schedules, productivity measurement, investment portfolio management, sports league scheduling, and ranking models) affords a practical mechanism for achieving that goal.

Moreover, the use of spreadsheets to obtain solutions to a diverse array of examples offers a reader-friendly way of addressing a topic (optimization) that can sometimes be viewed as intimidating. Many people are readily familiar with spreadsheets and how they work, yet are apt to be unaware of the incredible power of Excel for solving some rather complex optimization problems.

Another important contribution of the book is that it provides coverage of the mechanics of some common (yet sophisticated) statistical methods, which are often opaque to many users of such methods. Traditionally, methods such as regression, discriminant analysis, factor analysis, and cluster analysis were implemented using software packages such as SAS or SPSS. More recently, platforms such as R and Python have gained popularity for implementing the methods. However, with all of these packages and platforms, the tendency is to implement the method using simple commands and then examine the results. The details of the processes used to obtain the results typically must be garnered from other resources. By contrast, implementing the multivariate statistical methods in an Excel spreadsheet affords a powerful way of enhancing one's understanding of the methods. Specifically, the Excel spreadsheet approach provides clarity with respect to the optimization models that underlie the statistical methods, as well as the computational mechanics necessary to obtain solutions to the optimization models.

This is a book for someone who likes spreadsheets and wants to learn about a powerful way in which one can solve a diverse array of practical problems in business and statistics. However, it is also a book for someone who is very interested in optimization problems in business and statistics and desires a resource that shows how such problems can be clearly modeled and solved in a flexible and easy-to-understand manner using Excel spreadsheets. Accordingly, the book is not limited to a particular age group, course type, or arbitrary characteristic of readership.

1.4 Types of Optimization Problems

There are a variety of different ways to classify optimization problems and optimization methods. For example, a few possible classification schemes for optimization problems would be (i) continuous (smooth) or discrete, (ii) single variable or multivariable, (iii) constrained or unconstrained, and (iv) linear or nonlinear.

All of the examples that we consider in this book are multivariable in nature. A vast majority also have explicit constraints that must be satisfied, but there are a few problems (e.g., ordinary least squares regression) that are unconstrained. Many of the problems exhibit only linear functions of the variables in the objective function and constraints, but several nonlinear optimization problems are considered as well. We consider some examples that are purely smooth optimization problems, some that are purely discrete, and some that have both smooth and discrete components.

A common classification scheme (particularly in the context of discrete optimization) for optimization methods is whether they are *exact* (guaranteed to provide a *global* optimum) or *heuristic* (not guaranteed to provide a global optimum) in nature. For example, smooth optimization problems are commonly tackled using differential calculus. For some problems, taking first derivatives (or first partial derivatives) can be used to establish closed-form expressions for obtaining optimal solutions, whereas in other instances numerical estimation algorithms are required. Second derivative information can be used to support whether the solution is a *local* minimum or *local* maximum (or, possibly, a saddle point). However, a determination as to whether or not a local minimum or maximum is a global minimum or maximization requires further argument regarding the functional form of the optimization problem. Some smooth optimization problems have a host of local optima but verifying a global optimum (i.e., the best solution among the local optima) can be extremely difficult.

For discrete optimization, a local minimum or maximum is commonly framed with respect to some type of neighborhood search operation. For example, a clustering solution can be locally optimal with respect to all possible relocations of an item from its current cluster to one of the other clusters. Verifying a globally optimal solution requires explicit or implicit evaluation of all possible ways to cluster the objects into the specified number of clusters.

To put it simply, there are many examples in this book where we can be very confident that we have obtained the globally optimal solution, but there are other problems where we are unsure. To a great extent, the degree of confidence is based on whether the relationships are linear or nonlinear but other factors, such as problem size, can also play a role.

Linear programming models, which are the foundation for most of the examples in Chapters 2–11, are smooth, multivariable, constrained, and linear in nature. The goal is to find the values of the multiple variables (referred to as *decision variables*) that will minimize or maximize some linear function (referred to as the *objective function*) of those variables, subject to a set of *constraints* that are also linear functions of the decision variables. Values that are known in the linear programming model, such as coefficients for the variables in the objective function and constraints, are commonly referred to as *constants* or *parameters*. The statement of a linear programming model in terms of its decision variables, objective function, and constraints is known as a linear programming *formulation*. The solution of a linear programming formulation requires finding the values of the decision variables that minimize or maximize the objective function. This is most commonly accomplished using variants of the simplex method originally developed by Dantzig *et al.* (1954).

Integer programming can be viewed as an extension of linear programming where some (or all) of the variables are restricted to be integer values. There are many practical problems where, to be meaningful, decision variables must be integers (e.g., we cannot run a fractional number of advertisements or assign a fractional number of employees to a shift). For some problems, such as certain classes of network problems, decision variables will naturally assume integer values given integer constraint coefficients. For other problems, where decision variables take on large values, forcing integer restrictions is probably not necessary (e.g., a production result of 2173.4 units is perfectly fine, as the difference between 2173 vs. 2174 is pragmatically meaningless). There are other problems, however, where we will explicitly enforce the integer restrictions. Most notably, we have a number of applications that require *binary* decision variables (which we will specify via the definition of the variables), and the binary restrictions are always enforced on the variables. [In fact, we consider some binary integer linear programs where all of the variables are binary.] Integer linear programs are also typically solved using variants of the simplex method; however, the algorithmic process is augmented using methods such as branch and bound (Land & Doig, 1960) and cutting planes (Gomory, 1958) to ensure the integrality restrictions are preserved.

When the objective function and one more of the constraints are no longer linear functions of the decision variables, then we move into the realm of nonlinear programming, which can be much more challenging. Most smooth nonlinear programs are solved using numerical algorithms, such as Newton's method (also known as the Newton–Raphson method). Unlike the simplex method, however, global optimality can be more difficult to ensure depending on the particular functional form of the nonlinearity. Nonlinear problems with integer variables can be especially challenging and heuristic methods are common for such problems.

1.5 Optimization Using Excel

Most of the optimization problems in this book will be solved using the Excel Solver. The Solver is accessible from the 'Data' menu option. [If the Solver does not appear when clicking on the 'Data' menu, then it might be necessary to enable it by going to File -> Options -> Add-ins.] The Excel Solver uses a dialog box that allows users to build a formulation by telling the Solver where the model formulation information is located in the Excel worksheet. For example, the worksheet cell containing the objective function formula should be entered in the 'Set Objective' area of the Solver dialog box. The user will then tick the label for either 'Maximize' or 'Minimize'. The cells in the worksheet that correspond to the decision variables should be entered in the 'By Changing Cell Values' portion of the worksheet. The necessary constraints should be entered in the 'Subject to the Constraints' portion of the worksheet. When adding a constraint, functions of decision variables are entered on the left, the appropriate restriction type is entered in the middle, and constants are typically on the right side of the constraints. The middle portion of the 'Add' constraint box can also be used to place integer 'int' or binary 'bin' restrictions on the variables without any right-hand side.

There are three Solver engines available: (i) Simplex LP, (ii) GRG Nonlinear, and (iii) Evolutionary. The 'Simplex LP' engine is, by far, the one most frequently used in Chapters 2–11. It will obtain globally optimal solutions for standard linear programs. The 'Simplex LP' engine is also used for linear programs with integer or binary constraints on the decision variables. For such problems, greater care is needed when considering optimality, and it is useful to click on the 'Options' link in the Solver dialog box to check the settings for optimality tolerance, allowable time, and number of iterations. It is also important to ensure that the 'Ignore Integer Constraints' box is not checked if there are integers or binary variables in the model.

The 'GRG Nonlinear' engine is commonly used for smooth nonlinear optimization problems. We use this engine for some business applications, such as product life cycle analysis via the Bass model in Chapter 9 and the Markowitz portfolio optimization model in Chapter 10. We also use it for maximum likelihood estimation problems in Chapters 13, 15, and 16. Perhaps the most important setting for the 'GRG Nonlinear' engine is the convergence criterion, which we sometimes make smaller than the default to get greater precision.

The Evolutionary engine is useful for non-smooth problems with integer or binary variables. It is the most computationally demanding of the three search engines. By default, it requires upper and lower bounds for the decision variables to help narrow the search space. The Evolutionary engine is the least frequently used engine in our book. However, it performed very well for the multisource continuous facility location problems in Chapter 4 and is also useful for l_1-regularized logistic regression (i.e., the Lasso) in Chapter 13.

1.6   Organization of the Remainder of the Book

Part I of the book begins in Chapter 2 with coverage of production planning examples. These include basic product mix and blending problems that are commonly used to introduce linear programming in management science courses. Aggregate production planning and production scheduling examples are also included. Chapter 3 is similar in structure to Chapter 2; however, the examples in Chapter 3 correspond to aggregate workforce planning and workforce scheduling within the context of service operations.

Chapter 4 presents examples that pertain to continuous location theory and draws heavily from the work of Brusco (2022a). Chapter 5 focuses on discrete location examples, several of which draw from the work of Brusco (2022b). Some of the examples in Chapters 4 and 5 are suitable for the location of supply facilities that serve demand locations, whereas others are geared toward the location of emergency service facilities. Chapter 6 presents examples related to two important routing problems: (i) the shortest route problem, and (ii) the traveling salesperson problem.

Chapter 7 focuses on facility layout. Examples in this chapter include the use of integer linear programming to solve assembly line balancing problems for product layout, as well as the use of a pairwise interchange heuristic to obtain solutions for a process layout departmental arrangement problem. Chapter 8 examines the problem of 'crashing' the activities of a project so as to complete the project within some prespecified time but at minimum cost. This chapter draws heavily from the work of Huse and Brusco (2021).

Chapter 9 considers several examples that are at the interface of marketing and operations. These include allocation of an advertising budget, revenue management models, product life cycle analysis, and conjoint models for product design. Chapter 10 focuses on financial planning examples such as payout models, investment portfolio optimization, and productivity measurement. Chapter 11 examines sports-related examples such as league scheduling, ranking models, and mathematical elimination models.

Part II of the book begins in Chapter 12 with the coverage of ordinary least squares regression. The normal equations are derived and implemented in Excel using matrix functions. Chapter 13 describes the logistic regression model and implements the maximum likelihood estimation using the Excel Solver. Both computational and model selection issues are addressed in Chapters 12 and 13.

Chapter 14 examines Fisher's linear discriminant analysis model. The solution approach is based on a formidable generalized eigenproblem that, although challenging, can be implemented using Excel matrix capabilities. We consider an example for the simpler two-group classification problem, as well as a second example for a classic three-group problem. Chapter 15 focuses on factor analysis. The first section of Chapter 15 examines the foundational method for exploratory factor analysis, that is, principal component analysis. Like discriminant analysis, eigendecomposition is accomplished by matrix operations. The second section of Chapter 15 considers maximum likelihood-based confirmatory factor analysis, which is tackled using the GRG Nonlinear engine of the Solver. Finally, Chapter 16 looks at two fundamental problems in cluster analysis. The first is the assignment of items to clusters so as to minimize within-cluster sum-of-squared distance from items to their cluster centroids. This problem is addressed using a VBA implementation of K-means clustering. The second problem is a maximum-likelihood-based latent class mixture modeling problem that is solved using both the GRG engine and a worksheet implementation of the E–M algorithm.

Part I

Optimization Examples in Prescriptive Analytics

Chapter 2

Production Planning

There are a variety of different types of production planning applications of linear and nonlinear optimization. In this chapter, we focus on three categories of problems. In Section 2.1, we consider standard, single-period product mix and blending examples. Section 2.2 focuses on multiperiod aggregate planning examples in a production (or manufacturing) context. Here, the goal is to establish production, inventory and (possibly) backorder levels for a planning horizon of six months with the goal of minimizing costs.

In Section 2.3, we focus on shorter-term production scheduling examples where the goal is to generate a production schedule for a multiweek planning horizon. Examples in this section will capture real-world issues such as the costs of 'changing over' the production line to produce different contexts, as well as 'cutting stock' problems that arise in the paper and textile industries.

2.1 Product Mix and Blending

2.1.1 *Example* 2.1 — *Product mix*

A firm produces two types of decorative wooden boats. Type 1 boats sell for $200 and Type 2 boats sell for $210. The three primary resources for the weekly production of the boats are labor, high-quality oak, and high-quality pine. The production of each Type 1 boat requires three hours of labor, four square feet of oak, and nine square feet of pine. The production of each Type 2 boat requires three hours of labor, eight square feet of oak, and three square feet of pine.

The labor required to manufacture the boats costs $30 per hour. Oak costs $10 per square foot and pine costs $3 per square foot. Each week, there are 1260 hours of labor available to manufacture the boats. In addition, the availability of oak each week is limited to 2880 square feet and the availability of pine each week is limited to 2700 square feet.

The firm would like to determine how many Types 1 and 2 boats to manufacture each week so as to maximize weekly profit. The firm assumes that it will be able to sell all the Types 1 and 2 boats at the stated selling prices regardless of the number of boats of each type that are produced.

Linear programming can be used to solve the problem faced by the firm. The decision variables (unknowns) for the problem are production quantities for the two types of boats. Relevant constants for the model include limits on available labor, oak, and pine, as well as the quantities of labor, oak, and pine required to produce each unit of Types 1 and 2 boats.

The selling prices for the two types of boats and the costs of the resources must be translated into constants that indicate the contribution to profit for each unit of Types 1 and 2 boats:

Contribution to profit for Type 1 is
$200 - 3(30) - 4(10) - 9(3) = \$43/$unit,
Contribution to profit for Type 2 is
$210 - 3(30) - 8(10) - 3(3) = \$31/$unit.

Decision variables

$X1$ = the weekly production of Type 1 boats,
$X2$ = the weekly production of Type 2 boats.

Objective function

Maximize Profit = $Z = 43X1 + 31X2$.

Subject to constraints

$$3X1 + 3X2 <= 1260 \text{ (Labor)},$$

$$4X1 + 8X2 <= 2880 \text{ (Oak)},$$

$$9X1 + 3X2 <= 2700 \ (\text{Pine}),$$

$$X1 <= 0, \quad X2 >= 0.$$

The first step in the formulation is to obtain the contribution margins for each product by subtracting the variable costs from the selling prices. This results in objective function coefficients of 43 and 31 for Types 1 and 2 boats, respectively. These coefficients appear in cells B3:C3 of the Excel worksheet for Example 2.1, which is displayed in Figure 2.1. The constraint coefficients appear in cells B7:C9 and the resource limits (i.e., the constraint right-hand sides) appear in cells F7:F9.

The decision variables are cells B5:C5 in Figure 2.1. The objective function in cell F3 is the sumproduct of cells B5:C5 and B3:C3. The sumproduct of cells B5:C5 and the constraint coefficients produce the constraint left-hand sides in cells E7:E9. The Solver dialog box is displayed in Figure 2.2. The 'Set Objective' is cell F3, which contains the objective function. The decision variables in cells B5:C5 are the 'By Changing Variable Cells' in Figure 2.2, and the constraint left sides in cells E7:F9 are constrained to be equal to or less than the constraint rights sides in cells F7:F9.

As shown in Figure 2.1, the optimal solution is to produce X1 = 240 Type 1 boats and X2 = 180 Type 2 boats, which will yield a maximum profit of \$15,900. The labor and pine resources are used

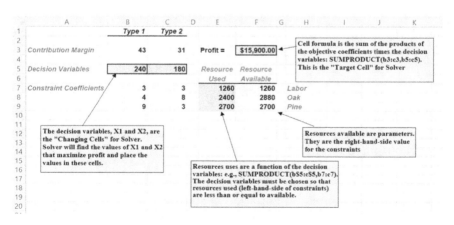

Figure 2.1.   Excel worksheet for Example 2.1.

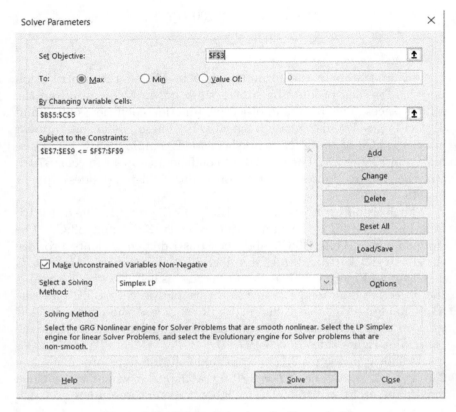

Figure 2.2.   Solver dialog box for Example 2.1.

to their maximum limits; however, 480 square feet of available oak are not used.

2.1.2   *Example 2.2 — Blending*

Bolger's House Coffee manufactures two coffees (regular and mountain) by blending three types of coffee beans. Regular coffee sells for $4.50 per pound and mountain coffee sells for $5.00 per pound. The cost per pound of beans 1, 2, and 3 are $3.25, $3.75, and $2.25, respectively. The available amount of beans 1, 2, and 3 in pounds during the coming planning period is 800, 600, and 600, respectively. There are also conditions that must be met to control the mix of beans in the coffee products. To meet flavor standards, regular coffee must consist of at least 20% bean 2 and no more than 40% bean 3. Likewise, mountain coffee must consist of at least 60% bean 2 and no more than 30% bean 3. The following is a linear programming formulation for maximum-profit blends that will provide at least 900 pounds of regular coffee, at least 600 pounds of mountain coffee, and exactly 1650 total (regular + mountain) pounds of coffee product:

Decision variables

R_j = the number of pounds of bean j in the regular coffee product; for $j = 1, 2, 3$;

M_j = the number of pounds of bean j in the mountain coffee product; for $j = 1, 2, 3$;

Objective function

Maximize: $\$1.25R_1 + \$0.75R_2 + \$2.25R_3 + \$1.75M_1 + \$1.25M_2 + \$2.75M_3$.

Subject to constraints

$$R_1 + M_1 \leq 800, \quad R_2 + M_2 \leq 600, \quad R_3 + M_3 \leq 600,$$

$$R_1 + R_2 + R_3 \geq 900, \quad M_1 + M_2 + M_3 \geq 600,$$

$$R_1 + R_2 + R_3 + M_1 + M_2 + M_3 = 1650,$$

$$R_2 \geq 0.2(R_1 + R_2 + R_3), \quad M_2 \geq 0.6(M_1 + M_2 + M_3),$$

$$R_3 \leq 0.4(R_1 + R_2 + R_3), \quad M_3 \leq 0.3(M_1 + M_2 + M_3).$$

Figure 2.3 displays the Excel worksheet for Example 2.2, and the corresponding Solver dialog box is shown in Figure 2.4. The profit coefficients for one pound of each bean in each product are contained in cells B8:G8 of Figure 2.3. For example, the contribution of one pound of bean 3 in mountain coffee is $2.75, which is equal to the selling price per pound for mountain coffee ($5.00) minus the cost per pound of bean 3 ($2.25). The constraint coefficients are in cells B12:G21 and the constraint right-hand sides are in cells J12:J21.

The decision variables are in cells B10:G10 and their sumproduct with the profit coefficients in B8:G8 is the objective function in cell I1 (the 'Set Objective' cell in the Solver dialog box in Figure 2.4). The sumproduct of B10:G10 with the constraint coefficients produces the constraint left-hand sides in cells I12:I21. These left-hand side cells

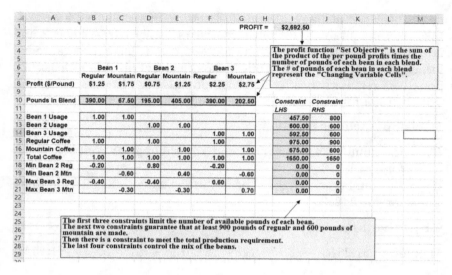

Figure 2.3.   Excel worksheet for Example 2.2.

are linked by the appropriate relational operator to the right-hand sides in cells J12:J21 in the 'Subject to the constraints' section of the Solver dialog box.

The constraints in rows 12–14 limit the availability of each of the three types of beans. For example, row 12 implements the constraint $R_1 + M_1 \leq 800$ to restrict the use of bean 1 to no more than 800 pounds. The constraints in rows 15–17 ensure that the production requirements for the coffee products are met. For example, row 15 implements the constraint $R_1 + R_2 + R_3 \geq 900$. The constraints in rows 18–21 enforce the mix restrictions. For example, row 18 implements the constraint $R_2 \geq 0.2(R_1 + R_2 + R_3)$. Note that, for implementation in Excel, all variable terms are moved to the left side of the inequality, leaving the constant zero on the right side of the inequality. Thus, the constraint in row 18 of Figure 2.3 is expressed in the form $-0.2R_1 + 0.8R_2 - 0.2R_3 \geq 0$.

The 1650 pounds of coffee produced yield a maximum profit of $2692.50. The 975 pounds of regular coffee comprised 40% bean 1 (390/975), 40% bean 3 (390/975), and 20% bean 2 (195/975). The 675 pounds of mountain coffee comprised 60% bean 2 (405/675), 30% bean 3 (202.5/675), and 10% bean 1 (67.5/675). Thus, all four mix

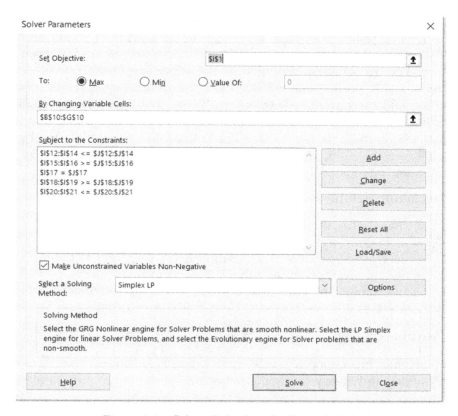

Figure 2.4. Solver dialog box for Example 2.2.

constraints were met exactly. Only bean 2 was used to full capacity but bean 3 was used to near capacity.

2.2 Aggregate Production Planning

2.2.1 *Example 2.3*

You have been asked to provide a linear programming formulation for aggregate planning over a six-month planning horizon (January to June). The goal is to determine the production, inventory, and backorder levels that will minimize the total costs associated with increasing and decreasing production from one month to the next, inventory holding costs, and backorder costs. The cost of increasing production from one month to the next is \$2.50 per unit. The cost of decreasing production from one month to the next is \$1.75 per unit. The cost to hold inventory is \$1.70 per unit based on ending inventory. The cost of backordering units is \$10 per unit. As of December, the monthly production was 3000 units, ending inventory was 1000 units, and there were no backorders. There should also be no backordered units when the planning horizon ends in June. The demand forecast (in units) is Jan = 4000, Feb = 2500, March = 2900, April = 2300, May = 3100, and June = 3800. The linear programming formulation for this problem is as follows:

Decision variables

P_j = the number of units produced in month j, for $j = 1$ to 6,
I_j = increase in production from month $j-1$ to j for $j = 1$ to 6,
D_j = decrease in production from month $j-1$ to j for $j = 1$ to 6,
E_j = ending inventory in month j for $j = 1$ to 6,
B_j = the number of units backordered at the end of month j for $j = 1$ to 6.

Parameters

$P_0 = 3000$, $E_0 = 1000$, $B_0 = 0$,
Q_j = demand in month j for $j = 1$ to 6.

Objective function

Minimize: $2.50(I_1 + I_2 + I_3 + I_4 + I_5 + I_6) + 1.75(D_1 + D_2 + D_3 + D_4 + D_5 + D_6) + 1.70(E_1 + E_2 + E_3 + E_4 + E_5 + E_6) + 10(B_1 + B_2 + B_3 + B_4 + B_5 + B_6)$.

Subject to constraints

$$E_0 + P_1 = 4000 + B_0 - B_1 + E_1 \qquad P_1 = P_0 + I_1 - D_1 \qquad B_6 = 0,$$

$$E_1 + P_2 = 2500 + B_1 - B_2 + E_2 \qquad P_2 = P_1 + I_2 - D_2,$$

$$E_2 + P_3 = 2900 + B_2 - B_3 + E_3 \qquad P_3 = P_2 + I_3 - D_3,$$

$$E_3 + P_4 = 2300 + B_3 - B_4 + E_4 \qquad P_4 = P_3 + I_4 - D_4,$$

$$E_4 + P_5 = 3100 + B_4 - B_5 + E_5 \qquad P_5 = P_4 + I_5 - D_5,$$

$$E_5 + P_6 = 3800 + B_5 - B_6 + E_6 \qquad P_6 = P_5 + I_6 - D_6.$$

	A	B	C	D	E	F	G	H
1	Production Increase Costs ($/unit)	$2.50						
2	Production Decrease Costs ($/unit)	$1.75						
3	Inventory Holding Costs ($/unit)	$1.70						
4	Backorder Costs ($/unit)	$10.00						
5								
6		December	January	February	March	April	May	June
7	Raw Demand Forecast		4000	2500	2900	2300	3100	3800
8	Backordered Units (current month)	0	0	0	0	0	0	0
9								
10	Ending Inventory	1000	0	200	0	400	550	0
11	Production Level	3000	3000	2700	2700	2700	3250	3250
12	Increase in Production		0	0	0	0	550	0
13	Decrease in Production		0	300	0	0	0	0
14								
15	Demand/Capacity Balance		0	0	0	0	0	0
16								
17								
18	Total Cost	$3,855.00						

Figure 2.5. Excel worksheet for Example 2.3.

Alternative for writing constraints

$$E_{j-1} + P_j = Q_j + B_{j-1} - B_j + E_j \quad \text{for all } j = 1, 2, 3, 4, 5, 6,$$
$$P_j = P_{j-1} + I_j - D_j \quad \text{for all } j = 1, 2, 3, 4, 5, 6, B_6 = 0.$$

The Excel worksheet for Example 2.3 is displayed in Figure 2.5, and the corresponding Solver dialog box is shown in Figure 2.6. The cost parameters are located in cells B1:B4 of Figure 2.5 and the demand parameters are in cells C7:H7.

The backorder decision variables are in cells C8:H8 and the ending inventory variables are in cells C10:H10. The production increase variables are located in cells C12:H12 and the production decrease variables are in cells C13:H13. These four sets of variables comprise the 'By Changing Variable Cells' portion of the Solver dialog box Figure 2.6.

The production variables are not explicitly included as changing cells in the Solver dialog box because they are linked directly to the increase and decrease variables by equality constraints. For example, the variable P_1 is equal to the parameter $P_0 = 3000$ plus the increase in month 1 (I_1) minus the decrease in month 1 (D_1). This relationship is captured in cell C11, which contains the formula B11 + C12–C13.

The objective function is located in cell B18 of Figure 2.5 and is computed by multiplying the cost coefficients in cells B1:B4 by their appropriate decision variable sums across the planning horizon. Cells C15:H15 contain formulas that capture the demand/capacity balance relationships. For example, the formula in cell D15 is

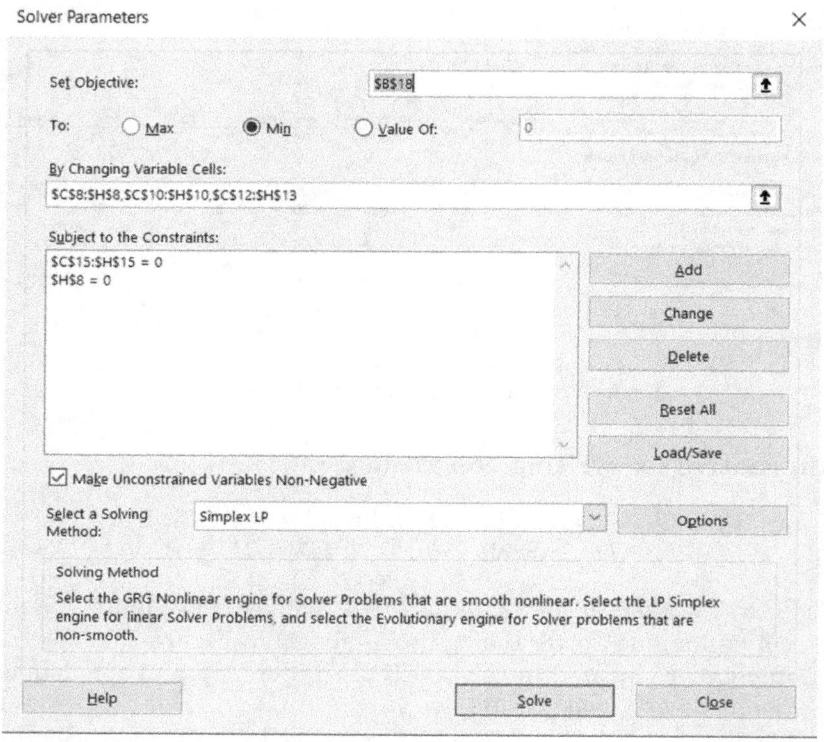

Figure 2.6. Solver dialog box for Example 2.3.

$C10 + D11 - D10 - D7 - C8 + D8$. When mapping this formula to the parameters and variables defined earlier, this translates to $E_1 + P_2 - E_2 - 2500 - B_1 + B_2$. Clearly, this is equivalent to the constraint in the second row of the constraint set. More specifically, subtracting $2500 + B_1 - B_2 + E_2$ from both sides of the constraint yields the cell formula and leaves zero on the right side. Thus, cells C15:H15 are simply constrained to equal zero to ensure the demand/capacity balance.

The optimal production plan in Figure 2.5 shows that production decreased only once (by 300 units in February) and increased only once (by 550 units in May) over the course of the six-month planning horizon. A modest amount of inventory (200 units) is held at the end of February but considerably more is held at the end of April and May (400 and 550 units, respectively) to help meet the peak requirements

	A	B	C	D	E	F	G	H
1	Production Increase Costs ($/unit)	$2.50						
2	Production Decrease Costs ($/unit)	$1.75						
3	Inventory Holding Costs ($/unit)	$1.70						
4	Backorder Costs ($/unit)	$1.00						
5								
6		December	January	February	March	April	May	June
7	Raw Demand Forecast		4000	2500	2900	2300	3100	3800
8	Backordered Units (current month)	0	250	0	150	0	0	0
9								
10	Ending Inventory	1000	0	0	0	300	500	0
11	Production Level	3000	2750	2750	2750	2750	3300	3300
12	Increase in Production		0	0	0	0	550	0
13	Decrease in Production		250	0	0	0	0	0
14								
15	Demand/Capacity Balance		0	0	0	0	0	0
16								
17								
18	Total Cost	$3,572.50						

Figure 2.7. Excel worksheet for Example 2.3 after reducing backorder costs to $1/unit.

in May and June. No backorders are held, which is not surprising given the appreciably higher cost of backordering. The cost of the plan is $3855.

If backordering costs are decreased from $10/unit to $1/unit, then the optimal plan does use backordering and costs are reduced to $3572.50 as shown in Figure 2.7.

2.2.2 Example 2.4

You have been asked to provide a linear programming formulation for aggregate planning over a six-month planning horizon (January to June). The goal is to determine the regular-time production, overtime production, inventory, and backorder levels that will minimize the total costs associated with regular-time production costs, overtime production costs, inventory holding costs, and backorder costs. The cost of regular-time production is $10 per unit. The cost of overtime production is $15 per unit. The cost to hold inventory is $2.00 per unit based on ending inventory. The cost of backordering units is $1.50 per unit. Maximum regular time production in any month is 3300 units. As of December, ending inventory was 1000 units, and there were no backorders. There should also be no backordered units when the planning horizon ends in June. The demand forecasts (in units) are Jan = 4000, Feb = 3500, March = 3900, April = 3300, May = 2900, and June = 3600. The LP formulation is as follows:

Decision variables

P_j = the number of units produced in regular time in month j, for $j = 1$ to 6,

T_j = the number of units produced in overtime in month j, j for $j = 1$ to 6,

E_j = ending inventory in month j for $j = 1$ to 6,

B_j = the number of units backordered at the end of month j for $j = 1$ to 6.

Parameters

$E_0 = 1000$, $B_0 = 0$,

Q_j = demand in month j for $j = 1$ to 6.

Objective function

Minimize: $10(P_1 + P_2 + P_3 + P_4 + P_5 + P_6) + 15(T_1 + T_2 + T_3 + T_4 + T_5 + T_6) + 2.00(E_1 + E_2 + E_3 + E_4 + E_5 + E_6) + 1.50 (B_1 + B_2 + B_3 + B_4 + B_5 + B_6)$.

Subject to constraints

$$E_0 + P_1 + T_1 = 4000 + B_0 - B_1 + E_1 \quad P_1 \le 3300 \quad B_6 = 0,$$
$$E_1 + P_2 + T_2 = 3500 + B_1 - B_2 + E_2 \quad P_2 \le 3300,$$
$$E_2 + P_3 + T_3 = 3900 + B_2 - B_3 + E_3 \quad P_3 \le 3300,$$
$$E_3 + P_4 + T_4 = 3300 + B_3 - B_4 + E_4 \quad P_4 \le 3300,$$
$$E_4 + P_5 + T_5 = 2900 + B_4 - B_5 + E_5 \quad P_5 \le 3300,$$
$$E_5 + P_6 + T_6 = 3600 + B_5 - B_6 + E_6 \quad P_6 \le 3300.$$

Alternative for writing constraints

$E_{j-1} + P_j + T_j = Q_j + B_{j-1} - B_j + E_j$ for all $j = 1, 2, 3, 4, 5, 6$,

$P_j \le 3300$ for all $j = 1, 2, 3, 4, 5, 6$,

$B_6 = 0$.

The Excel worksheet for Example 2.4 is displayed in Figure 2.8, and the corresponding Solver dialog box is shown in Figure 2.9. The cost parameters are located in cells B1:B4 of Figure 2.8, the limit on regular time production is in cell B5, and the demand parameters are in cells C8:H8.

	A	B	C	D	E	F	G	H
1	Regular Production Costs ($/unit)	$10.00						
2	OT Production Costs ($/unit)	$15.00						
3	Backordering Costs ($/unit)	$1.50						
4	Inventory Holding Costs ($/unit)	$2.00						
5	Maximum Regular Prod. Rate	3,300						
6								
7		December	January	February	March	April	May	June
8	Raw Demand Forecast		4000	3500	3900	3300	2900	3600
9	Backordered Units (current month)	0	0	0	100	100	0	0
10								
11	Ending Inventory	1000	300	100	0	0	300	0
12	Regular Production	3000	3300	3300	3300	3300	3300	3300
13	Overtime Production		0	0	400	0	0	0
14								
15	Demand/Capacity Balance		0	0	0	0	0	0
16								
17								
18	Total Cost	$205,700.00						

Figure 2.8. Excel worksheet for Example 2.4.

The backorder decision variables are in cells C9:H9 and the ending inventory variables are in cells C11:H11. The regular time production variables are located in cells C12:H12 and the overtime production variables are in cells C13:H13. These four sets of variables comprise the 'By Changing Variable Cells' portion of the Solver dialog box Figure 2.9.

The objective function is located in cell B18 of Figure 2.8 and is computed by multiplying the cost coefficients in cells B1:B4 by their appropriate decision variable sums across the planning horizon. Cells C15:H15 contain formulas that capture the demand/capacity balance relationships in a manner similar to what was done for Example 2.3 in Figure 2.5. For example, the formula in cell D15 is C11 + D12 + D13 − D11 − D8 − C9 + D9. When mapping this formula to the parameters and variables defined earlier, this translates to $E_1 + P_2 + T_2 - E_2 - 3500 - B_1 + B_2$. Clearly, this is equivalent to the constraint in the second row of the constraint set. More specifically, subtracting $3500 + B_1 - B_2 + E_2$ from both sides of the constraint yields the cell formula and leaves zero on the right side. Thus, cells C15:H15 are simply constrained to equal zero to assure the demand/capacity balance. Additional constraints require the regular time production variables in C12:H12 to be equal to or less than the maximum production rate in cell B5 and to force H9 = 0 so that backorders in June will be 0.

The optimal production plan in Figure 2.8 shows that production is kept level at the maximum rate of 3300 units for the entire six-month planning horizon. Overtime production is used only once over

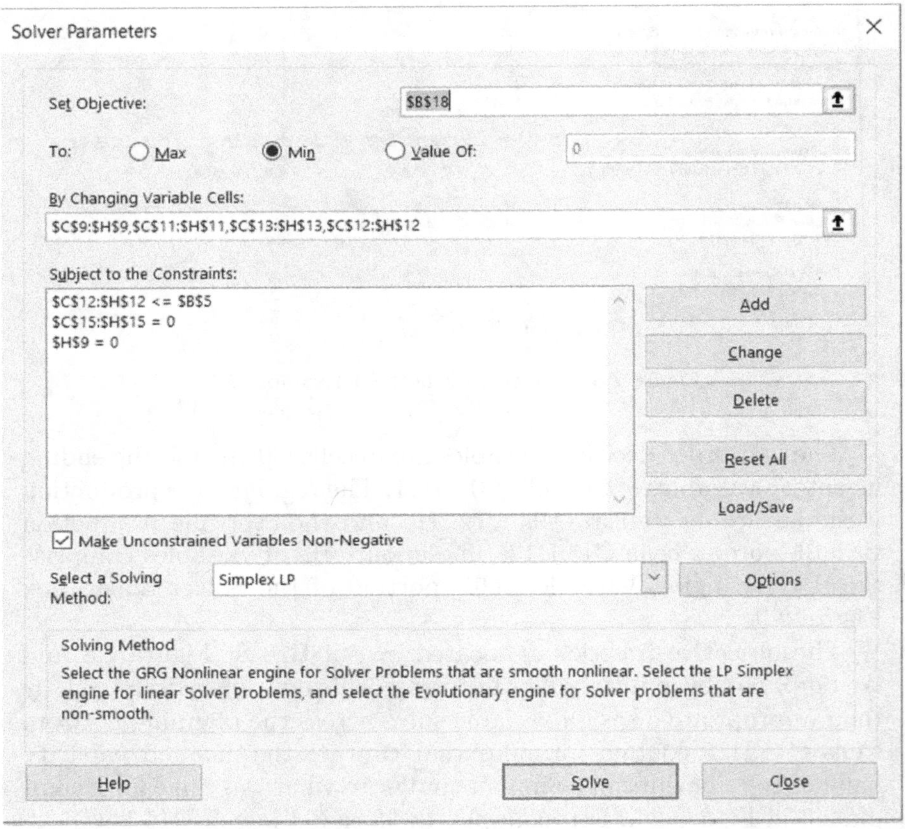

Figure 2.9.   Solver dialog box for Example 2.4.

the course of the six-month planning horizon, specifically in March at a level of 400 units. A modest amount of inventory is held at the end of January (300 units), February (100 units), and May (300 units). Modest levels of backordering are used in March and April (100 units in each of these months). The cost of the plan is $205,700.

2.3   Production Scheduling

Example 2.5 is a production scheduling example with changeover costs and is based on a case study described by Anderson *et al.* (2019, Ch. 7). A subtle difference in our example is that changeover costs

---

are not symmetric, that is, the cost of changing over the production line depends on the product last produced.

2.3.1 Example 2.5 — Line changeovers

Adroit manufacturing produces two types of transmissions for pickups: TMA and TMB. Production costs are \$625/unit for TMAs and \$740/unit for TMBs. The per unit cost of holding a TMA in inventory for one week is \$3 and the per unit cost of holding a TMB in inventory for one week is \$4. Each week, the production line can be set up to produce either TMAs or TMBs, but not both. If the line is set up to produce TMAs, then the production capacity for TMAs is 275 units. If the line is set up to produce TMBs, then the production capacity for TMBs is 180 units. The production line setup can be changed over the weekend. The cost of changing over the production line from TMAs to TMBs is \$2275 and the cost of changing over the line from TMBs to TMAs is \$1485. Currently, there are 145 TMAs in inventory and 115 TMBs in inventory. In the most recent week, the line was set up to produce TMBs. The following table provides the demand requirements for TMAs and TMBs for the next 5 weeks. At the end of the 5 weeks, there should be at least 50 TMAs and 35 TMBs in inventory. An integer linear programming model that will produce a production schedule for the next five weeks so as to meet the requirements while minimizing the sum of production costs, holding costs, and changeover costs is shown in the following.

	Week 1	Week 2	Week 3	Week 4	Week 5
TMA requirements	116	181	135	72	104
TMB requirements	64	145	86	114	47

Parameters

a_j = the number of TMAs required in week j; $j = 1$ to 5,
b_j = the number of TMBs required in week j; $j = 1$ to 5,
$p_0 = 145$, $h_0 = 115$, $s_0 = 0$.

Decision variables

x_j = the number of TMAs produced in week j; $j = 1$ to 5,
y_j = the number of TMBs produced in week j; $j = 1$ to 5,
p_j = the number of TMAs in ending inventory in week j; $j = 1$ to 5,
h_j = the number of TMBs in ending inventory in week j; $j = 1$ to 5,
$s_j = 1$ if the line is set up to produce TMAs or 0 if set up for TMBs in week j; $j = 1$ to 5,
$c_j = 1$ if the line is switched from TMAs in week $j - 1$ to TMBs in week j; $j = 1$ to 5,
$d_j = 1$ if the line is switched from TMBs in week $j - 1$ to TMAs in week j; $j = 1$ to 5.

Objective function

$$\text{Minimize}: \sum_{j=1}^{5} (625x_j + 740y_j + 3p_j + 4h_j + 2275c_j + 1485d_j).$$

$$(2.1)$$

Subject to constraints

$$p_{j-1} + x_j - a_j = p_j \text{ and } h_{j-1} + y_j - b_j = h_j \text{ for } j = 1, 2, 3, 4, 5,$$

$$(2.2)$$

$$x_j \leq 275s_j \quad \text{and} \quad y_j \leq 180(1 - s_j) \text{ for } j = 1, 2, 3, 4, 5, \quad (2.3)$$

$$d_j \geq s_j - s_{j-1} \quad \text{and} \quad c_j \geq s_{j-1} - s_j \text{ for } j = 1, 2, 3, 4, 5, \quad (2.4)$$

$$p_5 \geq 50 \quad \text{and} \quad h_5 \geq 35. \tag{2.5}$$

The Excel worksheet for Example 2.5 is displayed in Figure 2.10, and the corresponding Solver dialog box is shown in Figure 2.11.

The demand requirements in the table for TMAs and TMBs are established as parameters a_j and b_j, respectively. These a_j and b_j values are found in cells B6:F6 and B15:F15 of Figure 2.10, respectively. Sets of production variables x_j (cells B5:F5) and y_j (cells B14:F14) and ending inventory variables p_j (cells B7:F7) and h_j (B16:F16) are also established for TMAs and TMBs. The setup variables s_j (cells B22:F22) for each week are also defined. Changeover variables must be defined so as to capture the direction of the changeover.

	Week 1	Week 2	Week 3	Week 4	Week 5
Beginning Inventory	145	29	86	226	154
Production Level (TMA)	0	238	276	0	0
Demand	116	181	136	72	104
Ending Inventory	29	86	226	154	50
Capacity	275	275	276	275	275
Setup constraints	0	37	0	0	0
Beginning Inventory	115	231	86	0	0
Production Level (TMB)	180	0	0	114	82
Demand	64	145	86	114	47
Ending Inventory	231	86	0	0	35
Capacity	180	180	180	180	180
Setup constraints	0	0	0	66	98
Setup Variables	0	1	1	0	0
Changeover Variables TMA to TMB	0	0	0	1	0
Changeover Variables TMB to TMA	0	1	0	0	0
Changeover constraints TMA to TMB		1	0	0	0
Changeover constraints TMB to TMA	0	0	0	1	0

	p-heads	h-heads	Total
Production Costs	$320,625	$278,240	$598,865
Holding Costs	$1,635	$1,408	$3,043
TMA to TMB changeovers			$2,275
TMB to TMA changeovers			$1,485
Total Cost =			$605,668

Figure 2.10. Excel spreadsheet for Example 2.5.

For example, c_j (cells B23:F23) will pick up changeovers from TMAs to TMBs, whereas d_j (cells B24:F24) will pick up changeovers from TMBs to TMAs.

Equation (2.1) is the objective function, which consists of production, inventory, and changeover costs. These costs are independently collected in cells L5:L8 of the worksheet displayed in Figure 2.10 and their sum in cell L10 is the objective function.

Constraint set (2.2) includes production/demand/inventory balance constraints. They are implemented in cells B7:F7 and B16:F16 for TMAs and TMBs, respectively. For example, the B7 cell formula is B4 + B5 − B6, that is, beginning inventory plus production minus demand equal ending inventory for TMAs in week 1. Constraint set (2.3) limits maximum production based on the setup variables and this is accomplished for TMAs and TMBs in cells B11:F11 and B20:F20, respectively. Constraint set (2.4) picks up the changeovers in cells B26:F27. Row 26 picks up changeovers from TMAs to TMBs and, because we know that the line was set up to produce TMBs the week before the planning horizon began, then there is no possible switch from TMAs to TMBs at week 1 (hence the absence of a constraint for cell B26). Constraint (2.5) ensures minimum inventory levels at the end of the planning horizon which is accomplished via the restrictions on cells F7 and F16 in the worksheet.

The results in the worksheet in Figure 2.10 show that the production line was changed over twice: (i) once from TMBs to TMAs over the weekend between weeks 1 and 2, and (ii) once from TMAs to

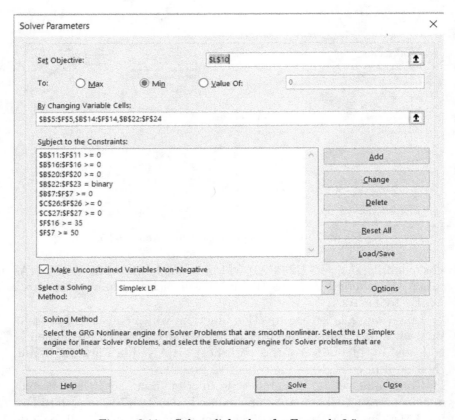

Figure 2.11.   Solver dialog box for Example 2.5.

TMBs over the weekend between weeks 3 and 4. Production of TMAs was near maximum in week 2 and at the maximum level in week 3. A substantial amount of TMA inventory was carried in weeks 2–4 to meet demand for the remainder of the week. Maximum production of TMBs was scheduled in week 1 and significant TMB inventory was carried in weeks 1 and 2 to meet requirements in weeks 2 and 3. There was no TMB ending inventory in weeks 3 and 4, and TMB production levels in weeks 4 and 5 were modest.

2.3.2   *Example 2.6 — Cutting stock*

Lindy lumber company produces one-foot *rulers* and three-foot *yardsticks*. They are produced by cutting 20-foot sticks of wood. Each

20-foot stick costs Lindy $1.00 in week 1, but the price is expected to increase by 20% in week 2, and again by 20% in week 3, and again by 20% in week 4. Lindy must cut the 20-foot sticks into widths of one foot (rulers) or three feet (yardsticks). Currently, in inventory, there are seven hundred rulers and no yardsticks. Currently, no rulers are backordered but three hundred yardsticks are backordered. The following table shows the number of rulers and yardsticks that are required for the next four weeks. The weekly cost of holding a ruler in inventory is $0.03 and the corresponding cost for a yardstick is $0.035. The weekly cost of backordering a ruler is $0.01 and the corresponding cost for a yardstick is $0.015. Each week, there are three possible cutting alternatives that produce different numbers of rulers and yardsticks. This information is also shown in the table. Lindy would like for you to develop an LP formulation for a cutting stock plan for the next four weeks that will meet the requirements for rulers and yardsticks. No rulers or yardsticks should be backordered at the end of the planning horizon in week 4. The goal is to minimize the total cost (i.e., the sum of the cost of the 20-foot sticks needed to meet the requirements, the inventory holding costs, and the backorder costs).

	Units per 20′ stick cut			Weekly requirements			
	Alt #1	Alt #2	Alt #3	Week 1	Week 2	Week 3	Week 4
Rulers	11	8	2	12,300	14,700	17,600	17,100
Yardsticks	3	4	6	9,100	2,300	9,500	1,500

Indices

i = the index for cutting alternatives ($i = 1, 2, 3$ for alternative #1, #2, and #3, respectively),
j = the index for products ($j = 1$ for rulers and $j = 2$ for yardsticks),
k = the index for weeks ($k = 1, 2, 3, 4$).

Parameters

R_{jk} = requirement for product type j in week k (for $j = 1, 2$ and $k = 1, 2, 3, 4$).

T_{ij} = units of product j from cutting a $20'$ stick using alternative i
(for $i = 1, 2, 3$ and $j = 1, 2, 3$).
$E_{10} = 700$, $E_{20} = 0$, $B_{10} = 0$, $B_{20} = 300$.

Decision variables

X_{ik} = the number of $20'$ sticks cut using alternative i in week k
(for $i = 1, 2, 3$ and $k = 1, 2, 3, 4$).
E_{jk} = the number of products of type j in inventory in week k
(for $j = 1, 2$ and $k = 1, 2, 3, 4$).
B_{jk} = the number of products of type j backordered in week k
(for $j = 1, 2$ and $k = 1, 2, 3, 4$).

Objective function

$$\text{Minimize: } \sum_{k=1}^{4} 1(1.2)^{(k-1)} \sum_{i=1}^{3} X_{ik} + \sum_{k=1}^{4}$$
$$(0.03E_{1k} + 0.035E_{2k} + 0.01B_{1k} + 0.015B_{2k}). \tag{2.6}$$

Subject to constraints

$$E_{j(k-1)} + \left(\sum_{i=1}^{3} T_{ij} X_{ik} \right) - R_{jk} - B_{j(k-1)}$$
$$= E_{jk} - B_{jk} \quad \text{for } j = 1, 2 \text{ and } k = 1, 2, 3, 4, \tag{2.7}$$
$$B_{14} = 0 \quad \text{and} \quad B_{24} = 0. \tag{2.8}$$

The Excel worksheet for Example 2.6 is displayed in Figure 2.12, and
the corresponding Solver dialog box is shown in Figure 2.13.

The demand requirements in the table for rulers ($j = 1$) and
yardsticks ($j = 2$) in week k are established as the R_{jk} parameters.
These parameter values are located in cells C10:F10 and C16:F16 of
Figure 2.12 for rulers and yardsticks, respectively. Sets of production variables X_{ik} indicating the number of $20'$ sticks cut using each
alternative each week are found in cells C3:F5. Ending inventory variables E_{jk} for rulers and yardsticks are located in cells C11:F11 and
C17:F17, respectively. Likewise, backorder variables B_{jk} for rulers
and yardsticks are located in cells C8:F8 and C14:F14, respectively.

Equation (2.6) is the objective function, which consists of production, inventory, and backorder costs. These costs are independently
collected in cells I1:I5 of the worksheet displayed in Figure 2.12, and
their sum in cell I7 is the objective function.

	A	B	C	D	E	F	G	H	I
1		Week 0	Week 1	Week 2	Week 3	Week 4		Production Costs	$8,254.27
2								Inv. Costs - Rulers	$348.00
3	Sticks Cut Using Alt #1		852.42	1336.36	2654.55	500.00		Inv. Costs - Yardsticks	$53.77
4	Sticks Cut Using Alt #2		0.00	0.00	0.00	0.00		Backorder Costs - Rulers	$0.00
5	Sticks Cut Using Alt #3		1111.67	0.00	0.00	0.00		Backorder Costs - Yardsticks	$2.59
6									
7								Total Cost	$8,658.64
8	Rulers Backordered	0	0.00	0.00	0.00	0.00			
9	Rulers Produced		11600.00	14700.00	29200.00	5500.00			
10	Demand Rulers		12300.00	14700.00	17600.00	17100.00			
11	Ending Inv. Rulers	700	0.00	0.00	11600.00	0.00			
12	Rulers - Balance		0.00	0.00	0.00	0.00			
13									
14	Yardsticks Backordered	300	172.73	0.00	0.00	0.00			
15	Yardsticks Produced		9227.27	4009.09	7963.64	1500.00			
16	Demand Yardsticks		9100.00	2300.00	9500.00	1500.00			
17	Ending Inv. Yardsticks	0	0.00	1536.36	0.00	0.00			
18	Yardsticks - Balance		0.00	0.00	0.00	0.00			

Figure 2.12.   Excel spreadsheet for Example 2.6.

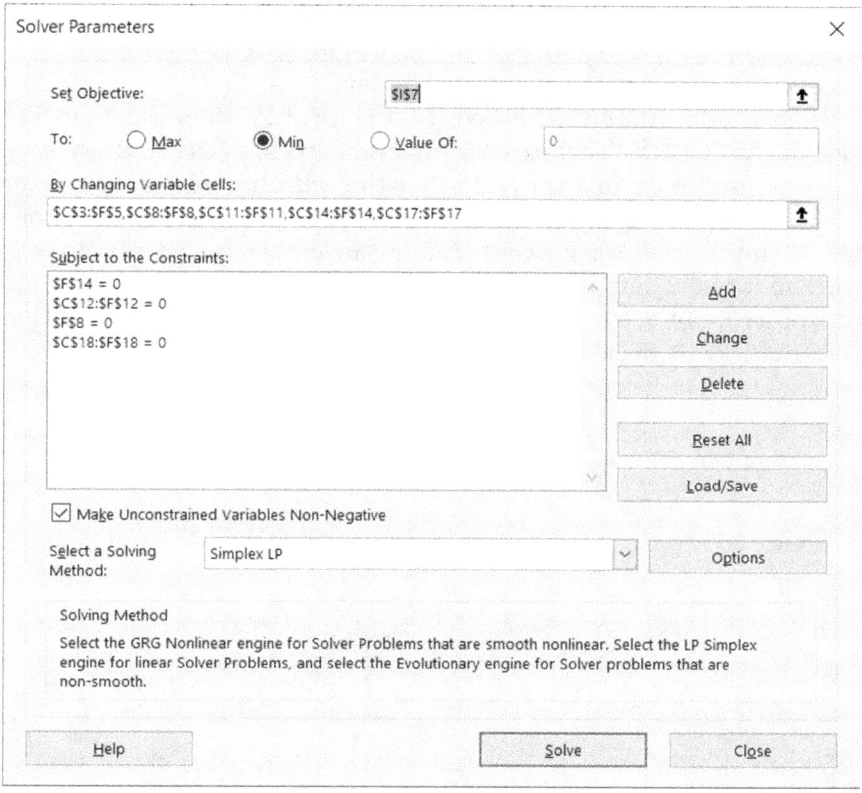

Figure 2.13.   Solver dialog box for Example 2.6.

Constraint set (2.7) includes production/demand/inventory/ backorder balance constraints. They are implemented in cells C12:F12 and C18:F18 for rulers and yardsticks, respectively. The production cells in rows 9 and 15 are linked to the cutting alternative decisions in rows 3–5. For example, cell C9 contains the formula 11*C3 + 8*C4 + 2*C5 to reflect the total number of rulers produced by the three cutting alternatives. Cell C12 is the balance constraint cell for rulers in week 1 and contains the formula B11 + C9 – C10 – B8 + C8 – C11. This represents beginning inventory in week 1 (i.e., ending inventory from week 0), plus production in week 1, minus demand in week 1, minus backorders from previous week (week 0), plus backorders for week 1, minus ending inventory for week 1.

Cells C12:F12 and C18:F18 are simply constrained to equal zero to ensure the production/demand/inventory/backorders balance each week for rulers and yardsticks, respectively. Additional constraints force F8 = 0 and F14 = 0 to ensure that there are no backorders at the end of the planning horizon.

The results in the worksheet in Figure 2.12 show that cutting alternative #1 was used extensively every week. Cutting alternative #3 was used only in week 1 and cutting alternative #2 was never used. Ending inventory for rulers was used extensively in week 3 but not in any of the other weeks. Ending inventory for yardsticks was used in week 2 but not in any of the other weeks. There was no backordering of rulers in the schedule and only modest backordering of yardsticks in week 1.

Chapter 3

Workforce Planning

Workforce planning applications of linear and nonlinear optimization problems can take many different forms. In this chapter, we focus on examples that can broadly be defined by two categories: (i) aggregate planning and (ii) scheduling. In Section 3.1, we consider aggregate planning examples, whereby the goal is to plan workforce staffing levels for a planning horizon of six months. These examples are based on nurse staff planning problems that have been studied for several decades (Kao & Queyranne, 1985; Brusco & Showalter, 1993). The goal is to establish regular-time and overtime nurse staffing hours for multiple nursing units over the planning horizon, so as to satisfy requirements at minimum cost. Supplemental sources of staffing, such as flex pools and float are also considered in the examples.

In Section 3.2, we examine shorter-term scheduling problems where the goal is to assign workers to schedules. Three types of workforce scheduling problems are considered based on Baker's (1976) classification scheme: (i) shift, (ii) days-off, and (iii) tour. Shift scheduling problems have a planning horizon of one day, and it is necessary to assign workers to shifts defined by a starting time, finishing time, and (possibly) a meal break. Days-off scheduling problems require the assignment of workers to work days and non-work days over a planning horizon that is commonly one week. Tour scheduling problems combine the shift and days-off problems, whereby workers are assigned to both work days and shifts for each work day for a weekly planning horizon.

3.1 Nurse Staff Planning

3.1.1 *Example 3.1 — With flex pool*

A hospital division consists of three large nursing units, A, B, and C. The registered nurses (RNs) in units A, B, and C earn $35, $40, and $38.75 per hour, respectively. Any overtime is paid at time and a half. In addition to nurses assigned to each unit, a flex pool of nurses is used by the division to meet peak requirements in the nursing units. These flex pool nurses are paid a premium of $47.50 per hour since they are not permanently assigned to any unit. The division wants to develop a staffing plan for January–June of 2025. The labor hours required in unit A are Jan = 2700, Feb = 2300, Mar = 2100, Apr = 2600, May = 3100, and June = 3300. The labor hours required in unit B are Jan = 2400, Feb = 2800, Mar = 2900, Apr = 2600, May = 2400, and June = 2300. The labor hours required in unit C are Jan = 3000, Feb = 2600, Mar = 2200, Apr = 2000, May = 2400, and June = 2800. The goal is to determine the number of regular-time, overtime, and flex nursing hours used in each nursing unit each month with the objective of minimizing total staffing costs for the division during the six-month period. The number of regular-time hours for any individual unit should be the same in each of the 6 months (i.e., regular-time hours in unit A might be more than unit B; however, the number of regular-time hours in A in January must be the same number in A for Feb, March, etc.). Overtime can be used as needed in any unit but cannot exceed 20% of regular-time hours in any month. Flex staff can also be used as needed, but the number of flex hours used across all three units must be the same for each month (i.e., if 1000 flex hours are used in Jan, then 1000 must be used in February and so on; however, the distribution of those hours across the units can vary.). The linear programming formulation for this problem is as follows:

Parameters and Indices

The index for the units is j: $j = $ A, B, C;
The index for the months is k: $k = 1 = $ Jan, $2 = $ Feb, $3 = $ Mar, $4 = $ Apr, $5 = $ May, $6 = $ Jun;
$D_{jk} = $ the number of hours required in unit j in month k ($j = $ A, B, C and $k = 1, 2, 3, 4, 5, 6$).

Decision variables

R_j = the number of regular-time hours used in unit j each month; $j =$ A, B, and C;

T_{jk} = the number of overtime hours used in unit j in month k; $j =$ A, B, C and $k = 1, \ldots, 6$;

F_{jk} = the number of flex pool hours used in unit j in month k; $j =$ A, B, C and $k = 1, \ldots, 6$;

TF = the total number of flex pool hours used each month.

Objective function

$$\text{Minimize: } 210R_A + 240R_B + 232.50R_C + 285TF$$

$$+ \sum_{k=1}^{6} [52.50T_{Ak} + 60T_{Bk} + 58.125T_{Ck}]. \qquad (3.1)$$

Subject to constraints

$$R_j + T_{jk} + F_{jk} \geq D_{jk} \quad \text{for} \quad j = \text{A, B, C} \quad \text{and} \quad k = 1, \ldots, 6, \qquad (3.2)$$

$$T_{jk} \leq .2R_j \quad \text{for} \quad j = \text{A, B, C} \quad \text{and} \quad k = 1, \ldots, 6, \qquad (3.3)$$

$$F_{Ak} + F_{Bk} + F_{Ck} = TF \quad \text{for} \quad k = 1, \ldots, 6. \qquad (3.4)$$

The index j in the formulation denotes the three units (A, B, and C) and the index k denotes the months of the planning horizon where $k = 1 =$ January, $2 =$ February, $3 =$ March, $4 =$ April, $5 =$ May, and $6 =$ June. The regular-time hour variables (R_j) do not require a k subscript because the number of regular-time hours in a given unit is the same each month. However, overtime and flex pool hours are differentiated by both unit and month.

The objective function in Equation (3.1) is to minimize the total staffing costs for the six-month planning horizon. The objective function coefficients for the regular-time hours for each unit, as well as the total flex hours (TF), are six times the hourly wage rates. For example, if $R_A = 2000$ hours, then each hour costs \$35 for each of the six months, so the objective coefficient is \$35(6) = \$210. The coefficients for the overtime variables are 1.5 times the hourly rate, so the cost of an hour of overtime in unit A is 1.5(\$35) = \$52.50.

Constraint set (3.2) ensures that, for each unit each month, the number of hours provided from regular time, overtime, and flex pool equals or exceeds the labor hours required. Constraint set (3.3) guarantees that, for each unit each month, the number of overtime hours does not exceed 20% of the number of regular-time hours. Constraint set (3.4) requires that the total number of flex pool hours assigned is the same each month.

Two Excel spreadsheets were prepared to solve this formulation. The first one uses a standard design where the columns correspond to the decision variables, there is a row containing the objective function coefficients for each decision variable, and the remaining rows pertain to the constraints. A partial screenshot of this worksheet is displayed in Figure 3.1, and a screenshot of the Solver dialog box is shown in Figure 3.2. The decision variables are cells B3:AO3 and the objective coefficients are in cells B5:AO5. The constraint coefficients span cells B7:AO48. The constraint right-hand side parameters are in cells AR7:AR48. The constraint left-hand sides, which are the sumproducts of the constraint coefficients and the decision variables, are in cells AQ7:AQ48. Cell AS1 contains the maximum overtime limit (currently 0.2). This cell can be changed to evaluate other overtime limits. The objective function is in cell AS2.

	A	B	C	D	E	F	G	H	I	J	K	L	M	N	O	P	Q	R
1	Decision Variable Labels	Ra	Rb	Rc	Ta1	Ta2	Ta3	Ta4	Ta5	Ta6	Tb1	Tb2	Tb3	Tb4	Tb5	Tb6	Tc1	Tc2
2																		
3	Decision Variable Values	2500.00	2500.00	2283.33	400.00	0.00	0.00	0.00	416.67	460.00	0.00	0.00	0.00	0.00	0.00	0.00	216.67	216.67
4																		
5	Objective Coefficients	210	240	232.5	52.5	52.5	52.5	52.5	52.5	52.5	60	60	60	60	60	60	58.125	58.125
6																		
7	Demand Unit A, Month 1	1			1													
8	Demand Unit A, Month 2	1				1												
9	Demand Unit A, Month 3	1					1											
10	Demand Unit A, Month 4	1						1										
11	Demand Unit A, Month 5	1							1									
12	Demand Unit A, Month 6	1								1								
13	Demand Unit B, Month 1		1								1							
14	Demand Unit B, Month 2		1									1						
15	Demand Unit B, Month 3		1										1					
16	Demand Unit B, Month 4		1											1				
17	Demand Unit B, Month 5		1												1			
18	Demand Unit B, Month 6		1													1		
19	Demand Unit C, Month 1			1													1	
20	Demand Unit C, Month 2			1														1
21	Demand Unit C, Month 3			1														
22	Demand Unit C, Month 4			1														
23	Demand Unit C, Month 5			1														
24	Demand Unit C, Month 6			1														
25	OT, Unit A, Month 1	-0.2			1													
26	OT, Unit A, Month 2	-0.2				1												
27	OT, Unit A, Month 3	-0.2					1											
28	OT, Unit A, Month 4	-0.2						1										
29	OT, Unit A, Month 5	-0.2							1									
30	OT, Unit A, Month 6	-0.2								1								
31	OT Unit B, Month 1		-0.2								1							
32	OT Unit B, Month 2		-0.2									1						
33	OT Unit B, Month 3		-0.2										1					
34	OT Unit B, Month 4		-0.2											1				
35	OT Unit B, Month 5		-0.2															

Figure 3.1.   Partial worksheet for Example 3.1 — standard layout.

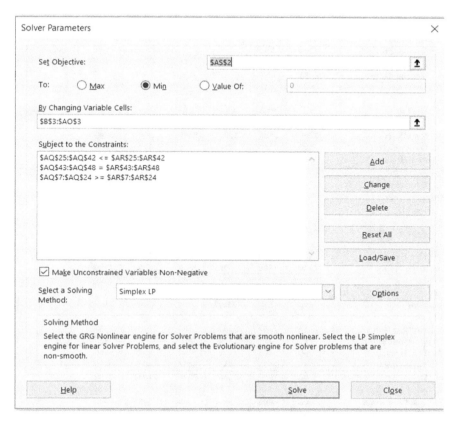

Figure 3.2.   Solver dialog box for Example 3.1 — standard layout.

The second Excel spreadsheet uses a more visually appealing design. A partial screenshot of the worksheet is provided in Figure 3.3, and a screenshot of the Solver dialog box is in Figure 3.4. The overtime limit parameter is cell B1. The R_A, R_B, R_C, and TF decision variables are in cells B2, B3, B4, and B5, respectively. Separate tables are used to illustrate coverage of requirements in the three nursing units. For example, row 12 collects the regular-time (row 9), overtime (row 10), and flex (row 11) hours in unit A and this total must equal or exceed the requirements for unit A in row 13. Row 31 is used to ensure that the total flex hours used each month are

	A	B	C	D	E	F	G	H
1	OT LIMIT	0.2						
2	Regular Hours in A	2300				Total Staffing Cost		$1,855,631.25
3	Regular Hours in B	2300						
4	Regular Hours in C	2283.33						
5	Total flex hours / mo.	600						
6								
7	Dept A	Jan	Feb	Mar	Apr	May	Jun	
8								
9	Regular Hours	2300	2300	2300	2300	2300	2300	
10	Overtime Hours	400	0	0	0	417	460	
11	Flex Hours	0	0	0	300	383	540	
12	Total Hours	2700	2300	2300	2600	3100	3300	
13	Demand	2700	2300	2100	2600	3100	3300	
14								
15	Dept B	Jan	Feb	Mar	Apr	May	Jun	
16								
17	Regular Hours	2300	2300	2300	2300	2300	2300	
18	Overtime Hours	0	0	0	0	0	0	
19	Flex Hours	100	500	600	300	100	0	
20	Total Hours	2400	2800	2900	2600	2400	2300	
21	Demand	2400	2800	2900	2600	2400	2300	
22								
23	Dept C	Jan	Feb	Mar	Apr	May	Jun	
24								
25	Regular Hours	2283	2283	2283	2283	2283	2283	
26	Overtime Hours	217	217	0	0	0	457	
27	Flex Hours	500	100	0	0	117	60	
28	Total Hours	3000	2600	2283	2283	2400	2800	
29	Demand	3000	2600	2200	2000	2400	2800	
30								
31	Total flex usage	600	600	600	600	600	600	
32	OT Constraints - A	60	460	460	460	43	0	
33	OT Constraints - B	460	460	460	460	460	460	
34	OT Constraints - C	240	240	457	457	457	0	

Figure 3.3.   Worksheet for Example 3.1 — efficient layout.

equal to the value of TF in cell B5. Rows 32, 33, and 34 handle the overtime conditions for units A, B, and C, respectively.

The solutions obtained by the two worksheets are the same. Units A, B, and C require 2300, 2300, and 2283.3 regular-time hours, respectively. The flex pool consists of 600 hours each month. Overtime is not used at all in unit B, which has the highest hourly wage rate of the three units. Units A and C each use overtime in three of the six months. Moreover, units A and C both use the maximum level of overtime in the month of June. The distribution of the 600 flex hours varies widely across the planning horizon. In January,

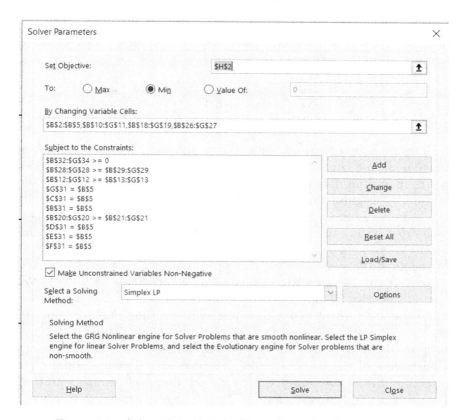

Figure 3.4.   Solver dialog box for Example 3.1 — efficient layout.

500 hours are used in unit C and the remaining 100 used in unit B. In March, all 600 hours are used in unit B. In June, 540 hours are used in unit A and the remaining 60 used in unit C.

3.1.2   *Example 3.2 — With float*

Example 3.2 is similar to Example 3.1; however, instead of a flex pool, there is the potential for floating nursing hours from unit B to unit A.

A hospital division consists of two large nursing units, A and B. The registered nurses (RNs) in units A and B earn $30 and $36 per hour, respectively. Overtime is paid at time and a half. There is also the possibility of assigning some of the hours from nurses assigned to unit B to meet requirements in unit A. The division wants to develop a six-month staffing plan. The minimum number of labor hours required in each unit each month is as follows:

	Jan	Feb	Mar	Apr	May	June
Unit A	2700	2300	2100	2600	3100	3300
Unit B	2400	2800	2900	2600	2400	2300

The number of regular-time hours allocated for any individual unit should be the same in each of the 6 months (i.e., regular-time hours in unit A might be more or less than unit B; however, the number of regular-time hours assigned for A in January must be the same number in A for Feb, March, etc.). Overtime can be used as needed in any unit but cannot exceed 15% of regular-time hours in any month. In any given month, up to 18% of the total number of regular-time plus overtime hours available from nurses assigned to unit B can be used to meet nursing requirements in unit A. We want to develop a linear programming formulation that will deter-mine the assignment of regular-time and overtime nursing hours to each nursing unit each month. The objective of the model is to mini-mize total staffing costs for the division during the six-month period. Constraints of the model should guarantee that nursing hours meet or exceed the labor hours required in each unit each month, and that the overtime and substitution (i.e., using hours assigned to B to meet requirements in A) conditions are satisfied.

Parameters and indices

The index for the units is $j : j =$ A, B;
The index for the months is $k : k = 1 =$ Jan, $2 =$ Feb, $3 =$ Mar, $4 =$ Apr, $5 =$ May, $6 =$ Jun;
$D_{jk} =$ the number of hours required in unit j in month k ($j =$ A, B and $k = 1, 2, 3, 4, 5, 6$).

Decision variables

$R_j =$ the number of regular-time nursing hours in unit j ($j =$ A, B);
$T_{jk} =$ the number of overtime nursing hours in unit j in month k ($j =$ A, B and $k = 1, 2, 3, 4, 5, 6$);
$S_k =$ the number of nursing hours from unit B that are allocated to unit A in month k ($k = 1, 2, 3, 4, 5, 6$).

Objective function

Minimize: $180R_\text{A} + 216R_\text{B} + 45(T_\text{A1} + T_\text{A2} + T_\text{A3} + T_\text{A4} + T_\text{A5}$

$$+ T_\text{A6}) + 54(T_\text{B1} + T_\text{B2} + T_\text{B3} + T_\text{B4} + T_\text{B5} + T_\text{B6}). \quad (3.5)$$

Subject to constraints

$$R_\text{A} + T_{\text{A}k} + S_k \geq D_{\text{A}k} \quad \text{for } k = 1, 2, 3, 4, 5, 6, \quad (3.6)$$

$$R_\text{B} + T_{\text{B}k} - S_k \geq D_{\text{B}k} \quad \text{for } k = 1, 2, 3, 4, 5, 6, \quad (3.7)$$

$$S_k \leq .18(R_\text{B} + T_{\text{B}k}) \quad \text{for } k = 1, 2, 3, 4, 5, 6, \quad (3.8)$$

$$T_{jk} \leq .15R_j \quad \text{for } j = \text{A, B} \quad \text{and} \quad k = 1, 2, 3, 4, 5, 6. \quad (3.9)$$

The parameters and decision variables for this problem are similar to those in Example 3.1. However, the flex pool variables are not relevant for this problem but the S_k variables are needed to indicate the number of hours from unit B assigned to unit A each month. The objective function in Equation (3.5) consists of regular-time and over-time costs for the two units across the planning horizon. Constraint set (3.6) ensures that the labor requirements are satisfied for unit A each month, whereas constraint set (3.7) accomplishes a similar result for unit B. Note that S_k is added to the left side of constraint set (3.6) but subtracted from the left side of constraint set (3.7) because the hours are moved from unit B to unit A. Constraint set (3.8) guarantees that, each month, no more than 18% of the total labor hours in unit B are allocated to unit A. Constraint set (3.9) limits the use of overtime in each unit each month to no more than 15% of the regular-time hours.

Once again, two Excel spreadsheets were prepared to solve this formulation. The first one uses a standard design where the columns correspond to the decision variables, there is a row containing the objective function coefficients for each decision variable, and the remaining rows pertain to the constraints. The second one is a more concise, reader-friendly worksheet. Although both worksheets are available with the online materials, we focus only on the second worksheet here. A screenshot of this worksheet is displayed in Figure 3.5. and a screenshot of the Solver dialog box is shown in Figure 3.6.

Cells B1 and B2 in Figure 3.5 contain the parameters that control the allowable proportions for overtime and allocation from B

	A	B	C	D	E	F	G	H
1	OT LIMIT	0.15						
2	B to A Allocation Limit	0.18						
3	Regular Hours in A	2430.54				Total Staffing Cost		$1,076,217.71
4	Regular Hours in B	2769.46						
5								
6	Dept A	Jan	Feb	Mar	Apr	May	Jun	
7								
8	Regular Hours	2431	2431	2431	2431	2431	2431	
9	Overtime Hours	0	0	0	0	300	365	
10	Hours from B	369	0	0	169	369	505	
11	Total Hours	2800	2431	2431	2600	3100	3300	
12	Demand	2700	2300	2100	2600	3100	3300	
13								
14	Dept B	Jan	Feb	Mar	Apr	May	Jun	
15								
16	Regular Hours	2769	2769	2769	2769	2769	2769	
17	Overtime Hours	0	31	131	0	0	35	
18	Hours to A	-369	0	0	-169	-369	-505	
19	Total Hours	2400	2800	2900	2600	2400	2300	
20	Demand	2400	2800	2900	2600	2400	2300	
21								
22								
23	OT Constraints - A	365	365	365	365	65	0	
24	OT Constraints - B	415	385	285	415	415	380	
25	B to A Constraint	129	504	522	329	129	0	
26								

Figure 3.5. Worksheet for Example 3.2.

to A, respectively. Cells B3 and B4 are the decision variables corresponding to regular-time hours for units A and B, respectively, and these cell values are copied to rows 8 and 16, respectively. Rows 9 and 17 contain the overtime decision variables for units A and B, respectively. Row 10 contains the decision variables for allocation of hours from unit B to unit A. Row 18 contains the negative of row 10 because the reallocated hours are subtracted from unit B. The total hours of coverage for units A and B are contained in rows 11 and 19, respectively, and these rows are constrained to equal or exceed the labor requirements (demand) in rows 12 and 20, respectively. Rows 23 and 24 handle the overtime constraints and row 25 accommodates the restriction on using hours from unit B to meet requirements in unit A.

The optimal number of regular-time hours for units A and B is approximately 2431 and 2769, respectively. Overtime is used sparingly in unit B, where it is more costly. Overtime is used extensively in unit A in May and used to its maximum level in unit A in June.

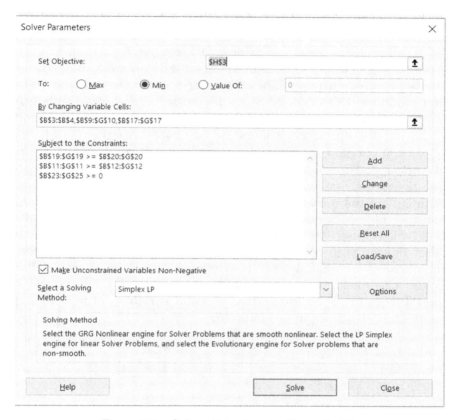

Figure 3.6.   Solver dialog box for Example 3.2.

Hours are allocated from B to A in four of the six months, with maximum allocation occurring in the month of June.

3.2   Workforce Scheduling

3.2.1   *Example 3.3 — Shift scheduling (minimizing labor hours)*

The first workforce scheduling example is a shift scheduling problem where the goal is to find an 'ideal' schedule that minimizes the total labor hours necessary to cover requirements in each hour of the day. Accordingly, the number of full-time and part-time workers is assumed to be unknown and will be determined by the solution to the problem.

A call center is open for business 11 hours per day (from 8 am to 7 pm). The call center management uses hourly planning intervals for staff scheduling purposes. The call center is staffed by full-time (FT) and part-time (PT) agents. Shifts for both FT and PT agents can start at the beginning of any hour, provided that they do not extend past 7 pm. FT shifts span 9 hours and can consist of either (1) 4 hours of work, a one-hour break, and 4 more hours of work or (2) 5 hours of work, a one-hour break, and 3 more hours of work. PT shifts span five hours with no break. Part-time agents cannot comprise more than 30% of the total workforce. The call center management would like to build an integer linear programming shift scheduling model that will be used to develop a schedule that will provide enough labor in each hour of the day (requirements shown in the following table), meet the staffing mix restriction, and minimize total labor hours scheduled (assume 8 productive hours for FT agents and 5 for PT agents). The formulation (decision variables, objective, and constraints) is shown in the following:

Hour	1	2	3	4	5	6	7	8	9	10	11
Req	10	15	23	32	37	40	33	26	19	13	8

Decision variables

X_{ij} = the number of FT agents who start their shift at the beginning of hour i and take a break after j hours of work (for $i = 1, 2, 3$ and $j = 4, 5$).

Y_i = the number of PT agents who start their shift at the beginning of hour i (for $i = 1, 2, 3, 4, 5, 6, 7$).

Objective function

$$\text{Minimize:} \sum_{i=1}^{3}\sum_{j=4}^{5} 8X_{ij} + \sum_{i=1}^{7} 5Y_i = 8X_{14} + 8X_{24} + 8X_{34} + 8X_{15}$$

$$+ 8X_{25} + 8X_{35}; \tag{3.10}$$

$$+ 5Y_1 + 5Y_2 + 5Y_3 + 5Y_4 + 5Y_5 + 5Y_6 + 5Y_7.$$

Subject to constraints

$$\sum_{i=1}^{7} Y_i \le 0.3 \left[\sum_{i=1}^{3} \sum_{j=4}^{5} X_{ij} + \sum_{i=1}^{7} Y_i \right]. \qquad (3.11)$$

X_{14}			$+X_{15}$			$+Y_1$							$\ge$	10
X_{14}	$+X_{24}$		$+X_{15}$	$+X_{25}$		$+Y_1$	$+Y_2$						$\ge$	15
X_{14}	$+X_{24}$	$+X_{34}$	$+X_{15}$	$+X_{25}$	$+X_{35}$	$+Y_1$	$+Y_2$	$+Y_3$					$\ge$	23
X_{14}	$+X_{24}$	$+X_{34}$	$+X_{15}$	$+X_{25}$	$+X_{35}$	$+Y_1$	$+Y_2$	$+Y_3$	$+Y_4$				$\ge$	32
	$+X_{24}$	$+X_{34}$	$+X_{15}$	$+X_{25}$	$+X_{35}$	$+Y_1$	$+Y_2$	$+Y_3$	$+Y_4$	$+Y_5$			$\ge$	37
X_{14}		$+X_{34}$		$+X_{25}$	$+X_{35}$		$+Y_2$	$+Y_3$	$+Y_4$	$+Y_5$	$+Y_6$		$\ge$	40
X_{14}	$+X_{24}$		$+X_{15}$		$+X_{35}$			$+Y_3$	$+Y_4$	$+Y_5$	$+Y_6$	$+Y_7$	$\ge$	33
X_{14}	$+X_{24}$	$+X_{34}$	$+X_{15}$	$+X_{25}$					$+Y_4$	$+Y_5$	$+Y_6$	$+Y_7$	$\ge$	26
X_{14}	$+X_{24}$	$+X_{34}$	$+X_{15}$	$+X_{25}$	$+X_{35}$					$+Y_5$	$+Y_6$	$+Y_7$	$\ge$	19
	$+X_{24}$	$+X_{34}$		$+X_{25}$	$+X_{35}$						$+Y_6$	$+Y_7$	$\ge$	13
		$+X_{34}$			$+X_{35}$							$+Y_7$	$\ge$	8

The X_{ij} and Y_i decision variables pertain to the assignment of agents to FT and PT shifts, respectively. FT shifts are defined by both a start time (hours $i = 1, 2$, or 3) and break placement (after $j = 4$ or 5 hours of work). PT shifts are defined only by start time (hours 1, 2, 3, 4, 5, 6, or 7). The objective function in Equation (3.10) seeks the minimization of the total number of labor hours scheduled. The objective function coefficients for the FT and PT shift decision variables are 8 and 5, respectively.

Constraint (3.11) ensures that the number of PT agents is no more than 30% of the total workforce. The constraints in the table guarantee that the number of agents working in each hour equals or exceeds the number of agents required in that hour.

The Excel spreadsheet for this problem is displayed in Figure 3.7 and the Solver dialog box is shown in Figure 3.8.

The objective function coefficients are in cells B3:N3 of the worksheet displayed in Figure 3.7, whereas the constraint coefficients are in cell B5:N16. The decision variables are in cells B18:N18. The objective function, which is the sumproduct of rows 3 and 18, is in cell H20. The number of agents working in each hour (cells P5:P15) is computed as the sumproduct of the constraint coefficients and the decision variables. Cells P5:P15 must equal or exceed the number of agents needed in cells Q5:Q15. The staffing mix constraint is enforced by requiring P16 to equal or exceed cell Q16. It is also evident from

	A	B	C	D	E	F	G	H	I	J	K	L	M	N	O	P	Q
1	Shift Variables	X14	X24	X34	X15	X25	X35	Y1	Y2	Y3	Y4	Y5	Y6	Y7			
2																	
3	Hours in Shift	8	8	8	8	8	8	5	5	5	5	5	5	5		Working	Needed
4																	
5	Hour 1	1			1			1								10	10
6	Hour 2	1	1		1	1		1	1							15	15
7	Hour 3	1	1	1	1	1	1	1	1	1						35	23
8	Hour 4	1	1	1	1	1	1	1	1	1	1					43	32
9	Hour 5		1	1	1	1	1	1	1	1	1	1				38	37
10	Hour 6	1		1		1	1		1	1	1	1	1			40	40
11	Hour 7	1	1		1		1			1	1	1	1	1		33	33
12	Hour 8	1	1	1	1	1					1	1	1	1		27	26
13	Hour 9	1	1	1	1	1		1				1	1	1		32	19
14	Hour 10		1	1		1	1						1	1		22	13
15	Hour 11			1			1			1				1		18	8
16	Staff Mix	0.3	0.3	0.3	0.3	0.3	0.3	-0.7	-0.7	-0.7	-0.7	-0.7	-0.7	-0.7		0.2	0
17																	
18	# of Agents Assigned	6	0	5	3	4	13	1	1	2	8	1	0	0			
19																	
20	Total Labor Hours =						313										
21																	
22	Total FT Agents	31															
23	Total PT Agents	13															

Figure 3.7.   Worksheet for Example 3.3.

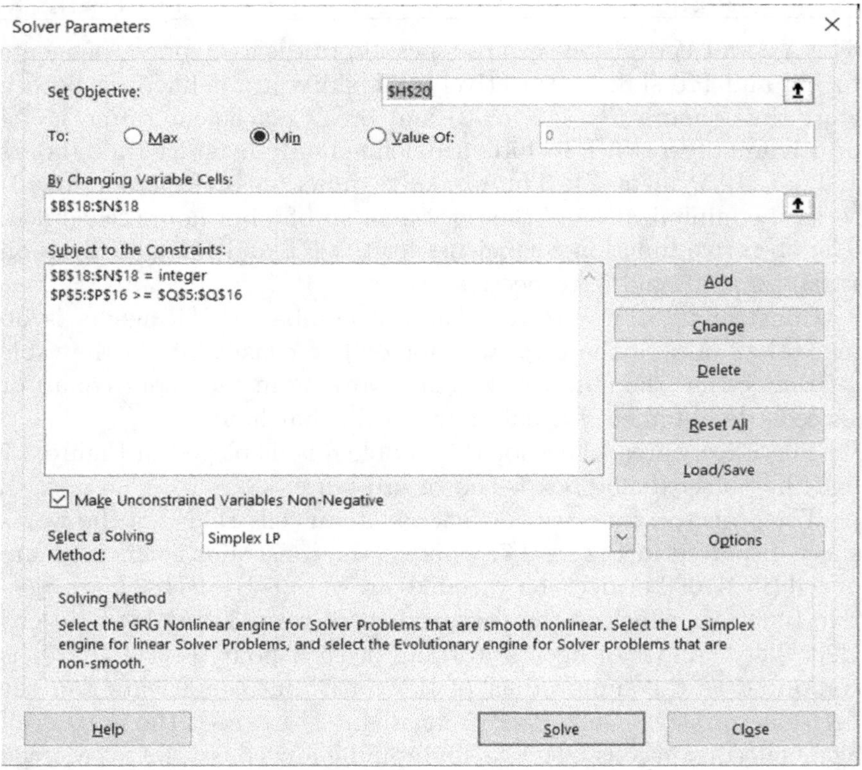

Figure 3.8.   Solver dialog box for Example 3.3.

the Solver dialog box in Figure 3.8 that constraints are added to ensure that the shift assignment decision variables assume integer values.

The optimal solution displayed in Figure 3.7 consists of 31 FT and 13 PT agents, which results in a minimum number of labor hours of $8(31) + 5(13) = 313$. Notice that $13/(31 + 13) = .294$, which is less than 0.3, so the staffing mix constraint is satisfied. There are some hours where the number of agents working is exactly equal to the number required (i.e., hours 1, 2, 6, 7) but there are other hours where the number of agents greatly exceeds the number required (i.e., 3, 4, 9, 10, 11). It should be noted, however, that the reported solution is not unique and there might be alternative optimal solutions with better distribution of surplus labor. Additional scheduling flexibility will also reduce surplus labor and the worksheet can be modified easily to accommodate such flexibility. Perhaps the simplest approach to reducing surplus labor is to increase the allowable percentage of PT agents.

3.2.2 *Example 3.4 — Shift scheduling (minimizing maximum shortage)*

Next, we present a similar shift scheduling example but now assume that the numbers of FT and PT agents are known. Accordingly, the total number of labor hours is known, so it is no longer a relevant objective criterion. Instead, we seek to schedule the pre-specified number of agents as efficiently as possible, so as to keep large shortages (which are detrimental to service quality) as small as possible.

Suppose that the call center in Example 3.3 only has 28 FT agents and 10 PT agents, which is not enough to meet the labor requirements for each hour. We want to revise the formulation so that, instead of minimizing labor hours, we schedule the fixed workforce to *minimize the maximum shortage* across all hours of the day. The new formulation is as follows:

Decision variables

X_{ij} = the number of FT agents who start their shift at the beginning of hour i and take a break after j hours of work (for $i = 1, 2, 3$ and $j = 4, 5$).

Y_i = the number of PT agents who start their shift at the beginning of hour i (for i = 1, 2, 3, 4, 5, 6, 7).

M = the maximum shortage across all hours of the day.

Objective function

$$\text{Minimize: } M. \tag{3.12}$$

Subject to constraints

$$\sum_{i=1}^{3} \sum_{j=4}^{5} X_{ij} = 28, \tag{3.13}$$

$$\sum_{i=1}^{7} Y_i = 10. \tag{3.14}$$

X_{14}			$+X_{15}$			$+Y_1$							$+M$	$\geq$	10
X_{14}	$+X_{24}$		$+X_{15}$	$+X_{25}$		$+Y_1$	$+Y_2$						$+M$	$\geq$	15
X_{14}	$+X_{24}$	$+X_{34}$	$+X_{15}$	$+X_{25}$	$+X_{35}$	$+Y_1$	$+Y_2$	$+Y_3$					$+M$	$\geq$	23
X_{14}	$+X_{24}$	$+X_{34}$	$+X_{15}$	$+X_{25}$	$+X_{35}$	$+Y_1$	$+Y_2$	$+Y_3$	$+Y_4$				$+M$	$\geq$	32
	$+X_{24}$	$+X_{34}$	$+X_{15}$	$+X_{25}$	$+X_{35}$	$+Y_1$	$+Y_2$	$+Y_3$	$+Y_4$	$+Y_5$			$+M$	$\geq$	37
X_{14}		$+X_{34}$		$+X_{25}$	$+X_{35}$		$+Y_2$	$+Y_3$	$+Y_4$	$+Y_5$	$+Y_6$		$+M$	$\geq$	40
X_{14}	$+X_{24}$		$+X_{15}$		$+X_{35}$			$+Y_3$	$+Y_4$	$+Y_5$	$+Y_6$	$+Y_7$	$+M$	$\geq$	33
X_{14}	$+X_{24}$	$+X_{34}$	$+X_{15}$	$+X_{25}$					$+Y_4$	$+Y_5$	$+Y_6$	$+Y_7$	$+M$	$\geq$	26
X_{14}	$+X_{24}$	$+X_{34}$	$+X_{15}$	$+X_{25}$	$+X_{35}$					$+Y_5$	$+Y_6$	$+Y_7$	$+M$	$\geq$	19
	$+X_{24}$	$+X_{34}$		$+X_{25}$	$+X_{35}$						$+Y_6$	$+Y_7$	$+M$	$\geq$	13
		$+X_{34}$			$+X_{35}$							$+Y_7$	$+M$	$\geq$	8

The X_{ij} and Y_i decision variables are the same as those in Example 3.3. The variable M, which represents the maximum shortage across all hours of the day, is the only additional variable. The objective function in Equation (3.12) seeks the minimization of the maximum shortage (M).

Constraint (3.13) and (3.14) ensure that we schedule only the 28 FT and 10 PT agents that we have available. The constraints in the table guarantee that the number of agents working in each hour *plus the maximum shortage* equals or exceeds the number of agents required in that hour. If it were possible to meet all of the requirements with the current workforce, then the maximum shortage M would be zero when the problem is solved. However, in this example, this is not the case.

The Excel spreadsheet for this problem is displayed in Figure 3.9 and the Solver dialog box is shown in Figure 3.10.

The constraint coefficients are in cells B3:O15 of Figure 3.9 and the decision variables are in cells B17:O17. The objective function cell is O17 (which contains M). The number of agents working in each hour (cells Q3:Q13) is computed as the sumproduct of the constraint coefficients and the decision variables. Cells R3:R13 are similar to Q3:Q13 but also include the value of M, and these cells must equal or exceed the number of agents needed in cells S3:S13. Requiring cells R14:R15 to equal S14:S15 ensures that exactly 28 FT and 10 PT agents are scheduled.

The solution in Figure 3.9 reveals that the 28 FT and 10 PT agents can be assigned to keep the maximum shortage to $M = 4$ agents. This maximum shortage occurs in hours 1, 5, 7, and 8. It is important to note, however, that the shortage of four agents in hour 1 is apt to be much more detrimental to service quality than the shortage of four agents in hour 5. That is, having only 6 agents working when you need 10 is arguably worse than having 33 agents when you need 37. Accordingly, rather than just considering raw shortage values, it might be preferable to consider the ratio of shortage to demand in the objective function. We will examine this possibility in the next example.

Shift Variables	X14	X24	X34	X15	X25	X35	Y1	Y2	Y3	Y4	Y5	Y6	Y7	M	Working	Working +M	Needed
Hour 1	1			1			1							1	6	10	10
Hour 2	1	1		1	1		1	1						1	14	18	15
Hour 3	1	1	1	1	1	1	1	1	1					1	29	33	23
Hour 4	1	1	1	1	1	1	1	1	1	1				1	38	42	32
Hour 5		1	1	1	1	1	1	1	1	1	1			1	33	37	37
Hour 6	1		1		1	1		1	1	1	1	1		1	37	41	40
Hour 7	1	1		1		1			1	1	1	1	1	1	29	33	33
Hour 8	1	1	1	1	1					1	1	1	1	1	22	26	26
Hour 9	1	1	1	1	1	1					1	1	1	1	28	32	19
Hour 10		1	1		1	1						1	1	1	23	27	13
Hour 11			1			1							1	1	15	19	8
FT EMP	1	1	1	1	1	1									28		28
PT EMP							1	1	1	1	1	1	1		10		10
Assignments	5	0	0	0	8	15	1	0	0	9	0	0	0	4			

Figure 3.9. Worksheet for Example 3.4.

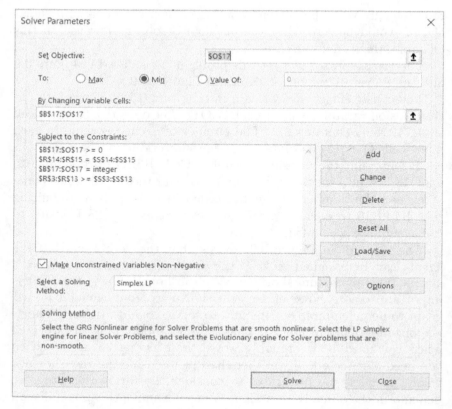

Figure 3.10.   Solver dialog box for Example 3.10.

3.2.3   *Example 3.5 — Days-off scheduling (minimizing sum of shortage ratios)*

In this example, we consider a days-off scheduling problem where the number of nurses is not sufficient to cover all requirements. Rather than focusing on the maximum shortage (as we did in Example 3.4), here we focus on the ratio of shortage to demand.

Daily requirements for a 10-hour nursing shift are shown in the following table. Nurses work four consecutive days per week and there are currently 10 nurses on staff. Due to a nursing shortage, there are not enough nurses available to cover all labor requirements. We seek a linear integer programming formulation that will minimize the sum (across all days) of the following ratio: (# of nurses short for the day/nursing requirement for the day). The formulation is provided in the following:

Day	Mo	Tu	We	Th	Fr	Sa	Su
Req	7	8	8	8	7	4	4

Decision variables

X_j = the number of nurses who begin their consecutive 4 days of work on day j (j = 1 = Monday, 2 = Tuesday, 3 = Wednesday, 4 = Thursday, 5 = Friday, 6 = Saturday, and 7 = Sunday).
S_j = the shortage of nurses on day j (j = 1 = Monday, 2 = Tuesday, 3 = Wednesday, 4 = Thursday, 5 = Friday, 6 = Saturday, and 7 = Sunday).

Objective function

Minimize: $S_1/7 + S_2/8 + S_3/8 + S_4/8 + S_5/7 + S_6/4 + S_7/4.$ (3.15)

Subject to constraints

$$X_1 + X_2 + X_3 + X_4 + X_5 + X_6 + X_7 = 10. \quad (3.16)$$

$+X_1$				$+X_5$	$+X_6$	$+X_7$	$+S_1$	$\geq$	7
$+X_1$	$+X_2$				$+X_6$	$+X_7$	$+S_2$	$\geq$	8
$+X_1$	$+X_2$	$+X_3$				$+X_7$	$+S_3$	$\geq$	8
$+X_1$	$+X_2$	$+X_3$	$+X_4$				$+S_4$	$\geq$	8
	$+X_2$	$+X_3$	$+X_4$	$+X_5$			$+S_5$	$\geq$	7
		$+X_3$	$+X_4$	$+X_5$	$+X_6$		$+S_6$	$\geq$	4
			$+X_4$	$+X_5$	$+X_6$	$+X_7$	$+S_7$	$\geq$	4

The X_j decision variables are the number of nurses who begin their workweek on a particular day j, whereas the S_j variables are the number of nurses short on a particular day j. The objective function in Equation (3.15) is the sum of ratios of daily shortage to daily requirements. If all days could be covered without a shortage, then all ratios would be zero and the objective function would be zero. The key aspect of this objective function is that any shortage greater than zero will have a more adverse effect on the objective function the smaller the requirement. For example, a shortage of one nurse on Tuesday would incur a penalty of only 1/8 in the objective function, whereas a shortage of one nurse on Sunday would incur a stronger penalty of $1/4$.

Constraint (3.16) ensures that all 10 nurses are assigned to work-weeks. The table ensures that nursing coverage plus the shortage variable equals or exceeds the requirement for each day. The Excel spreadsheet for this problem is displayed in Figure 3.11 and the Solver dialog box is shown in Figure 3.12.

The objective function coefficients are in cells I3:O3 of the worksheet displayed in Figure 3.11, whereas the constraint coefficients are in cells B5:O12. The decision variables are in cells B14:O14. The objective function, which is the sumproduct of rows 3 and 14, is in cell B16. The number of nurses working on each day (cells Q5:Q11) is computed as the sumproduct of the constraint coefficients and the decision variables. Cells R5:R11 are similar to Q5:Q11 but also include the daily shortage variable. Cells R5:R11 must exceed the number of nurses needed in cells S5:S11. The requirement to schedule all 10 nurses is accomplished by requiring R12 to equal cell S12. It is also evident from the Solver dialog box in Figure 3.8 that constraints are added to ensure that the days-off assignment decision variables assume integer values.

The solution in Figure 3.11 reveals that the 10 nurses are assigned to cover requirements efficiently but there are three days (Tuesday, Wednesday, and Thursday) that have a shortage of two nurses, resulting in an objective function value of $3(2/8) = 0.75$. The shortages occur on the three highest demand days, which is greatly preferable to a shortage of 2 nurses on the weekend when the requirement is only 4.

A	B	C	D	E	F	G	H	I	J	K	L	M	N	O	P	Q	R	S
Days off Pattern	P1	P2	P3	P4	P5	P6	P7	S1	S2	S3	S4	S5	S6	S7				
Obj Coefficients								.143	.125	.125	.125	.143	.250	.250			Shortage +	
																Working	Working	Needed
Monday	1				1	1	1	1								7	7	7
Tuesday	1	1				1	1		1							6	8	8
Wednesday	1	1	1			1				1						6	8	8
Thursday	1	1	1	1							1					6	8	8
Friday		1	1	1	1							1				7	7	7
Saturday			1	1	1	1							1			4	4	4
Sunday				1	1	1	1							1		4	4	4
Number of Nurses	1	1	1	1	1	1	1										10	10
Assigned to Pattern	3	3	0	0	4	0	0	0	2	2	2	0	0	0				
Shortage Cost	0.75																	

Figure 3.11.   Worksheet for Example 3.5.

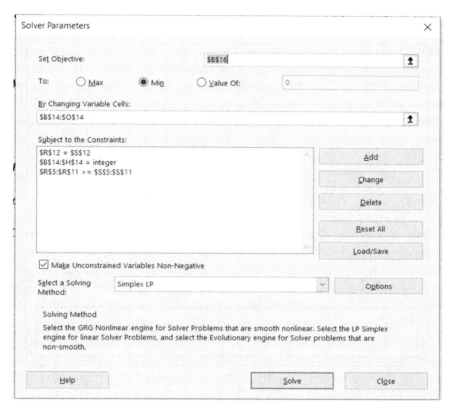

Figure 3.12. Solver dialog box for Example 3.5.

3.2.4 *Example 3.6 — Tour scheduling (minimizing cost)*

The next example in this chapter focuses on the tour scheduling problem, which requires the assignments of call center agents to both work days and non-work days, as well as shifts on each work day. This integration of the shift and days-off problem results in a tour scheduling problem that is appreciably larger and more complex than the two subproblems.

A call center is open 7 days per week from 8:00 am to 7:00 pm and uses hourly planning intervals. The minimum number of call center agents required in each hour of each day is shown in the following table. The call center will be staffed using full-time and part-time agents who each work five consecutive days per week. All seven possible patterns of five consecutive workdays are permissible. Each agent will begin their shift at the same time on each day of their work week. Full-time shifts span nine hours with a one-hour meal break after the fourth hour (i.e., 4 hours of work, a 1-hour break, then 4 more hours of work). Part-time shifts span 5 consecutive hours with no meal break. Full-time agents are paid $720 per week and part-time agents are paid $375 per week. Although both FT and PT agents can be used, there is a limit that no more than 30% of the agents can be part-time employees. The call center would like to construct a weekly work schedule that will minimize the weekly staffing costs, while assuring that there are a sufficient number of agents present to meet the minimum requirements in each hour of each day of the week. The integer linear programming formulation of this problem is shown in the following.

	8–9	9–10	10–11	11–12	12–1	1–2	2–3	3–4	4–5	5–6	6–7
Monday	17	27	33	20	18	26	31	25	23	19	14
Tuesday	20	31	37	26	21	25	31	32	24	20	16
Wednesday	24	34	43	32	27	30	36	34	29	25	21
Thursday	21	29	39	29	24	29	33	26	23	20	17
Friday	18	26	35	23	16	24	34	37	31	24	15
Saturday	11	14	16	11	9	15	20	15	11	8	6
Sunday	8	8	10	6	8	13	14	12	9	9	5

Parameters and sets

$I =$ the set of 77 hourly planning periods for the week (7 days of week beginning with Monday times 11 hours per work day beginning at 8am), $I = \{1, 2, \ldots, 77\}$.

$F =$ the set of 21 full-time tours to which FT agents can be assigned (7 days-off combinations times 3 FT shift starting times per day), $F = \{1, 2, \ldots, 21\}$.

$P =$ the set of 49 part-time tours to which PT agents can be assigned (7 days-off combinations times 7 PT shift starting times per day). $P = \{22, 23, \ldots, 70\}$.

$T =$ the set of all 70 full-time and part-time tours to which agents can be assigned: $T = \{1, 2, \ldots, 70\}$.

$D_i =$ the minimum requirement for hour i of the planning horizon (for all $i \in I$).

$A_{ij} = 1$ if hour i is a work period in tour j for all $i \in I$ and $j \in T$.

Decision variables

X_j = the number of agents assigned to tour j for all $j \in T$.

Objective function

$$\text{Minimize: } 720 \sum_{j \in F} X_j + 375 \sum_{j \in P} X_j. \tag{3.17}$$

Subject to constraints

$$\sum_{j \in T} A_{ij} X_j \geq D_i \quad \text{for all } i \in I, \tag{3.18}$$

$$\sum_{j \in P} X_j \leq 0.3 \sum_{j \in T} X_j. \tag{3.19}$$

For a seven-day week and 11 hours per day, there are $7 \times 11 = 77$ hours in the planning horizon, which are contained in the set I. There are seven possible workweek patterns and three possible FT shift starting times for each of these patterns, resulting in a total of 21 FT tours in set F. Note that the assumption that agents have the same starting time on each work day of their schedule is critical. If starting times for individual agents are permitted to float (or vary) across the work days, then the number of possible tours can rapidly become enormous and better models for accommodating float will become necessary (Jacobs & Brusco, 1996; Brusco & Jacobs, 2000). Because there are seven possible PT shift starting times, there are $7 \times 7 = 49$ PT tours in set P. The set of all tours is $T = F \cup P$. The D_i parameters are the minimum requirements for each hour of the day on each day of the week. The A_{ij} parameters can be viewed as a 77×70 matrix with elements of 1 in the matrix indicating that the hour in the row is a work period in the tour corresponding to the column.

The objective function in Equation (3.17) is the total cost of the FT and PT agents. Constraint set (3.18) guarantees that, for each of the 77 hours of the planning horizon, the assignment of agents to tours will provide a number of agents working that is sufficient to equal or exceed the minimum requirement. Constraint (3.19) limits the number of PT agents to no more than 30% of the total workforce. Although not explicitly stated, we also impose integer restrictions on all of the X_j variables because fractional assignments of agents to tours are not useful.

A partial view of the Excel spreadsheet for this tour scheduling problem is provided in Figure 3.13 and the Solver dialog box is shown in Figure 3.14. Cell A2 allows the user to specify a minimum percentage for the PT workforce and this value is currently set to 0.3 in accordance with the problem description. Cells B6:BS6 contain the objective function coefficients. Cells B10:BS86 display the coefficients for constraint set (3.18), that is, the A_{ij} values. The decision variables span cells B8:BS8. The product of the decision variables and the constraint coefficients for each hour are found in cells BV10:BV86 and the values in these cells must equal or exceed the minimum requirements for each hour shown in cells BW10:BW86. Row 87 handles the PT staffing mix limit in the same manner as shown for the shift scheduling problem in Example 3.

The optimal solution displayed in Figure 3.13 reveals that there are 41 FT agents and 17 PT agents, for a total cost of 720(41) + 375(17) = \$35,895. Notice that 17/(41+17) = 0.293, so the part-time workforce is just under 30% of the total workforce, thus satisfying the constraint. As was observed in Example 3.3, there is significant overstaffing in some time periods. This can be mitigated by allowing more scheduling flexibility, which can be accomplished by increasing the allowable percentage for PT agents, allowing alternative break placements for FT agents, or other mechanisms.

3.2.5 *Example 3.7 — Shift scheduling with start-time restrictions*

All of the preceding workforce scheduling examples have not imposed any restrictions on the number of shifts and/or days-off patterns that can be used. However, in real-world shift and tour scheduling problems, it is not uncommon to encounter such restrictions. For example, when scheduling airport ground personnel at United Airlines, as reported by Brusco *et al.* (1995), there were 96 15-minute planning intervals in the day and, accordingly 96 possible shift starting times. If even 20% of these were used, there would be employees arriving, leaving, or taking breaks in many planning intervals of the day, which is administratively cumbersome. For this reason, a common constraint that was imposed at many of the airports was the

	A	B	C	D	E	F	G	H	I	J	K	L	M	N	O	P	Q	R	S	T	U	V	W	X	Y	Z	AA	AB	AC	AD	AE	AF	AG	AH
1	*PT percentage*		Sa-Su			Su-Mo			Mo-Tu			Tu-We			We-Th			Th-Fr			Fr-Sa			Sa-Su							Su-Mo			
2	0.3		(4-B-4)			(4-B-4)			(4-B-4)			(4-B-4)			(4-B-4)			(4-B-4)			(4-B-4)			(5-B-3)							(5-B-3)			
3																																		
4	*Tot Cost =*		35895		FT =	41		PT =	17.00																									
5																																		
6	*Weekly cost*	720	720	720	720	720	720	720	720	720	720	720	720	720	720	720	720	720	720	720	720	720	375	375	375	375	375	375	375	375	375	375	375	375
7																																		
8	*Num. Scheduled*	12.00	7.00	7.00	1.00	1.00	1.00	3.00	0.00	1.00	1.00	0.00	0.00	0.00	0.00	0.00	4.00	1.00	1.00	0.00	0.00	1.00	0.00	0.00	0.00	0.00	0.00	0.00	3.00	2.00	1.00	0.00	0.00	0.00
9																																		
10	*Monday 8-9*	1									1			1			1			1			1											
11	*Monday 9-10*	1	1								1	1		1	1		1	1		1	1		1	1										
12	*Monday 10-11*	1	1	1							1	1	1	1	1	1	1	1	1	1	1	1	1	1	1									
13	*Monday 11-12*	1	1	1							1	1	1	1	1	1	1	1	1	1	1	1	1	1	1	1								
14	*Monday 12-1*		1	1								1	1		1	1		1	1		1	1	1	1	1	1	1							
15	*Monday 1-2*	1		1							1		1	1		1	1		1	1		1		1	1	1	1	1						
16	*Monday 2-3*	1	1								1	1		1	1		1	1		1	1				1	1	1	1	1					
17	*Monday 3-4*	1	1	1							1	1	1	1	1	1	1	1	1	1	1	1				1	1	1	1					
18	*Monday 4-5*	1	1	1							1	1	1	1	1	1	1	1	1	1	1	1					1	1	1					
19	*Monday 5-6*		1	1								1	1		1	1		1	1		1	1						1	1					
20	*Monday 6-7*			1									1			1			1			1							1					
21	*Tuesday 8-9*	1			1									1			1			1			1							1				
22	*Tuesday 9-10*	1	1		1	1								1	1		1	1		1	1		1	1						1	1			
23	*Tuesday 10-11*	1	1	1	1	1	1							1	1	1	1	1	1	1	1	1	1	1	1					1	1	1		
24	*Tuesday 11-12*	1	1	1	1	1	1							1	1	1	1	1	1	1	1	1	1	1	1	1				1	1	1	1	
25	*Tuesday 12-1*		1	1		1	1								1	1		1	1		1	1	1	1	1	1	1			1	1	1	1	1
26	*Tuesday 1-2*	1		1	1		1							1		1	1		1	1		1		1	1	1	1	1			1	1	1	1
27	*Tuesday 2-3*	1	1		1	1								1	1		1	1		1	1				1	1	1	1	1			1	1	1
28	*Tuesday 3-4*	1	1	1	1	1	1							1	1	1	1	1	1	1	1	1				1	1	1	1				1	1
29	*Tuesday 4-5*	1	1	1	1	1	1							1	1	1	1	1	1	1	1	1					1	1	1					1

Figure 3.13. A partial view of the worksheet for Example 3.6.

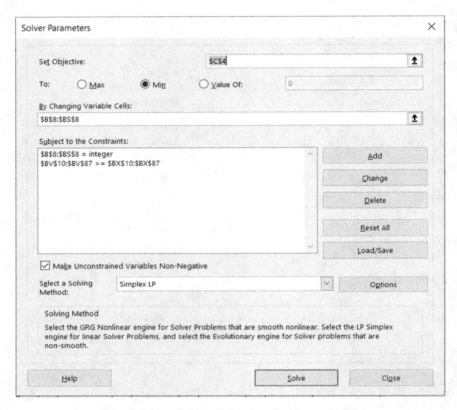

Figure 3.14. Solver dialog box for Example 3.6.

limitation that no more than seven starting times could be used for full-time employees and no more than two starting times could be used for part-time employees. Our final example in this chapter is rooted in the frequent need for these types of constraints.

A telephone call center operates from 7 am to 6 pm (Monday-Friday) and uses hourly planning intervals for its scheduling decisions. The call center uses FT shifts that span a total of nine hours (eight hours of work and a one-hour meal break). The meal breaks in a FT shift can occur after the 3rd hour of work (3 hours of work, a one-hour break, and 5 hours of work) or after the fourth hour of work (4 hours of work, a one-hour break, and 4 hours of work). The call center also uses PT shifts that consist of 4 consecutive hours of work with no break. Potential starting times for full-time and part-time shifts are at the beginning of any hour, provided that the shift would not extend past the end of the operating day. FT employees are paid \$190 per day, whereas PT employees are paid \$80 per day. The number of PT employees cannot exceed 35% of the total number of employees scheduled for the day. An additional constraint is that no more than three shift starting times can be used for the PT employees (note: three of the eight possible PT starting times should not be arbitrarily selected but, instead, they should be determined by the solution to the model). The following is an integer linear programming model for scheduling FT and PT employees, so as to minimize total daily cost, meet the PT staffing constraint, and guarantee that the number of employees scheduled each hour is sufficient to meet the requirements shown in the following table.

Hour	1	2	3	4	5	6	7	8	9	10	11
Req	15	26	35	31	30	26	35	37	36	24	11

Decision variables

X_{ij} = the number of FT agents who start their shift at the beginning of hour i and take a break after j hours of work (for $i = 1, 2, 3$ and $j = 3, 4$).

Y_i = the number of PT agents who start their shift at the beginning of hour i (for $i = 1, 2, 3, 4, 5, 6, 7, 8$).

$S_i = 1$ if hour i is selected as a PT shift starting time, 0 otherwise (for $1 \leq i \leq 8$).

Objective function

$$\text{Minimize:} \quad \sum_{i=1}^{3} \sum_{j=3}^{4} 190 X_{ij} + \sum_{i=1}^{8} 80 Y_i \quad (3.20)$$

$$= 190 X_{14} + 190 X_{24} + 190 X_{34}$$

$$+ 190 X_{15} + 190 X_{25} + 190 X_{35} + 80 Y_1 + 80 Y_2$$

$$+ 80 Y_3 + 80 Y_4 + 80 Y_5 + 80 Y_6 + 80 Y_7 + 80 Y_8. \quad (3.21)$$

Subject to constraints

$$\sum_{i=1}^{8} Y_i \leq 0.35 \left[\sum_{i=1}^{3} \sum_{j=3}^{4} X_{ij} + \sum_{i=1}^{8} Y_i \right], \quad (3.22)$$

$$Y_i \leq 37S_i \quad (\text{for } 1 \leq i \leq 8), \qquad (3.23)$$

$$\sum_{i=1}^{8} S_i \leq 3. \qquad (3.24)$$

X_{14}			$+X_{13}$			$+Y_1$								≥ 15
X_{14}	$+X_{24}$		$+X_{13}$	$+X_{23}$		$+Y_1$	$+Y_2$							≥ 26
X_{14}	$+X_{24}$	$+X_{34}$	$+X_{13}$	$+X_{23}$	$+X_{33}$	$+Y_1$	$+Y_2$	$+Y_3$						≥ 35
X_{14}	$+X_{24}$	$+X_{34}$	$+X_{13}$	$+X_{23}$	$+X_{33}$	$+Y_1$	$+Y_2$	$+Y_3$	$+Y_4$					≥ 31
	$+X_{24}$	$+X_{34}$	$+X_{13}$	$+X_{23}$	$+X_{33}$		$+Y_2$	$+Y_3$	$+Y_4$	$+Y_5$				≥ 30
X_{14}		$+X_{34}$		$+X_{23}$	$+X_{33}$			$+Y_3$	$+Y_4$	$+Y_5$	$+Y_6$			≥ 26
X_{14}	$+X_{24}$		$+X_{13}$		$+X_{33}$				$+Y_4$	$+Y_5$	$+Y_6$	$+Y_7$		≥ 35
X_{14}	$+X_{24}$	$+X_{34}$	$+X_{13}$	$+X_{23}$						$+Y_5$	$+Y_6$	$+Y_7$	$+Y_8$	≥ 37
X_{14}	$+X_{24}$	$+X_{34}$	$+X_{13}$	$+X_{23}$	$+X_{33}$						$+Y_6$	$+Y_7$	$+Y_8$	≥ 36
	$+X_{24}$	$+X_{34}$		$+X_{23}$	$+X_{33}$							$+Y_7$	$+Y_8$	≥ 24
		$+X_{34}$			$+X_{33}$								$+Y_8$	≥ 11

The X_{ij} and Y_i decision variables pertain to the assignment of agents to FT and PT shifts, respectively. FT shifts are defined by both a start time (hours $i = 1, 2$, or 3) and break placement (after $j = 3$ or 4 hours of work). PT shifts are defined only by start time (hours 1, 2, 3, 4, 5, 6, 7, or 8). The objective function in Equation (3.20) seeks the minimization of the total labor cost. The objective function coefficients for the FT and PT shift decision variables are 190 and 80, respectively.

Constraint (3.22) ensures that the number of PT agents is no more than 30% of the total workforce. Constraint set (3.23) guarantees that PT agents are only assigned to shifts for selected starting times (if $S_i = 0$, then no PT agents can start in hour i). The coefficient of 37 for the S_i variables was selected because 37 is the maximum labor requirement across the day and we would, therefore, never assign more than 37 agents to any shift. However, any sufficiently large constant will work. Constraint (3.24) allows no more than three starting times to be selected for PT agents. The constraints in the following table guarantee that the number of agents working in each hour equals or exceeds the number of agents required in that hour.

Two Excel spreadsheets for this problem are displayed in Figures 3.15 and 3.16. They differ only in that Figure 3.15 corresponds to a solution where the limitation on the number of starting times is removed (by changing the right side of constraint (3.24) from

	FT1	FT2	FT3	FT4	FT5	FT6	PT1	PT2	PT3	PT4	PT5	PT6	PT7	PT8	S1	S2	S3	S4	S5	S6	S7	S8	Working	Needed
Hour 1	1	1	1	1																			15	15
Hour 2	1	1	1	1	1			1															26	26
Hour 3	1	1	1	1	1	1	1	1	1														35	35
Hour 4	1	1	1	1	1	1	1	1	1	1													32	31
Hour 5		1	1	1	1	1	1	1	1	1	1												30	30
Hour 6	1	1	1	1	1	1	1	1	1	1	1	1											26	26
Hour 7	1	1	1	1	1	1				1	1	1	1										36	35
Hour 8	1	1	1	1	1	1					1	1	1	1									37	37
Hour 9	1	1	1	1	1	1						1	1	1									36	36
Hour 10	1		1	1		1							1	1									24	24
Hour 11				1										1									11	11
Staff Mix	0.35	0.35	0.35	0.35	0.35	0.35	-0.65	-0.65	-0.65	-0.65	-0.65	-0.65	-0.65	-0.65									1	0
PT Starts Used															1	1	1	1	1	1	1	1	6	8
PT Start 1															-37								-34	0
PT Start 2																-37							-36	0
PT Start 3									1								-37						0	0
PT Start 4										1								-37					-32	0
PT Start 5											1								-37				-36	0
PT Start 6												1	1							-37			0	0
PT Start 7																					-37		-34	0
PT Start 8														1								-37	-35	0
Assignments	4	6	4	8	4	5	3	1	0	5	1	0	3	2	2	1	0	1	1	0	1	1		
Total Cost																								
$7,090.00																								

Figure 3.15.   Worksheet for Example 3.7 (no restriction on PT starting times).

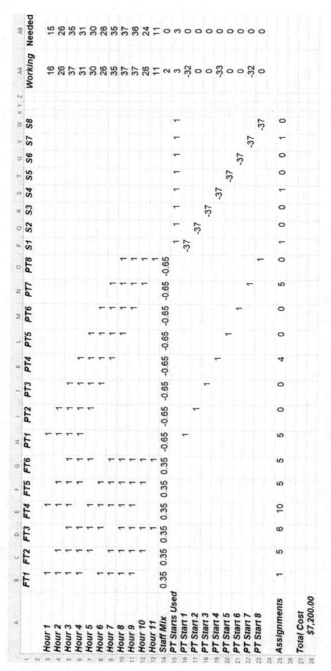

Figure 3.16. Worksheet for Example 3.7 (limit of three PT starting times).

Figure 3.17. Solver dialog box for Example 3.7.

3 to 8), whereas Figure 3.16 corresponds to the solution where the number of PT starting times is limited to three. The Solver dialog box is shown in Figure 3.17.

For the most part, the structure of the worksheets in Figures 3.15 and 3.16 is similar to the one Figure 3.7 for Example 3.3. The main differences pertain to rows 15–23. Cell AA15 contains the sumproduct of P15:W15 and P25:W25 (these are the binary S_i variables and are constrained as binary in the Solver dialog box in Figure 3.17) and, therefore, represents the total number of PT starting times used. Cell AA15 is constrained to be equal to or less than the upper limit on PT starting times in cell AB15. Rows 16–23 operationalize constraint set (3.23). Cells AA16:AA23 contain $Y_i - 37S_i$ and these cells

are constrained to be equal to or less than the values of zero in cells
AB16:AB23.

The results in Figure 3.15 reveal that, in the absence of limits on
the number of PT shift starting times, the optimal solution consists
of 31 FT and 15 PT agents for a total cost of 31($190) + 15($80) =
$7090. Moreover, PT agents are assigned to six of the eight shift
variables (i.e., six starting times are used for PT agents). Coverage
of the labor requirements is extremely efficient, with only one more
agent scheduled than required in hours 4 and 7. The question is as
follows: How much of a cost penalty (if any) will be realized from
adding the restriction that no more than three PT shift starting
times can be used?

The results in Figure 3.16 show that the cost penalty for enforcing
the limit of no more than three PT shift starting times is very modest.
The optimal solution consists of 32 FT and 14 PT agents for a total
cost of 32($190) + 14($80) = $7200. So, effectively, the start-time
constraints require the replacement of one PT employee with one
FT employee for a cost increase of $190 – $80 = $110 (from $7090 to
$7200). Coverage of the requirements is still efficient, with an excess
of one agent in hours 1 and 9 and an excess of two agents in hours 3
and 10.

Other types of constraints can also be imposed on starting times
to make schedules more manageable and/or to satisfy union restrictions. For example, another constraint that was incorporated in the
United Airlines application reported by Brusco *et al.* (1995) was that
selected PT shift starting times had to be separated by at least eight
hours (the span of an FT shift).

Chapter 4

Continuous Facility Location

In this chapter, we focus on continuous facility location problems. The chapter is divided into two main sections: (i) single-source continuous location and (ii) multisource continuous location. The principal focus in the single-source section is the Weber problem, which is arguably the most fundamental problem in continuous location theory. The center-of-gravity problem, which is commonly covered in operations and supply-chain management textbooks and is closely related to the Weber problem (see Morrison, 2010), is also discussed. In the multisource section, we emphasize the generalization of the Weber problem to the case of locating multiple supply sources.

4.1 Single-Source Continuous Location

4.1.1 *Historical background*

In the 1600s, Pierre de Fermat posed a problem that served as the foundation for continuous location problems in the plane (i.e., Euclidean 2-space). Given three points in the plane (P1, P2, P3), locate a fourth point (P4) such that the sum of the Euclidean distances from P4 to each of the other three points is minimized. An important generalization of this problem, typically referred to as the Weber problem (Weber, 1909), assumes n points in the plane with coordinates (x_j, y_j) for $1 \leq j \leq n$ and positive weights (r_j) for each point $1 \leq j \leq n$. The goal is to find a point (u, v) that minimizes the sum of the weighted Euclidean distances from the point (u, v) to the n weighted points:

$$\min_{u,v} Z = \sum_{j=1}^{n} r_j \sqrt{(x_j - u)^2 + (y_j - v)^2}. \qquad (4.1)$$

In location theory, the n points in the plane are commonly existing *demand sites* that each have some requirement (r_j) that must be satisfied from the *supply* facility to be located. Thus, the selection of the coordinates (u, v) for the supply facility will tend to be pulled toward those demand sites with greater requirements.

Nearly all operations and supply chain management textbooks have a chapter on facility location. Many of these textbook chapters cover an approach to single-facility location problems that is typically known as the center-of-gravity *method* (e.g., Heizer *et al.*, 2017; Krajewski *et al.*, 2016; Stevenson, 2018); however, here, we will use the term *center-of-gravity problem* for consistency with the Weber problem. It has been noted that there is significant variability across textbooks with respect to the accuracy and detail of the coverage of the center-of-gravity problem (Kuo & White, 2004). The coordinates obtained using formulas for the center-of-gravity problem do not provide an optimal solution to the Weber problem in Equation (4.1); however, they do provide an optimal solution to the problem associated with requirement-weighted *squared* Euclidean distance (Schärlig, 1973):

$$\min_{u,v} Z_S = \sum_{j=1}^{n} r_j \left[(x_j - u)^2 + (y_j - v)^2 \right]. \qquad (4.2)$$

There are three questions that commonly arise when it is explained that the center-of-gravity method minimizes the sum of requirement-weighted *squared* Euclidean distances. The first question is as follows: Why is the emphasis on squared distance? The distance between the supply facility and a demand site is easy to grasp, but the squared distance between them is somewhat less so. The second question that naturally follows is this: Can the *closed-form* center-of-gravity formulas be modified to minimize requirement-weighted Euclidean distance instead? Finally, the third question is as follows: Does the distinction between Euclidean distance and squared Euclidean distance make much of a difference in the location obtained?

The answers to the first two questions are intertwined. As we shall see in subsequent sections, the answer to the second question is 'no' because the solution to the Weber problem requires numerical estimation. This is likely, in part, the reason why center-of-gravity is favored in textbooks. That is, the center-of-gravity problem has a simple closed-form solution whereas the Weber problem does not. This is not to suggest that the center-of-gravity problem would never be preferred on substantive grounds (see, for example, the discussion by Kuo and White (2004), pertaining to fire loss). The answer to the third question is 'sometimes'.

4.1.2 *The center-of-gravity problem*

Taking the partial derivative of Z_S in Equation (4.2) with respect to u and setting it equal to zero yield the following results:

$$\frac{\partial Z_S}{\partial u} = \sum_{j=1}^{n} -2r_j(x_j - u) = 0, \tag{4.3}$$

$$-2\sum_{j=1}^{n} r_j x_j + 2u \sum_{j=1}^{n} r_j = 0, \tag{4.4}$$

$$2u \sum_{j=1}^{n} r_j = 2\sum_{j=1}^{n} r_j x_j, \tag{4.5}$$

$$u = \frac{\sum_{j=1}^{n} r_j x_j}{\sum_{j=1}^{n} r_j}. \tag{4.6}$$

Equation (4.6) is the center-of-gravity formula for finding the coordinate value for u that minimizes Equation (4.2). Taking the partial derivative of Z_S in Equation (4.2) with respect to v and applying identical algebra yield the following center-of-gravity formula for coordinate v:

$$v = \frac{\sum_{j=1}^{n} r_j y_j}{\sum_{j=1}^{n} r_j}. \tag{4.7}$$

Succinctly, the problem posed by Equation (4.2), finding a location that minimizes the sum of the requirement-weighted squared

Euclidean distances, has a closed-form solution that is given by the center-of-gravity formulas in Equations (4.6) and (4.7).

4.1.3   *The weber problem*

Taking the partial derivatives of Z in Equation (4.1) with respect to u and v and setting them equal to zero yield the following results:

$$\frac{\partial Z}{\partial u} = \sum_{j=1}^{n} \frac{-r_j(x_j - u)}{\sqrt{(x_j - u)^2 + (y_j - v)^2}} = 0, \qquad (4.8)$$

$$\frac{\partial Z}{\partial v} = \sum_{j=1}^{n} \frac{-r_j(y_j - v)}{\sqrt{(x_j - u)^2 + (y_j - v)^2}} = 0. \qquad (4.9)$$

Unfortunately, there is no closed-form solution to these equations. So, the center-of-gravity formulas cannot be modified to minimize the sum of the requirement-weighted Euclidean distances. Instead, it is necessary to resort to iterative numerical algorithms and two options are discussed here.

The first option is the use of Newton's method for optimization. Given estimates of the coordinates at iteration k (u^k and v^k), we can compute the values of first and second partial derivatives for these estimates and use them to produce updated estimates (u^{k+1} and v^{k+1}) as follows:

$$\begin{bmatrix} u^{k+1} \\ v^{k+1} \end{bmatrix} = \begin{bmatrix} u^k \\ v^k \end{bmatrix} - \frac{\lambda}{\frac{\partial^2 Z}{\partial u^2}\left(\frac{\partial^2 Z}{\partial v^2}\right) - \left(\frac{\partial^2 Z}{\partial u \partial v}\right)^2} \begin{bmatrix} \frac{\partial^2 Z}{\partial v^2} & \frac{-\partial^2 Z}{\partial u \partial v} \\ \frac{-\partial^2 Z}{\partial u \partial v} & \frac{\partial^2 Z}{\partial u^2} \end{bmatrix} \begin{bmatrix} \frac{\partial Z}{\partial u} \\ \frac{\partial Z}{\partial v} \end{bmatrix}.$$

$$(4.10)$$

The updated estimates (u^{k+1} and v^{k+1}) are obtained by taking the estimates at the current iteration (u^k and v^k) and subtracting the product of three terms: (a) the step size constant (λ), (b) the inverse of the Hessian matrix, and (c) the gradient vector. When $\lambda = 1$, the updating product is just the inverse Hessian times the gradient. However, the specification of $0 < \lambda < 1$ can lead to greater stability at the expense of slower convergence.

Although the use of Newton's method to solve the Weber problem is explored in the following, the potential for numerical instability associated with this approach to the problem is well known

(Cooper & Katz, 1981). For this reason, we also consider the Weiszfeld algorithm (Weiszfeld, 1937), which uses the following updating scheme:

$$u^{k+1} = \frac{\sum_{j=1}^{n} \frac{r_j x_j}{\sqrt{(x_j-u^k)^2+(y_j-v^k)^2}}}{\sum_{j=1}^{n} \frac{r_j}{\sqrt{(x_j-u^k)^2+(y_j-v^k)^2}}}, \tag{4.11}$$

$$v^{k+1} = \frac{\sum_{j=1}^{n} \frac{r_j y_j}{\sqrt{(x_j-u^k)^2+(y_j-v^k)^2}}}{\sum_{j=1}^{n} \frac{r_j}{\sqrt{(x_j-u^k)^2+(y_j-v^k)^2}}}. \tag{4.12}$$

The updating steps of the Weiszfeld algorithm are simpler and the algorithm is generally robust. However, it should be acknowledged that, if the estimated coordinates pass through a demand site (i.e., $x_j = u^k$ and $y_j = v^k$), then division by zero occurs in (4.11) and (4.12). To circumvent this problem, we add a small ε value in the square root terms of these equations. More formal treatments of convergence are provided in the literature (Brimberg, 1995; Cánovas et al., 2002; Kuhn & Kuenne, 1962; Vardi & Zhang, 2000).

4.1.4 Example 4.1 — A small numerical example

The worksheet displayed in Figure 4.1 enables a comparison of models and methods on a small problem with only $n = 4$ demand sites.

Euclidean / Squared Euclidean:

	$u =$	6.721	4.944	Lambda	1
	$v =$	5.524	5.155		

j	x_j	y_j	r_j	$r_j d_j$	$r_j d_j^2$	$r_j x_j$	$r_j y_j$
1	8	6	31	42.31	311.72	248	186
2	1	7	17	100.44	322.26	17	119
3	2	3	15	80.30	199.63	30	45
4	7	2	8	28.28	113.46	56	16
SUM	18	18	71	251.33	947.07	351	366

Center of gravity formula results = 4.944 5.155

Euclidean Distance Coordinates Using Newton's Method.

u	v	Hessian		Inv Hessian		Gradient
3.000	3.000	4.5625	-.3684	.2195	.0038	-11.7407895748
		-.3684	21.4956	.0038	.0466	-29.2143456650
5.687	4.405	7.0305	-3.8486	.1677	.0464	-.4664860191
		-3.8486	13.8972	.0464	.0848	-13.4670488557
6.390	5.569	4.3441	-4.8102	.3017	.0646	-1.9152316620
		-4.8102	22.4650	.0646	.0583	3.0816673261
6.769	5.513	6.2152	-8.3207	.2892	.0958	.3821033049
		-8.3207	25.1106	.0958	.0716	-.6739156008
6.723	5.525	5.8210	-7.7327	.2927	.0910	.0127086450
		-7.7327	24.8711	.0910	.0685	-.0150130489
6.721	5.524	5.8095	-7.7103	.2927	.0908	.0000121714
		-7.7103	24.8473	.0908	.0684	-.0000248514
6.721	5.524	5.8095	-7.7103	.2927	.0908	.0000000000
		-7.7103	24.8473	.0908	.0684	.0000000000
6.721	5.524	5.8095	-7.7103	.2927	.0908	.0000000000
		-7.7103	24.8473	.0908	.0684	.0000000000
6.721	5.524	5.8095	-7.7103	.2927	.0908	.0000000000
		-7.7103	24.8473	.0908	.0684	.0000000000
6.721	5.524	5.8095	-7.7103	.2927	.0908	.0000000000
		-7.7103	24.8473	.0908	.0684	.0000000000
6.721	5.524	5.8095	-7.7103	.2927	.0908	.0000000000
		-7.7103	24.8473	.0908	.0684	.0000000000
6.721	5.524	5.8095	-7.7103	.2927	.0908	.0000000000
		-7.7103	24.8473	.0908	.0684	.0000000000
6.721	5.524	5.8095	-7.7103	.2927	.0908	.0000000000
		-7.7103	24.8473	.0908	.0684	.0000000000

Euclidean Distance Coordinates Using Weiszfeld's Algorithm

u	v
3.000	3.000
3.500	4.181
4.103	4.704
4.612	4.960
5.014	5.074
5.327	5.140
5.571	5.190
5.761	5.233
5.912	5.271
6.034	5.303
6.132	5.332
6.214	5.356
6.281	5.377
6.338	5.395
6.386	5.410
6.427	5.424
6.463	5.435
6.493	5.446
6.519	5.455
6.542	5.462
6.562	5.469
6.580	5.475
6.596	5.481
6.609	5.485
6.621	5.490

Figure 4.1. Excel worksheet for Example 4.1 ($\lambda = 1.0$, starting $u = v = 3$).

The worksheet is displayed in Figure 4.1. The input data (cells A7:D10) are the x_j, y_j, and r_j values for the four demand sites. Cells E7:E10 contain the requirement-weighted Euclidean distance of each demand site from the supply facility for the Weber model. The sum of these values, which is Z in Equation (4.1) is stored in cell E12. Similarly, Cells F7:F10 contain the requirement-weighted squared Euclidean distance of each demand site from the supply facility for the center-of-gravity model, and the sum of these values, which is Z_S in Equation (4.2), is stored in cell F12.

The solution to the center-of-gravity problem was obtained in two different ways. First, Excel Solver was used to find values for cells C3:C4 (which contain u and v) so as to minimize cell F12. Second, we used Equations (4.6) and (4.7), whereby the $r_j x_j$ and $r_j y_j$ products were computed in columns G and H, respectively. Dividing the sum of these products (cells G12 and H12) by the sum of the r_j values in cell D12 results in the same optimal values of $u = 4.944$ and $v = 5.155$ as shown in cells G13 and H13.

The solution to the Weber problem was obtained in three different ways. First, Excel Solver was used to find values for cells B3:B4 (which contain u and v) so as to minimize cell E12. Second, Newton's method was applied using an initial solution of $u = v = 3$. Implementing Newton's method in the spreadsheet is somewhat tedious due to the complexity of the derivative formulas in the Hessian and gradient columns and the user must specify a step size in cell F3. Nevertheless, Newton's method converges to the same optimal coordinates (cells B3:B4) obtained by the Excel Solver within just five or six iterations. Third, the Weiszfeld algorithm was applied using an initial solution of $u = v = 3$. The results for the Weiszfeld algorithm show a much slower convergence than Newton's method. The algorithm does converge, but it took 80 iterations to match the solution of $u = 6.721$ and $v = 5.524$ obtained via Newton's method to three decimal places.

Based on the results in Figure 4.1, it would appear that Newton's method would be preferable to the Weiszfeld algorithm for the Weber problem, but such a conclusion is premature. Figure 4.2 presents results assuming an initial starting solution of $u = v = 1$ for the three methods. The inherent instability of Newton's approach arose when using $\lambda = 1$. Therefore, it was necessary to reduce the step size (note that $\lambda = .4$ in Figure 4.2) to ensure convergence, albeit at a slower rate. By contrast, the Weiszfeld algorithm has no problem with this starting solution and still converges in about 80 iterations. For simplicity and avoidance of the step size parameter, the Weisfeld algorithm is preferred.

Figure 4.3 provides a visual display of the demand sites and the optimal supply facility locations associated with the Weber

	Squared						
	Euclidean	Euclidean					
u =	6.721	4.944	Lambda	0.4			
v =	5.524	5.155					

j	x_j	y_j	r_j	$r_j d_j$	$r_j d_j^2$	$r_j x_j$	$r_j y_j$
1	8	6	31	42.31	311.72	248	186
2	1	7	17	100.44	322.26	17	119
3	2	3	15	80.30	199.63	30	45
4	7	2	8	28.28	113.46	56	16
SUM	18	18	71	251.33	947.07	351	366

Center of gravity formula results = 4.944 5.155

Euclidean Distance Coordinates Using Newton's Method.

u	v	Hessian		Inv Hessian		Gradient
1.000	1.000	9.4529	-4.6010	.1914	.1758	-39.8250929306
		-4.6010	5.0075	.1758	.3613	-49.7499837687
7.548	10.990	8.6622	-1.2054	.1251	.0694	20.7600266105
		-1.2054	2.1720	.0694	.4989	60.0265614739
4.842	-1.566	8.1425	-.2669	.1239	.0153	-1.2010690942
		-.2669	2.1572	.0153	.4655	-63.6990308885
5.290	10.300	7.9953	.7177	.1271	-.0225	1.4934221245
		.7177	4.0612	-.0225	.2502	58.1037676686
5.736	4.499	6.9682	-4.0819	.1726	.0496	-.4944539517
		-4.0819	14.2005	.0496	.0847	-12.3462009867
6.015	4.927	6.5220	-5.1719	.2048	.0650	-.5877325975
		-5.1719	16.3071	.0650	.0819	-7.1660617258
6.249	5.177	6.2045	-5.9486	.2332	.0751	-.4864315443
		-5.9486	18.4719	.0751	.0783	-4.1424705686
6.419	5.321	6.0185	-6.5185	.2543	.0814	-.3465267152
		-6.5185	20.3729	.0814	.0751	-2.4042678587
6.532	5.405	5.9189	-6.9285	.2686	.0851	-.2303435010
		-6.9285	21.8592	.0851	.0727	-1.4059911732
6.605	5.453	5.8676	-7.2114	.2779	.0874	-.1468702184
		-7.2114	22.9241	.0874	.0711	-.8285023105
6.650	5.482	5.8411	-7.3986	.2836	.0888	-.0913887852
		-7.3986	23.6407	.0888	.0701	-.4912000580
6.678	5.499	5.8272	-7.5184	.2872	.0896	-.0560375953
		-7.5184	24.1029	.0896	.0694	-.2924871426
6.695	5.509	5.8196	-7.5933	.2894	.0901	-.0340646892

Euclidean Distance Coordinates Using Weiszfeld's Algorithm

u	v
1.000	1.000
3.802	4.498
4.370	4.863
4.825	5.028
5.180	5.110
5.456	5.166
5.672	5.212
5.841	5.252
5.976	5.288
6.085	5.318
6.175	5.344
6.249	5.367
6.311	5.386
6.363	5.403
6.408	5.417
6.446	5.430
6.478	5.441
6.507	5.450
6.531	5.459
6.553	5.466
6.572	5.472
6.588	5.478
6.603	5.483
6.615	5.487
6.627	5.491

Figure 4.2. Excel worksheet for Example 4.1 (λ = 0.4.0, starting $u = v = 1$).

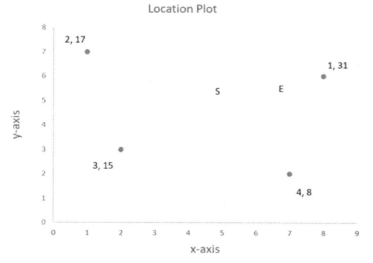

Figure 4.3. Plot of Weber (E) and center-of-gravity (S) solutions for Example 4.1.

(shown by E in Figure 4.3) and center-of-gravity (shown by S in Figure 4.3) solutions. For each demand site, the label is (j, r_j), that is, the site's label number and its requirement. The Weber and center-of-gravity solutions have similar y-coordinate values (i.e., similar values of v), but differ more significantly on the x-axis. In the Weber solution, the supply facility is strongly pulled toward demand site 1, which has (by far) the greatest requirement. However, this pulls the supply facility farther away from demand sites 2 and 3, which have a sum of requirement comparable to site 1's requirement. The results for the center-of-gravity method suggest that, when using the more punitive squared distance penalty, it is not optimal to push the supply facility to a point so distant from demand sites 2 and 3. For this reason, the supply facility is more centrally located on the x-axis in the center-of-gravity solution.

4.1.5   *VBA macros*

We have implemented Newton's method and the Weisfeld algorithm as VBA macros. The macro for Newton's algorithm runs one iteration at a time. Specifically, when the button 'Run One Iteration of Newton's Method' is clicked, the macro reads values of λ, u, v, and n from, respectively, cells B1, B2, B3, and B11 of the worksheet displayed in Figure 4.4. The relevant coordinates and requirements are also read from the worksheet. The macro then computes the gradient, Hessian, and inverse Hessian matrix and updates the u and v coordinates accordingly. This information is then written back to the worksheet. By successively clicking on the button, subsequent iterations are performed and the convergence can be evaluated by both the small changes in the coordinates and the vanishing gradient.

The macro for the Weisfeld algorithm can also be used to solve center-of-gravity and Weber problems of arbitrary size. The user would input the value of n in cell C1, along with the corresponding x_j, y_j, and r_j values in columns B, C, and D beginning at row 4. Clicking on the 'Run Location Analysis' button will execute a VBA macro that reads all of the input data in columns B, C, and D. The macro then uses Equations (4.6) and (4.7) to compute the center-of-gravity coordinates and writes these values of u and v to cells I4 and I5 of the worksheet. The VBA macro then uses the center-of-gravity coordinates as initial input for the Weiszfeld algorithm. Two hundred iterations of the algorithm are completed and the coordinates

	A	B	C	D	E	F	G	H	I	J	K	L
1	n =		22									
2											Weiszfeld Iterations	
3	j		x_j	y_j	r_j		Center of Gravity			1	20.31818	21.72727
4	1	Aachen	-57	28	1		u =	20.3182		2	16.69488	20.98140
5	2	Ausburg	54	-65	1		v =	21.7273		3	15.58790	20.70708
6	3	Bielefeld	0	71	1					4	15.24505	20.58605
7	4	Braunschweig	46	79	1		Weber			5	15.14019	20.52403
8	5	Bremen	8	111	1		u =	15.1033		6	15.10947	20.48822
9	6	Essen	-36	52	1		v =	20.4209		7	15.10147	20.46586
10	7	Freigburg	-22	-76	1					8	15.10013	20.45126
11	8	Hamburg	34	129	1					9	15.10052	20.44151
12	9	Hof	74	6	1		Run			10	15.10121	20.43492
13	10	Karlsruhe	-6	-41	1		Location Analysis			11	15.10180	20.43045
14	11	Kassel	21	45	1					12	15.10226	20.42741
15	12	Kiel	37	155	1					13	15.10258	20.42533
16	13	Koln	-38	35	1					14	15.10280	20.42391
17	14	Lubeck	50	140	1					15	15.10295	20.42295
18	15	Mannheim	-5	-24	1					16	15.10306	20.42229
19	16	Munchen	70	-74	1					17	15.10313	20.42184
20	17	Munster	-20	70	1					18	15.10318	20.42153
21	18	Nurnberg	59	-26	1					19	15.10321	20.42132
22	19	Passau	114	-56	1					20	15.10324	20.42118
23	20	Regensburg	83	-41	1					21	15.10325	20.42108
24	21	Saarbrucken	-40	-28	1					22	15.10326	20.42102
25	22	Wurzburg	21	-12	1					23	15.10327	20.42097
26	23									24	15.10328	20.42094
27	24									25	15.10328	20.42092

Figure 4.4.   Excel worksheet for Example 4.2 using the Weisfeld algorithm.

for each iteration are pasted in cells K3:L202. The coordinate values at iteration 200 are taken as the solution to the Weber problem and are written to cells I8 and I9.

4.1.6   *Example 4.2 — German towns' data*

We illustrate the macros using a larger test problem corresponding to coordinates for $n = 22$ German towns published by Späth (1980, pp. 43, 80). The towns and their coordinates are displayed in cells B4:D25 of the worksheet in Figure 4.4. Although Brusco (2022a) created some synthetic demand requirements for these towns, we will assume equal weighting by using 1.0 as the r_j value for each town.

We begin by running the macro for the Weisfeld algorithm for German towns' data. This is accomplished by clicking the 'Run Location Analysis' button, which produces the results shown in Figure 4.4. The optimal coordinates for the center-of-gravity problem are computed as $u = 20.3182$ and $v = 21.7273$. These are then used as the starting coordinates for the Weisfeld algorithm (see cells K3:L3). The Weisfeld algorithm converges rapidly to the optimal coordinates for the Weber problem, which are $u = 15.1033$ and $v = 20.4209$. So, in this particular example, the optimal y coordinates for the center-of-gravity and Weber solutions are very similar; however, the optimal x coordinate for the Weber problem is a good bit further west than the optimal x coordinate for the center-of-gravity problem.

Next, we can examine the convergence of Newton's algorithm using the Newton macro. We began by entering the center-of-gravity solution (i.e., $u = 20.3182$ and $v = 21.7273$) as the starting point for the algorithm in cells B2 and B3. Figure 4.5 displays the results

	A	B	C	D	E	F	G	H	I	J
1	Lambda	0.4								
2	u =	18.2415								
3	v =	21.2454					Run One Iteration of Newton's Method			
4					Gradient					
5	Hessian	.227	.021		.727					
6		.021	.107		.152					
7										
8	Inverse	4.479	-.882							
9	Hessian	-.882	9.512							
10										
11	n =	22								
12										
13	j	x_j	y_j	r_j						
14	Aachen	-57	28	1						
15	Ausburg	54	-65	1						
16	Bielefeld	0	71	1						
17	Braunschweig	46	79	1						
18	Bremen	8	111	1						
19	Essen	-36	52	1						
20	Freigburg	-22	-76	1						
21	Hamburg	34	129	1						
22	Hof	74	6	1						
23	Karlsruhe	-6	-41	1						
24	Kassel	21	45	1						
25	Kiel	37	155	1						
26	Koln	-38	35	1						
27	Lubeck	50	140	1						
28	Mannheim	-5	-24	1						
29	Munchen	70	-74	1						
30	Munster	-20	70	1						
31	Nurnberg	59	-26	1						
32	Passau	114	-56	1						
33	Regensburg	83	-41	1						
34	Saarbrucken	-40	-28	1						
35	Wurzburg	21	-12	1						
36										

Figure 4.5.   Excel worksheet for Example 4.2 after one iteration of Newton's algorithm.

after one iteration of Newton's algorithm, which is accomplished by clicking the 'Run One Iteration of Newton's Method' button. The new values in cells B2 and B3 are $u = 18.2415$ and $v = 21.2454$, respectively, which indicate that the movement toward the optimal Weber coordinates has begun.

As the button is repeatedly clicked, the coordinates begin to converge to the optimal Weber coordinates found by the Weisfeld algorithm and the gradient in cells E5:E6 converges to zero. The worksheet after tapping the button 20 or 30 times is shown in Figure 4.6.

	A	B	C	D	E	F	G	H	I	J
1	Lambda	0.4								
2	u =	15.1033								
3	v =	20.4209					Run One Iteration of Newton's Method			
4					Gradient					
5	Hessian	.225	.019		.000					
6		.019	.109		.000					
7										
8	Inverse	4.507	-.787							
9	Hessian	-.787	9.283							
10										
11	n =	22								
12										
13	j	x_j	y_j	r_j						
14	Aachen	-57	28	1						
15	Ausburg	54	-65	1						
16	Bielefeld	0	71	1						
17	Braunschweig	46	79	1						
18	Bremen	8	111	1						
19	Essen	-36	52	1						
20	Freigburg	-22	-76	1						
21	Hamburg	34	129	1						
22	Hof	74	6	1						
23	Karlsruhe	-6	-41	1						
24	Kassel	21	45	1						
25	Kiel	37	155	1						
26	Koln	-38	35	1						
27	Lubeck	50	140	1						
28	Mannheim	-5	-24	1						
29	Munchen	70	-74	1						
30	Munster	-20	70	1						
31	Nurnberg	59	-26	1						
32	Passau	114	-56	1						
33	Regensburg	83	-41	1						
34	Saarbrucken	-40	-28	1						
35	Wurzburg	21	-12	1						
36										

Figure 4.6. Excel worksheet for Example 4.2 after 30 iterations of Newton's algorithm.

The analyses in this section are useful for illustrating single-source continuous facility location. They highlight differences between the center-of-gravity problem and the Weber problem and also illustrate the mechanics of different estimation methods for the Weber problem. However, the key limitation in the section is the restriction to the case of a single-source facility.

4.2 Multisource Continuous Location

4.2.1 *Background*

Brusco (2022a) described a spreadsheet solution approach for multisource generalizations of the center-of-gravity and Weber problems. His description of the multisource problem uses some of the same notation described earlier for the single-source continuous location problem, namely, n demand locations in the plane with coordinates (x_j, y_j) for $1 \leq j \leq n$ and positive demand requirements (r_j) for each demand site $1 \leq j \leq n$. However, rather than a single-source facility, demand requirements are now to be met from a set of P *source* facilities with coordinates u_p and v_p for $1 \leq p \leq P$. Each of the n demand locations is assigned to its nearest source facility, which will satisfy the demand locations' requirements in their entirety. Like the single-source problem, the goal is to locate the source facilities to minimize the sum of requirement-weighted distance. Brusco (2022a) provided a general statement of the problem as follows:

$$\min_{u_p, v_p (1 \leq p \leq P)} Z = \sum_{j=1}^{n} \min_{(1 \leq p \leq P)} \left\{ r_j [|x_j - u_p|^\alpha + |y_j - v_p|^\alpha]^\beta \right\}. \quad (4.13)$$

The α and β parameters can define a wide range of distance measures but among the most common configurations are (i) $\alpha = \beta = 1$, for metropolitan distance (also known as Manhattan and rectilinear distance), (ii) $\alpha = 2$ and $\beta = 1$ for squared Euclidean distance, and (iii) $\alpha = 2$ and $\beta = 0.5$ for Euclidean distance. As noted by Kuo and White (2004, p. 220), the most suitable of these three distance measures depends on the application.

As its name implies, metropolitan distance is often suitable for travel within a large city. It may also be practical for location modeling within a warehouse facility. As noted previously, squared

Euclidean distance is typically harder to justify and Euclidean distance is arguably most common. Although the Excel spreadsheet we use for subsequent analyses in this section is built to accommodate metropolitan and squared Euclidean distance, we focus our attention solely on Euclidean distance. In the special case of $P = 1$ with $\alpha = 2$ and $\beta = 0.5$, Equation (4.13) reduces to the Weber problem. When $P > 1$, the problem is known as the *multisource Weber problem* (Brimberg *et al.*, 2000; Cooper, 1964), which is our focus here.

The Excel worksheet developed by Brusco (2022a) for multisource continuous location problems uses the Evolutionary engine of the Solver. Although the GRG Nonlinear engine might work sufficiently well for the special case of $P = 1$, the Evolutionary engine is generally more robust for the complex nonlinear functions associated with instances of $P > 1$. Of course, the price paid for selecting the Evolutionary engine over the GRG Nonlinear is greater computation time. A globally optimal solution to multisource location problems is not guaranteed from a single run of the Evolutionary engine and there is the potential for comparatively poor local optima. Although there remains no guarantee that a globally optimal solution will be found, we recommend rerunning the Evolutionary engine multiple times from different starting solutions to at least mitigate the chances of a poor local optimum.

4.2.2 *Example 4.3 — German towns' data*

Example 4.3 uses the same German towns' data in Example 4.2 but the focus is now on the use of $P > 1$ source facilities. The Excel worksheet for tackling the multisource problems is displayed in Figure 4.7 and the corresponding Solver dialog box is shown in Figure 4.8.

The worksheet is capable of solving multisource continuous location problems with up to five source facilities. The coordinates (u_j and v_j) for the five source facilities are located in cells B2:F3 of the worksheet displayed in Figure 4.7. These are the 'By Changing Variable Cells' in the Solver dialog box in Figure 4.8. Because the Evolutionary engine is selected as the Solver, it is best to set boundaries for the decision variables, which correspond to the minimum and maximum values of the x and y coordinates for the German towns. These boundaries are placed in cells G2:H3. For example, the

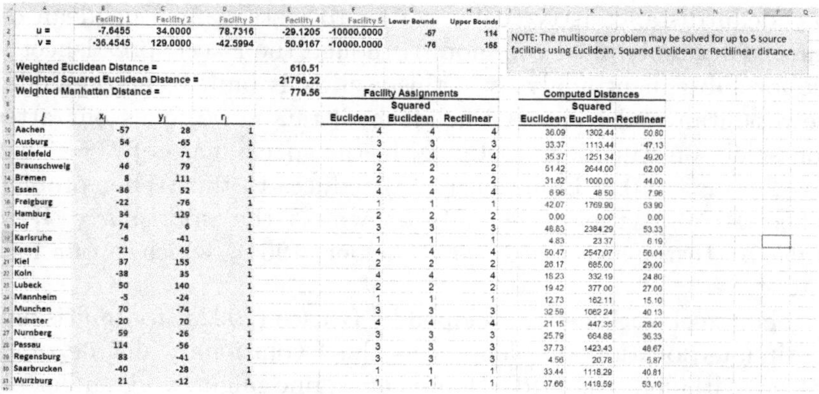

	Facility 1	Facility 2	Facility 3	Facility 4	Facility 5	Lower Bounds	Upper Bounds	
u =	-7.6455	34.0000	78.7316	-29.1205	-10000.0000	-67	114	NOTE: The multisource problem may be solved for up to 5 source
v =	-36.4545	129.0000	-42.5994	50.9167	-10000.0000	-76	155	facilities using Euclidean, Squared Euclidean or Rectilinear distance.

Weighted Euclidean Distance = 610.51
Weighted Squared Euclidean Distance = 21796.22
Weighted Manhattan Distance = 779.56

				Facility Assignments			Computed Distances		
					Squared			Squared	
i	x_i	y_i	r_i	Euclidean	Euclidean	Rectilinear	Euclidean	Euclidean	Rectilinear
Aachen	-57	28	1	4	4	4	36.09	1302.44	50.80
Ausburg	54	-65	1	3	3	3	33.37	1113.44	47.13
Bielefeld	0	71	1	4	4	4	35.37	1251.34	49.20
Braunschweig	46	79	1	2	2	2	51.42	2644.00	62.00
Bremen	8	111	1	2	2	2	31.62	1000.00	44.00
Essen	-36	52	1	4	4	4	6.96	48.50	7.96
Freigburg	-22	-76	1	1	1	1	42.07	1769.90	53.90
Hamburg	34	129	1	2	2	2	0.00	0.00	0.00
Hof	74	6	1	3	3	3	48.83	2384.29	53.33
Karlsruhe	-6	-41	1	1	1	1	4.83	23.37	6.19
Kassel	21	45	1	4	4	4	50.47	2547.07	56.04
Kiel	37	155	1	2	2	2	26.17	685.00	29.00
Koln	-38	35	1	4	4	4	18.23	332.19	24.80
Lubeck	50	140	1	2	2	2	19.42	377.00	27.00
Mannheim	-5	-24	1	1	1	1	12.73	162.11	15.10
Munchen	70	-74	1	3	3	3	32.59	1062.24	40.13
Munster	-20	70	1	4	4	4	21.15	447.35	28.20
Nurnberg	59	-26	1	3	3	3	25.79	664.88	36.33
Passau	114	-56	1	3	3	3	37.73	1423.43	48.67
Regensburg	83	-41	1	3	3	3	4.56	20.78	5.87
Saarbrucken	-40	-28	1	1	1	1	33.44	1118.29	40.81
Wurzburg	21	-12	1	1	1	1	37.66	1419.59	53.10

Figure 4.7.   Excel worksheet for Example 4.3 ($P = 4$ source facilities).

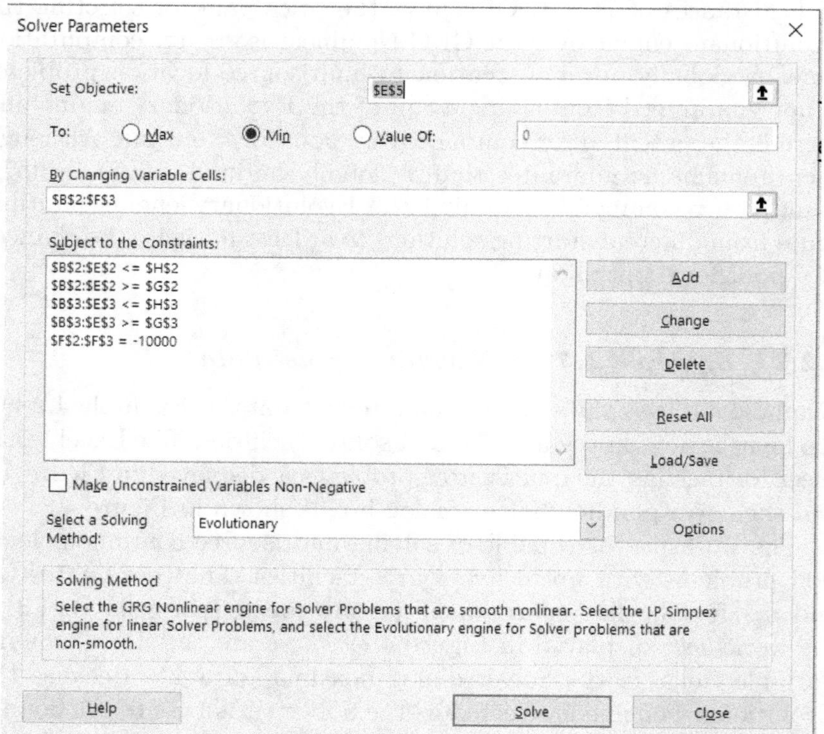

Figure 4.8.   Solver dialog box for Example 4.3 ($P = 4$ source facilities).

minimum x coordinate is –57 (Aachen) and the maximum x coordinate is 114 (Passau).

For implementation of the Evolutionary engine when using fewer than five facilities, the unused facilities are 'closed' by forcing their u_j and v_j values to be –10000 so that no demand locations could possibly be nearest (and therefore assigned) to those closed facilities. The Solver dialog box in Figure 4.8 is configured to run the analysis for $P = 4$ source facilities. Cells B2:E2 are constrained to fall within the boundaries for the x coordinates, whereas cells B3:E3 are constrained to fall within the boundaries for the y coordinates. Because source facility 5 should not be used, cells F2 and F3 are constrained to equal –10000. Cell E5, total weighted Euclidean distance, is the 'Set Objective' cell in the Solver dialog box.

We used the Evolutionary engine of the Excel to Solver to obtain solutions for all integer values of P on the interval $1 \leq P \leq 5$. The solution for $P = 1$ matched the solution obtained by Newton's method and the Weisfeld algorithm reported in Section 4.1.6. The minimum distance associated with that solution is 1722.13. Moving to $P = 2$ source facilities substantially reduces the travel distance to 1102.79. The $P = 3$ solution yielded a travel distance of 840.13. A further reduction of more than 200 was realized by increasing the number of source facilities to $P = 4$, where the travel distance was 610.51. Increasing to $P = 5$ source facilities only reduced the travel distance to 528.30.

We select the $P = 4$ source facility for interpretation based on several factors. First, there is a fairly strong reduction in travel distance when going from $P = 3$ to $P = 4$ facilities. Second, the reduction in travel distance when going from $P = 4$ to $P = 5$ facilities is comparatively modest. Third, the $P = 4$ source facility solution has an even distribution of towns to source facilities (each source facility gets either 5 or 6 towns). Fourth, we have observed that students have often selected the $P = 4$ solution when this problem has been used as a class exercise.

As noted previously, Figure 4.7 displays the $P = 4$ facility solutions. The assignment of towns to facilities is provided in cells F10:F31, and the Euclidean distances between each town and the source facility to which it is assigned are located in cells J10:J31. The sum of cells J10:J31 is 610.51, which is the value of the objective function in cell E5.

A visual display of the $P = 4$ source facility solution is provided in Figure 4.9. The approximate coordinates of the towns are shown as black dots with the name of the town adjacent to the dots. Lines are used to establish a rough sketch of a convex hull of the towns assigned to each source facility. The approximate coordinates for each source facility are labeled as S1, S2, S3, and S4. For example, source facility S1 is located at coordinates $u_1 = -7.6455$ and $v_1 = -36.4545$, and

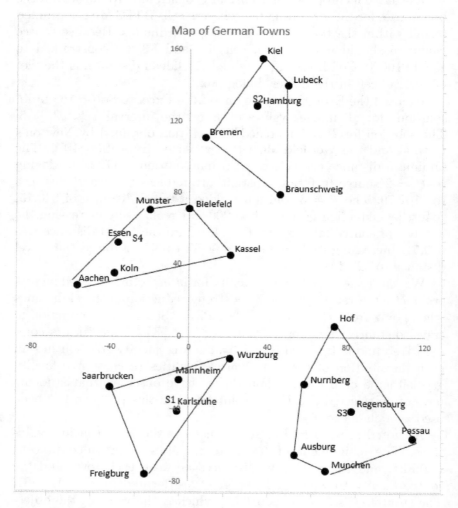

Figure 4.9.   The $P = 4$ source facility solution for Example 4.3.

the towns assigned to this source facility are Freigburg, Karlsruhe, Mannheim, Saarbrücken, and Würzburg.

The optimal location for source facility S1 is very close to the town of Karlsruhe ($x = -6, y = -41$). The optimal location for source facility S2 is identical to the coordinates for the town of Hamburg ($u_2 = x = 34, v_2 = y = 129$). The optimal location coordinates for source facility S3 ($u_3 = 78.7315, v_3 = -42.5995$) are close to the coordinates of the town of Regensburg ($x = 83, y = -41$). The optimal location coordinates for source facility S4 ($u_3 = -29.1205, v_3 = 50.9167$) are close to the coordinates of the town of Essen ($x = -36, y = 52$).

If the coordinates of the four source facilities are changed to the coordinates of their nearest town, then the total Euclidean distance increases by less than 1% to 615.10. This highlights an important point regarding the continuous location problem, namely, the goal is to establish an approximate location for facilities. The actual coordinates from the estimation process might be infeasible but modest adjustments often do not create a serious distance penalty.

Chapter 5

Discrete Facility Location

Chapter 4 focused on problems where source facilities could be located anywhere in the continuous two-dimensional space to service demand sites with explicit coordinates in the plane. In Chapter 5, the locations of source facilities are limited to discrete candidates (see Daskin, 2013 for a review of discrete location problems). In Section 5.1, we first describe the classic transportation problem (Hitchcock, 1941), where the goal is to minimize variable distribution costs. Although source facilities are predetermined in the basic transportation problem, we also consider the fixed-charge transportation problem. This problem introduces binary variables for *candidate* source facilities and seeks to trade off fixed and variable costs. Section 5.1 also considers a transshipment problem, whereby there are two stages of distribution (e.g., plants to regional distribution centers and distribution centers to local warehouses). These *two-echelon* problems are common in real-world settings (Camm *et al.*, 1997; Robinson *et al.*, 1993).

Section 5.2 describes two classic types of 'covering' location problems, which are common in the location of emergency services (Aktas *et al.*, 2013; Dekle *et al.*, 2005). The concept of 'covering' pertains to the proximity of a facility to a region (such as a zone or census tract within a county). For example, a zone might be defined as 'covered' by a facility if it is within three miles, or within five minutes of travel time. The first model considered in this section is the location set covering problem (Toregas & ReVelle, 1972; Toregas *et al.*, 1971), which seeks to find the minimum number of facilities necessary to ensure that all regions are covered by at least one facility. The second model is the maximal covering location problem (Church & ReVelle, 1974), which seeks to cover a maximum population given a fixed number of facilities.

Section 5.3 presents examples of three classic discrete plant location models: (i) the p-median problem (Hakimi, 1964, 1965), (ii) the p-centers problem (Hakimi, 1964), and (iii) the simple plant location problem (Balinski, 1965; Kuehn & Hamburger, 1963). These problems are important not only for their utility in facilities planning but also for their more general relevance to cluster analysis (Brusco & Steinley, 2015). We note that the number of variables and/or constraints for these three models may exceed the limits of the base system of the Excel Solver for even modestly-sized problems, at which point OpenSolver (Mason, 2012) or some other package with extended capabilities might be necessary.

5.1 Transportation and Transshipment Problems

5.1.1 *Example 5.1 — Transportation problem*

A firm has three locations for supply centers (SCs), 1, 2, and 3, that will ship units to eight existing demand centers (DCs), 1, 2, 3, 4, 5, 6, 7, and 8. Management has asked you to come up with a model that will determine the distribution of units from SCs to DCs. The optimal plan should satisfy the total demand at each DC exactly, not exceed the capacities of each SC, and minimize the sum of the fixed operating costs and variable costs of distribution.

	DC 1	DC 2	DC 3	DC 4	DC 5	DC 6	DC 7	DC 8	Capacity
SC 1	$1.80	$2.40	$2.10	$1.80	$2.10	$1.80	$1.70	$2.00	7200
SC 2	$1.30	$3.10	$1.60	$2.70	$1.90	$1.40	$1.60	$1.80	5200
SC 3	$1.50	$1.70	$1.80	$2.30	$2.20	$2.50	$1.90	$1.70	6100
Demand	2000	1700	2500	1300	2800	2700	1800	2200	

The linear programming formulation of this problem is as follows:

Indices

SC index goes from $i = 1$ to 3 and DC index goes from $j = 1$ to 8.

Parameters

a_{ij} = the variable cost of shipping from SC i to DC j;
p_i = the capacity of SC i;
d_j = the demand at DC j.

Decision variables

x_{ij} = the number of units shipped from SC i to DC j.

Objective function

$$\text{Minimize: } \sum_{i=1}^{3}\sum_{j=1}^{8} a_{ij}x_{ij}. \tag{5.1}$$

Subject to constraints

$$\sum_{j=1}^{8} x_{ij} \le p_i \quad \text{for } 1 \le i \le 3, \tag{5.2}$$

$$\sum_{i=1}^{3} x_{ij} = d_j \quad \text{for } 1 \le j \le 8. \tag{5.3}$$

The objective function in Equation (5.1) is the total variable cost of distribution, which is to be minimized. Constraint set (5.2) ensures that capacity is not exceeded for any SC, whereas constraint set (5.3) guarantees that demand will be satisfied for each DC. Figure 5.1 displays the Excel worksheet for Example 5.1, and the corresponding Solver dialog box is in Figure 5.2.

Figure 5.1 contains the cost parameters in cells B2:I4, the capacity parameters in cells L8:L10, and the demand parameters in cells B14:I14.

The decision variables, which are shown in the 'by Changing Cells' box in Figure 5.2, are located in cells B8:I10. The sumproduct of

	A	B	C	D	E	F	G	H	I	J	K	L
1		DC 1	DC 2	DC 3	DC 4	DC 5	DC 6	DC 7	DC 8			VARIABLE COST
2	SC 1	$1.80	$2.40	$2.10	$1.80	$2.10	$1.80	$1.70	$2.00			$28,690.00
3	SC 2	$1.30	$3.10	$1.60	$2.70	$1.90	$1.40	$1.60	$1.80			
4	SC 3	$1.50	$1.70	$1.80	$2.30	$2.20	$2.50	$1.90	$1.70			
5												
6											Actual	Potential
7		DC 1	DC 2	DC 3	DC 4	DC 5	DC 6	DC 7	DC 8		SC Supply	SC Capacity
8	SC 1	0	0	0	1,300	2,800	0	1,800	0		5,900	7,200
9	SC 2	0	0	2,500	0	0	2,700	0	0		5,200	5,200
10	SC 3	2,000	1,700	0	0	0	0	0	2,200		5,900	6,100
11												
12												
13	Supplied to	2,000	1,700	2,500	1,300	2,800	2,700	1,800	2,200			
14	Cust. Demand	2,000	1,700	2,500	1,300	2,800	2,700	1,800	2,200			

Figure 5.1. Excel worksheet for Example 5.1.

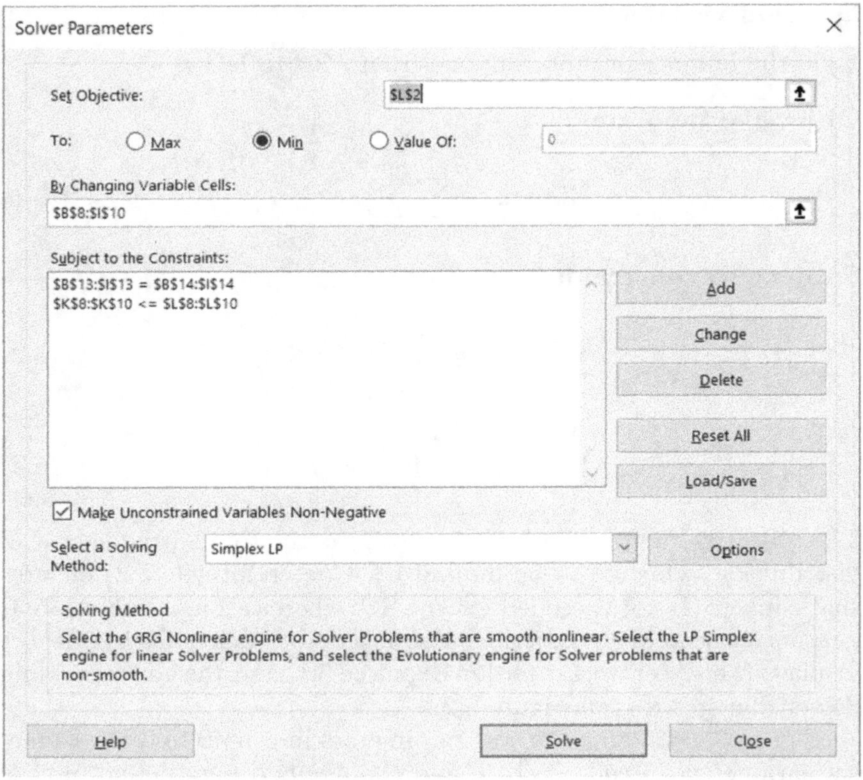

Figure 5.2.   Solver dialog box for Example 5.1.

the cost parameters and these decision variables is contained in cell L2, which is the 'Set Objective' cell in the Solver dialog box. Cells K8:K10 contain the row sums of the decision variables, representing the amount supplied by each SC. These cells are constrained to be equal to or less than the SC capacities in cells L8:L10. Cells B13:I13 contain the column sums of the decision variables, which represent the amount shipped to each DC. These cells are constrained to equal the DC requirements in cells B14:I14.

The optimal solution reveals that SC1 fully satisfies the requirements for DC4, DC5, and DC7. Similarly, SC2 fully satisfies the requirements for DC3 and DC6. Finally, SC3 fully satisfies the requirements for DC1, DC2, and DC8. The total variable cost of distribution for the shipment plan is $28,690.

5.1.2　*Example 5.2 — Fixed-charge transportation problem*

A firm has five *potential* locations for supply centers (SCs), 1, 2, 3, 4, and 5, that can ship units to eight existing demand centers (DCs), 1, 2, 3, 4, 5, 6, 7, and 8. The fixed operating costs for the potential SC locations 1, 2, 3, 4, and 5 are \$34,000, \$23,500, \$24,500, \$10,500, and \$27,500, respectively. Management has asked you to come up with a model that will determine which SCs to open and the distribution of units from SCs to DCs. The optimal plan should satisfy the total demand at each DC exactly, not exceed the capacities of each SC, and minimize the sum of the fixed operating costs and variable costs of distribution. An additional constraint is that if SC 2 is opened, then SC 4 must also be opened.

	DC 1	DC 2	DC 3	DC 4	DC 5	DC 6	DC 7	DC 8	Capacity
SC 1	\$1.80	\$2.40	\$2.10	\$1.80	\$2.10	\$1.80	\$1.70	\$2.00	7200
SC 2	\$1.30	\$3.10	\$1.60	\$2.70	\$1.90	\$1.40	\$1.60	\$1.80	5200
SC 3	\$1.50	\$1.70	\$1.80	\$2.30	\$2.20	\$2.50	\$1.90	\$1.70	6100
SC 4	\$1.90	\$2.00	\$2.30	\$1.60	\$2.40	\$2.00	\$1.50	\$2.40	3200
SC 5	\$2.40	\$2.50	\$3.10	\$2.90	\$1.40	\$2.20	\$1.90	\$2.20	5700
Demand	2000	1700	2500	1300	2800	2700	1800	2200	

Example 5.2 differs from Example 5.1 in two interrelated aspects. First, all of the SCs in Example 5.1 were assumed to be opened, whereas some of the SCs in Example 5.2 might be opened while others are not opened. Therefore, in Example 5.2, each SC requires a decision as to whether it is opened or not. Second, in addition to variable distribution costs, the fixed operating costs of the SCs are relevant and must be incorporated into the objective function. The integer linear programming formulation of the problem is as follows:

Indices

SC index goes from $i = 1$ to 5 and DC index goes from $j = 1$ to 8.

Parameters

a_{ij} = the variable cost of shipping from SC i to DC j;
p_i = the capacity of SC i;
f_i = the fixed operating cost of SC i;
d_j = the demand at DC j.

Decision variables

x_{ij} = the number of units shipped from SC i to DC j;
y_i = 1 if SC i is opened and 0 otherwise.

Objective function

$$\text{Minimize:} \quad \sum_{i=1}^{5} \sum_{j=1}^{8} a_{ij} x_{ij} + \sum_{i=1}^{5} f_i y_i. \tag{5.4}$$

Subject to constraints

$$\sum_{j=1}^{8} x_{ij} \leq p_i y_i \quad \text{for } 1 \leq i \leq 5. \tag{5.5}$$

$$\sum_{i=1}^{5} x_{ij} = d_j \quad \text{for } 1 \leq j \leq 8. \tag{5.6}$$

$$y_4 \geq y_2. \tag{5.7}$$

The objective function in Equation (5.4), which is to be minimized, is the total variable cost of distribution plus the fixed operating costs of the selected SCs. Constraint set (5.5) not only ensures that capacity is not exceeded for any opened SC but, via the inclusion of the y_i variable on the right side of the constraint, also ensures that the SCs that are not selected will not be allowed to supply any units to DCs. Constraint set (5.6) guarantees that demand will be satisfied for each DC. Constraint set (5.7) requires SC 4 to be opened if SC 2 is opened. Figure 5.3 displays the Excel worksheet for Example 5.2, and the corresponding Solver dialog box is in Figure 5.4.

Figure 5.3 contains the variable cost parameters in cells B2:I6, the fixed cost parameters in cells O10:O14, the *potential* capacity parameters in cells L10:L14, and the demand parameters in cells B18:I18. The x_{ij} variables are located in cells B10:I14 and y_i variables are in cells M10:M14. The total variable and fixed costs are in cell L2 and L3, respectively. The sum of cells L2 and L3 is in cell L4, which is the 'Set Objective' cell in the Solver dialog box. Cells K10:K14 contain the row sums of the decision variables, representing the amount supplied by each SC. These cells are constrained to be equal to or less than the *actual* SC capacities in cells N10:N14.

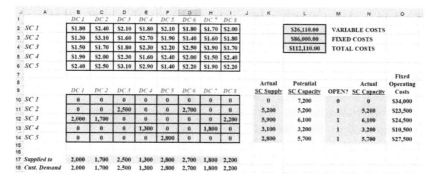

Figure 5.3.   Excel worksheet for Example 5.2.

Solver Parameters   ×

Se_t Objective: $L$4   ⬆

To: ○ _Max   ● Mi_n   ○ _Value Of: 0

_By Changing Variable Cells:

$B$10:$I$14,$M$10:$M$14   ⬆

Su_bject to the Constraints:

$K$10:$K$14 <= $N$10:$N$14
$M$10:$M$14 = binary   Add
$B$17:$I$17 = $B$18:$I$18
$M$13 >= $M$11   Change

                                                                   Delete

                                                                   Reset All

                                                                   Load/Save

☑ Mak_e Unconstrained Variables Non-Negative

S_elect a Solving   Simplex LP   Options
Method:

Solving Method

Select the GRG Nonlinear engine for Solver Problems that are smooth nonlinear. Select the LP Simplex
engine for linear Solver Problems, and select the Evolutionary engine for Solver problems that are
non-smooth.

        Help   _Solve   Cl_ose

Figure 5.4.   Solver dialog box for Example 5.2.

The actual capacities are the products of the potential capacities in L10:L14 and the binary SC selection variables in cells M10:M14 (in other words, potential capacity is only actualized if the SC is opened). B17:I17 contain the column sums of the x_{ij} variables, which represent the amount shipped to each DC. These cells are constrained to equal or exceed the DC requirements in cells B18:I18.

The optimal solution in Figure 5.4 reveals that four of the five SCs are opened (only SC 1 is not opened) and the total cost is $112,100. The requirements for each DC are met by exactly one of the SCs. Three of the four opened SCs supply at least two DCs; however, SC 5 only supplies DC 5. Therefore, there is substantial excess capacity (2,900 units) at SC 5. In light of this result, it might be interesting to evaluate the cost implications of adding constraints to ensure that the unused capacity for each of the opened SCs does not exceed 700 units. The necessary constraint set for this condition is shown in Equation 5.8 and the resulting solution after these constraints are appended is shown in Figure 5.5.

$$\sum_{j=1}^{8} x_{ij} \geq p_i y_i - 700 \quad \text{for } 1 \leq i \leq 5. \tag{5.8}$$

The addition of the capacity utilization constraints results in a cost increase of $3,510 to $115,620. Only three SCs are opened instead of four but fixed costs remain unchanged ($86,000) because SC 1 has replaced SC 2 and SC 4 in the solution and the sum for the fixed costs for SC 2 and SC 4 is equal to the fixed cost for SC 1

	A	B	C	D	E	F	G	H	I	J	K	L	M	N	O	P
1		DC 1	DC 2	DC 3	DC 4	DC 5	DC 6	DC 7	DC 8							
2	SC 1	$1.80	$2.40	$2.10	$1.80	$2.10	$1.80	$1.70	$2.00			$29,620.00	VARIABLE COSTS			
3	SC 2	$1.30	$3.10	$1.60	$2.70	$1.90	$1.40	$1.60	$1.80			$86,000.00	FIXED COSTS			
4	SC 3	$1.50	$1.70	$1.80	$2.30	$2.20	$2.50	$1.90	$1.70			$115,620.00	TOTAL COSTS			
5	SC 4	$1.90	$2.00	$2.30	$1.60	$2.40	$2.00	$1.50	$2.40							
6	SC 5	$2.40	$2.50	$3.10	$2.90	$1.40	$2.20	$1.90	$2.20							
7																
8											Actual	Potential		Actual	Fixed Operating	Actual
9		DC 1	DC 2	DC 3	DC 4	DC 5	DC 6	DC 7	DC 8		SC Supply	SC Capacity	OPEN?	SC Capacity	Costs	SC MinUse
10	SC 1	0	0	700	1,300	0	2,700	1,800	0		6,500	7200	1	7200	$34,000	6500
11	SC 2	0	0	0	0	0	0	0	0		0	5200	0	0	$23,500	0
12	SC 3	2,000	1,700	1,800	0	0	0	0	0		5,500	6100	1	6100	$24,500	5400
13	SC 4	0	0	0	0	0	0	0	0		0	3200	0	0	$10,500	0
14	SC 5	0	0	0	0	2,800	0	0	2,200		5,000	5700	1	5700	$27,500	5000
15																
16																
17	Supplied to	2,000	1,700	2,500	1,300	2,800	2,700	1,800	2,200							
18	Cust. Demand	2000	1700	2500	1300	2800	2700	1800	2200							

Figure 5.5. Excel worksheet for Example 5.2 with capacity utilization constraints.

(i.e., $23,500 + $10,500 = $34,000$). Despite the increase in variable costs, the solution in Figure 5.5 might be preferred to the one in Figure 5.3 because the one in Figure 5.5 uses fewer SCs with better capacity utilization.

5.1.3 *Example 5.3 — Two-echelon (transshipment) problem*

Consider a two-echelon supply chain consisting of three potential origins (plants), two transshipment nodes (distribution centers), and three destinations (warehouses). Plants (P1, P2, P3) can supply warehouses (W1, W2, W3) by shipping to the distribution centers (DC1, DC2) that will subsequently supply the warehouses. The plant capacities are 1800, 1400, and 1700 units for P1, P2, and P3, respectively. The distribution centers have no capacity limits. Demand at warehouses 1, 2, and 3 is 1200, 1000, and 800 units, respectively. Currently, there are 170 units stored at DC1 and 85 units stored at DC2. Management would like a distribution plan that leaves 120 units at both DCs after demand at each warehouse is satisfied. The fixed costs associated with plants P1, P2, and P3 are $3400, $2500, and $3600, respectively. The variable costs of shipping are shown in the following table. We want to find a distribution plan that meets all constraints and minimizes total variable distribution cost.

	DC1	DC2			W1	W2	W3
P1	17	13		DC1	6	3	9
P2	14	15		DC2	5	4	7
P3	12	18					

Indices

plant index goes from $i = 1, 2, 3$, DC index from $j = 1, 2$, and warehouse index from $k = 1, 2, 3$.

Parameters

a_{ij} = the variable cost of shipping from plant i to DC j;
b_{jk} = variable cost of shipping from DC j to warehouse k;
f_i = the fixed operating cost of plant i;
p_i = the capacity of plant i;
d_k = the demand at warehouse k.

Decision variables

x_{ij} = the number of units shipped from plant i to DC j;
v_{jk} = the number of units shipped from DC j to warehouse k;
y_i = 1 if plant i is opened and 0 otherwise.

Objective function

$$\text{Minimize:} \sum_{i=1}^{3}\sum_{j=1}^{2} a_{ij}x_{ij} + \sum_{j=1}^{2}\sum_{k=1}^{3} b_{jk}v_{jk} + \sum_{i=1}^{3} f_i y_i. \qquad (5.9)$$

Subject to constraints

$$\sum_{j=1}^{2} x_{ij} \leq p_i y_i \quad \text{for } 1 \leq i \leq 3. \qquad (5.10)$$

$$\sum_{j=1}^{2} v_{jk} = d_k \quad \text{for } 1 \leq k \leq 3. \qquad (5.11)$$

$$170 + x_{11} + x_{21} + x_{31} - y_{11} - y_{12} - y_{13} = 120. \qquad (5.12)$$

$$85 + x_{12} + x_{22} + x_{32} - y_{21} - y_{22} - y_{23} = 120. \qquad (5.13)$$

The objective function in Equation (5.9), which is to be minimized, is the total variable cost of distribution from plants to DCs, plus the variable cost of distribution from DCs to warehouses, plus the fixed operating costs of the selected plants. Constraint set (5.10) not only ensures that capacity is not exceeded for any opened plant but, via the inclusion of the y_i variable on the right side of the constraint, also ensures that plants that are not selected will not be allowed to supply units to DCs. Constraint set (5.11) guarantees that demand will be satisfied for each warehouse. Constraint sets (5.12) and (5.13) ensure that beginning inventory at the DCs plus whatever is shipped to them from plants is equal to whatever is shipped out of the DCs plus the desired ending inventory. Figure 5.6 displays the Excel worksheet for Example 5.3, and the corresponding Solver dialog box is in Figure 5.7.

Figure 5.6 contains the plant to DC cost parameters in cells B3:C5, the DC to warehouse cost parameters in cells B14:D15, the plant fixed cost parameters in cells E3:E5, the plant capacity parameters in cells H7:H9, and the demand parameters in cells B21:D21.

	A	B	C	D	E	F	G	H	I	J
1					Plant					
2		DC1	DC2		Fixed		Total Cost =			$59,455
3	P1	17	13		3400					
4	P2	14	15		2500					TRUE
5	P3	12	18		3600		Plant	Plant	Plant	Plant
6							Out	Cap	Open?	Cap
7	P1	0	1285				1285	1800	1	1800
8	P2	0	0				0	1400	0	0
9	P3	1700	0				1700	1700	1	1700
10										
11	DC in	1700	1285							
12										
13		W1	W2	W3						
14	DC1	6	3	9			TOT	TOT		
15	DC2	5	4	7			DC	DC		
16							OUT	IN		
17	DC1	750	1000	0			1870	1870		
18	DC2	450	0	800			1370	1370		
19										
20	Whse in	1200	1000	800						
21	Whse dem	1200	1000	800						

Figure 5.6. Excel worksheet for Example 5.3.

The x_{ij} variables are located in cells B7:C9, the v_{jk} variables in cells B17:D18, and the y_i variables in cells I7:I9. The total variable and fixed costs are in cell J2, which is the 'Set Objective' cell in the Solver dialog box. Cells G7:G9 contain the total number of units shipped out of each plant and these cells are constrained to be equal to or less than the true (or actual) plant capacities in cells J7:J9. The actual capacities are the products of the potential capacities in H7:H9 and the binary plant selection variables in cells I7:I9.

Cells B11:C11 contain the number of units shipped into the DCs. In cells H17:H18, these quantities are augmented by the beginning inventories to establish the number of units 'in' the DCs. Cells G17:G18 contain total shipments 'out' of the DCs plus the desired ending inventory levels of 120. As shown in the Solver dialog box in Figure 5.7, cells G17:G18 are constrained to equal cells H17:H18 so as to satisfy constraints (5.12) and (5.13). Cells B20:D20 contain the total shipments in the warehouses and these are constrained to equal warehouse requirements in cells B21:D21.

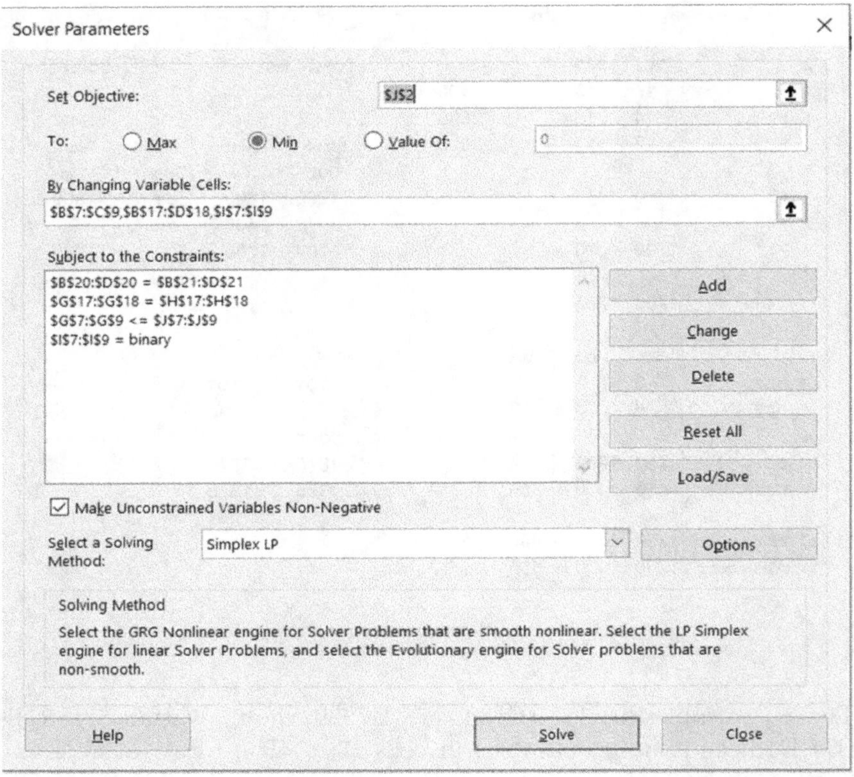

Figure 5.7. Solver dialog box for Example 5.3.

The optimal solution displayed in Figure 5.6 reveals that the two plants with the largest fixed costs (plants 1 and 3) are opened but plant 2 is not. Plant 1 ships 1285 units to DC 2, bringing the total number of available units at DC 2 to $1285 + 85 = 1370$. Plant 3 ships 1700 units to DC 1, bringing the total number of units available at DC 1 to $1700 + 170 = 1870$.

The demand of 1000 at warehouse 2 is met entirely from DC 1, whereas the demand of 800 at warehouse 3 is met entirely by DC 2. The demand of 1200 at warehouse 1 is met by 750 units from DC 1 and 450 units from DC 2. The total cost of the location distribution plan is \$59,455.

5.2 Covering Models

5.2.1 *Location set covering problem*

Parameters and sets

$I =$ a set of m 'demand' nodes, indexed $(1 \leq i \leq m)$;
$J =$ a set of n candidate location sites, indexed $(1 \leq j \leq n)$;
$a_{ij} =$ coverage parameters whereby $a_{ij} = 1$ if demand node i is within some threshold for travel time or distance from candidate location site j and $a_{ij} = 0$ otherwise, for all $i \in I$ and $j \in J$.

Decision variables

$y_j =$ binary decision variables such that $y_j = 1$ if candidate site j is selected as a facility location and $y_j = 0$ otherwise (for $1 \leq j \leq n$).

Objective function

$$\text{Minimize: } Z_{LSCP} = \sum_{j \in J} y_j. \tag{5.14}$$

Subject to constraints

$$\sum_{j \in J} a_{ij} y_j \geq 1 \quad \forall i \in I, \tag{5.15}$$

$$y_j \in \{0, 1\} \quad \forall j \in J. \tag{5.16}$$

The objective function in Equation (5.14) is to minimize the number of candidate sites selected. Constraint set (5.15) ensures that each demand node is 'covered' by at least one selected site. Constraint set (5.16) places binary restrictions on the decision variables. (Although constraint set (5.16) is implied by the variable definitions, it is not uncommon to see the binary restrictions explicitly presented in these types of formulations.)

5.2.2 *Example 5.4 — Location set covering problem example*

Here, we provide a specific example of a location set covering problem. Moreover, rather than the generic shorthand for the problem in

Section 5.2.1, here we will make use of a longhand presentation of the formulation to offer a different perspective.

A firm wishes to select from among eight potential locations for emergency service stations. The objective is to minimize the total number of stations opened, subject to constraints that ensure that all 12 county zones are covered. The goal is to prepare an integer linear programming formulation that will accomplish this task. The potential locations and the zones covered are shown in the following table.

Potential Emergency Service Location	Zones Covered
A	1, 2, 5, 9
B	2, 3, 4, 9, 10
C	1, 3, 5, 10
D	2, 6, 7, 11
E	3, 4, 10, 11
F	5, 8, 9, 12
G	6, 7, 8, 12
H	5, 6, 8, 9, 11

Decision variables

$y_j = 1$ if a station is opened at potential location j and 0 otherwise (for j = A, B, C, D, E, F, G, H).

Objective function

$$\text{Minimize: } y_A + y_B + y_C + y_D + y_E + y_F + y_G + y_H.$$

Subject to constraints

$$\text{Zone 1: } y_A + y_C \geq 1,$$

$$\text{Zone 2: } y_A + y_B + y_D \geq 1,$$

$$\text{Zone 3: } y_B + y_C + y_E \geq 1,$$

$$\text{Zone 4: } y_B + y_E + \geq 1,$$

$$\text{Zone 5: } y_A + y_C + y_F + y_H \geq 1,$$

$$\text{Zone 6: } y_D + y_G + y_H \geq 1,$$

$$\text{Zone 7: } y_D + y_G \geq 1,$$

	Loc A	Loc B	Loc C	Loc D	Loc E	Loc F	Loc G	Loc H		
	A	B	C	D	E	F	G	H	I	J	K
1		Loc A	Loc B	Loc C	Loc D	Loc E	Loc F	Loc G	Loc H		
2											# of Locations
3	Open?	1	0	0	0	1	0	1	0		3
4											
5											Covered
6	Zone 1	1		1							1
7	Zone 2	1	1		1						1
8	Zone 3		1	1		1					1
9	Zone 4	1				1					1
10	Zone 5	1		1			1		1		1
11	Zone 6					1		1	1		1
12	Zone 7					1		1			1
13	Zone 8						1	1	1		1
14	Zone 9	1	1				1		1		1
15	Zone 10		1	1		1					1
16	Zone 11				1	1			1		1
17	Zone 12						1	1			1
18											

Figure 5.8.   Excel worksheet for Example 5.4.

$$\text{Zone 8: } y_F + y_G + y_H \geq 1,$$

$$\text{Zone 9: } y_A + y_B + y_F + y_H \geq 1,$$

$$\text{Zone 10: } y_B + y_C + y_E \geq 1,$$

$$\text{Zone 11: } y_D + y_E + y_H \geq 1,$$

$$\text{Zone 12: } y_F + y_G \geq 1.$$

The Excel worksheet for this problem is displayed in Figure 5.8, and the corresponding Solver dialog box is in Figure 5.9.

Cells B3:I3 in Figure 5.8 contain the binary decision variables for the potential stations. The sum of these cells is in cell K3 and this is the objective function or 'Set Objective' cell in the Solver dialog box. The constraint coefficients indicating which zones are covered by which locations are in cells B6:I17. Cells K6:K17 contain the sumproduct of B3:I3 and the coefficients for each zone. These cells are constrained to equal or exceed one in the Solver dialog box to ensure that each zone is covered.

The optimal solution in Figure 5.6 reveals that a minimum of three sites is needed to cover all of the zones. The optimal set of candidate sites are (i) site A (covering zones 1, 2, 5, 9), (ii) site E (covering zones 3, 4, 10, 11), and (iii) site G (covering zones 6, 7, 8, 12). Two observations are noteworthy. First, the optimal solution in Figure 5.6 is unique; however, many location set covering problems

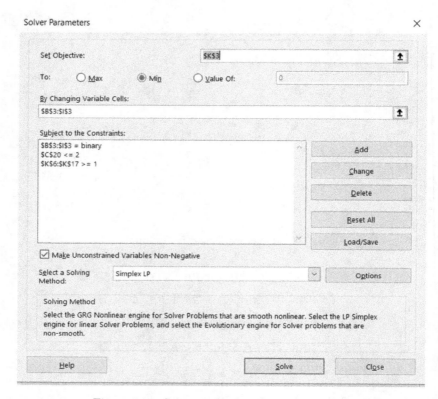

Figure 5.9.　Solver dialog box for Example 5.4.

will have multiple optimal solutions. Second, it is likely that this problem could be solved using trial and error within a reasonable amount of time. However, that is not an effective strategy for larger instances found in practice (cf. Dekle *et al.*, 2005).

5.2.3　*Maximal covering location problem*

Parameters and sets

$I = $ a set of m 'demand' nodes, indexed $(1 \leq i \leq m)$;
$J = $ a set of n candidate location sites, indexed $(1 \leq j \leq n)$;
$a_{ij} = $ coverage parameters whereby $a_{ij} = 1$ if demand node i is within some threshold for travel time or distance from candidate location site j and $a_{ij} = 0$ otherwise, for all $i \in I$ and $j \in J$;

g_i = a measure of demand at node i, for $(1 \leq i \leq m)$;
p = the number of facilities to be located.

Decision variables

y_j = binary decision variables such that $y_j = 1$ if candidate site j is selected as a facility location and $y_j = 0$ otherwise (for $1 \leq j \leq n$);

v_i = binary decision variables for demand node coverage, whereby $v_i = 1$ if demand node i is covered and 0 otherwise (for $i \in I$).

Objective function

$$\text{Maximize: } Z_{MCLP} = \sum_{i \in I} g_i v_i. \tag{5.17}$$

Subject to constraints

$$v_i - \sum_{j \in J} a_{ij} y_j \leq 0 \quad \forall i \in I, \tag{5.18}$$

$$\sum_{j \in J} y_j = p, \tag{5.19}$$

$$0 \leq v_i \leq 1 \quad \forall i \in I, \tag{5.20}$$

$$y_j \in \{0, 1\} \quad \forall j \in J. \tag{5.21}$$

The objective function in Equation (5.17) is to maximize the total population covered. Constraint set (5.18) ensures that the coverage variable for each demand node can only assume a value of one if the node is 'covered' by at least one selected site. Constraint (5.19) restricts the number of opened sites to a predetermined value, p. Constraint set (5.20) places bounds on the coverage variables and constraint set (5.21) places binary restrictions on the site open/close decision variables.

5.2.4 Example 5.5 — Maximal covering location problem example

A firm wishes to select two locations from among eight potential locations for emergency service stations. The objective is to maximize the total population covered across 12 zones. The zone populations (in thousands) are as follows: zone 1 — 11.4, zone 2 — 5.6, zone 3 — 13.8, zone 4 — 12.8, zone 5 — 12.7, zone 6 — 9.7, zone 7 — 14.1, zone 8 — 9.9, zone 9 — 3.6, zone 10 — 14.4, zone 11 — 4.9, and zone 12 — 14.2. We want to prepare an integer linear programming formulation that will accomplish this task. The potential locations and the zones covered are shown in the following table and a longhand version of the formulation follows.

Potential Emergency Service Location	Zones Covered
A	1, 2, 5, 9
B	2, 3, 4, 9, 10
C	1, 3, 5, 10
D	2, 6, 7, 11
E	3, 4, 10, 11
F	5, 8, 9, 12
G	6, 7, 8, 12
H	5, 6, 8, 9, 11

Decision variables

$y_j = 1$ if a station is opened at potential location j and 0 otherwise (for j = A, B, C, D, E, F, G, H).

$v_i = 1$ if zone i is covered and 0 otherwise (for $i = 1, 2, \ldots, 12$).

Objective function

Minimize: $11.4v_1 + 5.6v_2 + 13.8v_3 + 12.8v_4 + 12.7v_5 + 9.7v_6$

$$+ 14.1v_7 + 9.9v_8 + 3.6v_9 + 14.4v_{10} + 4.9v_{11} + 14.2v_{12}.$$

Subject to constraints

$$y_A + y_B + y_C + y_D + y_E + y_F + y_G + y_H = 2,$$

Zone 1: $y_A + y_C \geq v_1,$

Zone 2: $y_A + y_B + y_D \geq v_2,$

Zone 3: $y_B + y_C + y_E \geq v_3,$

Zone 4: $y_B + y_E + \geq v_4,$

Zone 5: $y_A + y_C + y_F + y_H \geq v_5,$

Zone 6: $y_D + y_G + y_H \geq v_6$,

Zone 7: $y_D + y_G \geq v_7$,

Zone 8: $y_F + y_G + y_H \geq v_8$,

Zone 9: $y_A + y_B + y_F + y_H \geq v_9$,

Zone 10: $y_B + y_C + y_E \geq v_{10}$,

Zone 11: $y_D + y_E + y_H \geq v_{11}$,

Zone 12: $y_F + y_G \geq v_{12}$.

The Excel worksheet for this problem is displayed in Figure 5.10, and the corresponding Solver dialog box is in Figure 5.11.

Cells B3:I3 in Figure 5.10 contain the binary decision variables for the potential stations. Cells L6:L17 contain the variables indicating whether or not a zone is covered. Cell K3, the objective function or 'Set Objective' cell, contains the total population covered, which is the sum product of L6:L17 and the zone populations in cells M6:M17. Cells K6:K17 are the sumproducts of the site selection variables in B3:I3 and the coverage constraint coefficients (cells B6:I17) for each zone. The coverage variables in cells L6:L17 are constrained to be equal to or less than K6:K17. The sum of the station selection variables is in cell M3, and this sum is constrained to equal 2, that is, the number of stations to select.

The optimal solution in Figure 5.10 reveals that the two stations that are opened, C and G, cover only eight of the 12 zones.

	A	B	C	D	E	F	G	H	I	J	K	L	M
1		Loc A	Loc B	Loc C	Loc D	Loc E	Loc F	Loc G	Loc H				
2											Pop Covered		Opened
3	Open?	0	0	1	0	0	0	1	0		100.2		2
4													
5											Covers	Covered?	Pop
6	Zone 1	1		1							1	1	11.4
7	Zone 2	1	1		1						0	0	5.6
8	Zone 3		1	1		1					1	1	13.8
9	Zone 4	1			1						0	0	12.8
10	Zone 5	1		1		1		1			1	1	12.7
11	Zone 6			1			1	1			1	1	9.7
12	Zone 7			1				1			1	1	14.1
13	Zone 8						1	1	1		1	1	9.9
14	Zone 9	1	1				1		1		0	0	3.6
15	Zone 10		1	1		1					1	1	14.4
16	Zone 11				1	1			1		0	0	4.9
17	Zone 12						1	1			1	1	14.2
18													

Figure 5.10. Excel worksheet for Example 5.5.

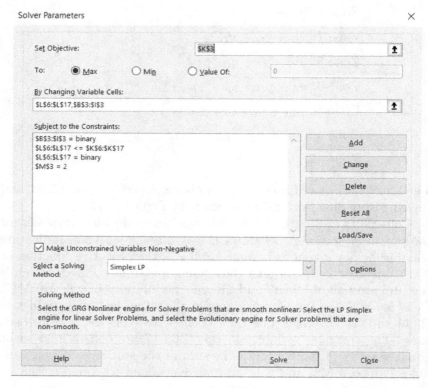

Figure 5.11.   Solver dialog box for Example 5.5.

Zones 2, 4, 9, and 11 are not covered. However, three of these four zones (Zone 4 being the exception) have very small population figures. Thus, although stations C and G cover only eight zones, they tend to be high-population zones resulting in a population cover of 100.2. By contrast, locations sites B and H will jointly cover 9 zones, but the total population covered is only 87.4.

5.3   Other Discrete Plant Location Models

The distances between the five source (SC1, SC2, SC3, SC4, SC5) and eight demand (DC1, DC2, DC3, DC4, DC5, DC6, DC7, DC8) facilities in the three examples in this section (Examples 5.6, 5.7, and 5.8) are Euclidean distances computed based on the following set of (X, Y) coordinates:

	SC1	SC2	SC3	SC4	SC5	DC1	DC2	DC3	DC4	DC5	DC6	DC7	DC8
X	22	37	36	91	91	12	60	65	6	55	82	80	37
Y	38	49	74	37	66	10	96	27	66	1	53	53	15

The Euclidean distances in the distance tables and formulations are rounded. However, the Excel worksheets and the reported results preserve the greater precision in the distances.

5.3.1 *p-median model*

Parameters and sets

$I =$ a set of m 'demand' nodes, indexed $(1 \le i \le m)$;

$J =$ a set of n candidate location sites for source facilities, indexed $(1 \le j \le n)$;

$d_{ij} =$ a measure of distance or cost between demand node i and candidate site j (for $1 \le i \le m$ and $1 \le j \le n$);

$g_i =$ a measure of demand at node i, for $(1 \le i \le m)$;

$h_j =$ a measure of capacity at candidate site j, for $(1 \le j \le n)$;

$p =$ the number of facilities to be located.

Decision variables

$y_j =$ binary decision variables such that $y_j = 1$ if candidate site j is selected as a facility location and $y_j = 0$ otherwise (for $1 \le j \le n$);

$x_{ij} =$ binary decision variables such that $x_{ij} = 1$ if demand node i is assigned to candidate site j and $x_{ij} = 0$ otherwise (for $1 \le i \le m$ and $1 \le j \le n$).

Objective function

$$\text{Minimize: } Z_{PMP} = \sum_{i \in I} \sum_{j \in J} d_{ij} x_{ij}. \tag{5.22}$$

Subject to constraints

$$\sum_{j \in J} x_{ij} = 1 \quad \forall i \in I, \tag{5.23}$$

$$x_{ij} \le y_j \quad \forall i \in I, \ j \in J, \tag{5.24}$$

$$\sum_{j \in J} y_j = p, \tag{5.25}$$

$$\sum_{i \in I} g_i x_{ij} \leq h_j \quad \forall j \in J, \tag{5.26}$$

$$x_{ij} \in \{0,1\} \quad \forall i \in I, \ j \in J, \tag{5.27}$$

$$y_j \in \{0,1\} \quad \forall j \in J. \tag{5.28}$$

The objective function of the p-median model in Equation (5.22) is to minimize the sum of the distances of the demand nodes to the source facility to which they are assigned. Constraint set (5.23) ensures that each demand node is assigned to exactly one source facility. Constraint set (5.24) guarantees that a demand node can only be assigned to a candidate source facility that has been selected (i.e., opened). Constraint (5.25) requires that exactly p source facilities are selected.

Constraint set (5.26) is not part of the p-median model in its most basic form. However, this constraint set is relevant for the *capacitated* p-median model where the supply capacities of the source facilities are limited. Specifically, constraint set (5.26) requires that the sum of the requirements of the demand nodes assigned to a particular source facility is less than the capacity of that source facility. Constraint sets (5.27) and (5.28) place binary restrictions on the decision variables.

5.3.2   *Example 5.6 — p-median example*

Here, we provide a specific example of a capacitated p-median problem. Rather than the generic shorthand for the problem in Section 5.3.1, here we will make use of a longhand presentation of the formulation to offer a different perspective.

A firm has five *potential* locations for supply centers (SCs), 1, 2, 3, 4, and 5, that can supply eight existing demand centers (DCs), 1, 2, 3, 4, 5, 6, 7, and 8. The distances (in miles) between each DC and each SC are shown in the following table, as are the capacities for the SCs and demand levels for the DCs. The management would like to select three SCs to open and assign each DC to one of the three opened SCs (the demand at each DC will be met entirely by the SC to which it is assigned). The goal is to minimize total distance. The optimal plan should ensure that the capacities of the SCs are not exceeded.

	DC 1	DC 2	DC 3	DC 4	DC 5	DC 6	DC 7	DC 8	Capacity
SC 1	30	69	44	32	50	62	60	27	7200
SC 2	46	52	36	35	51	45	43	34	5200
SC 3	68	33	55	31	75	51	49	59	6100
SC 4	83	67	28	90	51	18	19	58	4400
SC 5	97	43	47	85	74	16	17	74	4200
Demand	2000	1700	2500	1300	2800	2700	1800	2200	

Decision variables

$x_{jk} = 1$ if DC k is assigned to SC j and 0 otherwise, for $j = 1$ to 5 and $k = 1$ to 8;

$y_j = 1$ if SC j is opened and 0 otherwise (for $j = 1$ to 5).

Objective function

Minimize: $30x_{11} + 69x_{12} + 44x_{13} + 32x_{14} + 50x_{15} + 62x_{16} + 60x_{17}$

$$+ 27x_{18} + 46x_{21} + 52x_{22} + 36x_{23} + 35x_{24} + 51x_{25} + 45x_{26}$$

$$+ 43x_{27} + 34x_{28} + 68x_{31} + 33x_{32} + 55x_{33} + 31x_{34} + 75x_{35}$$

$$+ 51x_{36} + 49x_{37} + 59x_{38} + 83x_{41} + 67x_{42} + 28x_{43} + 90x_{44}$$

$$+ 51x_{45} + 18x_{46} + 19x_{47} + 58x_{48} + 97x_{51} + 43x_{52} + 47x_{53}$$

$$+ 85x_{54} + 74x_{55} + 16x_{56} + 17x_{57} + 74x_{58}$$

Subject to constraints

$y_1 + y_2 + y_3 + y_4 + y_5 = 3$

$x_{11} + x_{21} + x_{31} + x_{41} + x_{51} = 1 \quad x_{12} + x_{22} + x_{32} + x_{42} + x_{52} = 1$

$x_{13} + x_{23} + x_{33} + x_{43} + x_{53} = 1 \quad x_{14} + x_{24} + x_{34} + x_{44} + x_{54} = 1$

$x_{15} + x_{25} + x_{35} + x_{45} + x_{55} = 1 \quad x_{16} + x_{26} + x_{36} + x_{46} + x_{56} = 1$

$$x_{17} + x_{27} + x_{37} + x_{47} + x_{57} = 1 \quad x_{18} + x_{28} + x_{38} + x_{48} + x_{58} = 1$$

$$x_{jk} \le y_j \text{ for all } j = 1 \text{ to } 5 \text{ and } k = 1 \text{ to } 8.$$

$$2000x_{11} + 1700x_{12} + 2500x_{13} + 1300x_{14} + 2800x_{15} + 2700x_{16}$$
$$+ 1800x_{17} + 2200x_{18} \le 7200$$

$$2000x_{21} + 1700x_{22} + 2500x_{23} + 1300x_{24} + 2800x_{25} + 2700x_{26}$$
$$+ 1800x_{27} + 2200x_{28} \le 5200$$

$$2000x_{31} + 1700x_{32} + 2500x_{33} + 1300x_{34} + 2800x_{35} + 2700x_{36}$$
$$+ 1800x_{37} + 2200x_{38} \le 6100$$

$$2000x_{41} + 1700x_{42} + 2500x_{43} + 1300x_{44} + 2800x_{45} + 2700x_{46}$$
$$+ 1800x_{47} + 2200x_{48} \le 4400$$

$$2000x_{51} + 1700x_{52} + 2500x_{53} + 1300x_{54} + 2800x_{55} + 2700x_{56}$$
$$+ 1800x_{57} + 2200x_{58} \le 4200.$$

The Excel worksheet for this problem is displayed in Figure 5.12, and the corresponding Solver dialog box is in Figure 5.13.

Figure 5.12 contains the Euclidean distances between SCs and DCs in cells B2:I6, the SC capacity parameters in cells L10:L14, and the demand parameters in cells B17:I17. The DC to SC assignment variables are located in cells B10:I14 and the variables controlling the

	A	B	C	D	E	F	G	H	I	J	K	L	M	N
1		DC 1	DC 2	DC 3	DC 4	DC 5	DC 6	DC 7	DC 8					
2	SC 1	29.73	69.34	44.38	32.25	49.58	61.85	59.91	27.46			268.22	DISTANCE	
3	SC 2	46.32	52.33	35.61	35.36	51.26	45.18	43.19	34.00					
4	SC 3	68.35	32.56	55.23	31.05	75.43	50.57	48.75	59.01					
5	SC 4	83.49	66.65	27.86	89.81	50.91	18.36	19.42	58.31					
6	SC 5	96.83	43.14	46.87	85.00	74.30	15.81	17.03	74.28					
7														
8											Actual	Potential		
9		DC 1	DC 2	DC 3	DC 4	DC 5	DC 6	DC 7	DC 8		SC Supply	SC Capacity	OPEN?	
10	SC 1	1	0	0	0	1	0	0	1		7,000	7,200	1	
11	SC 2	0	0	0	0	0	0	0	0		0	5,200	0	
12	SC 3	0	1	0	1	0	1	0	0		5,700	6,100	1	
13	SC 4	0	0	1	0	0	0	1	0		4,300	4,400	1	
14	SC 5	0	0	0	0	0	0	0	0		0	4,200	0	
15														
16														
17	Cust. Demand	2,000	1,700	2,500	1,300	2,800	2,700	1,800	2,200				3	
18	DC Assigned	1	1	1	1	1	1	1	1					

Figure 5.12.   Excel worksheet for Example 5.6.

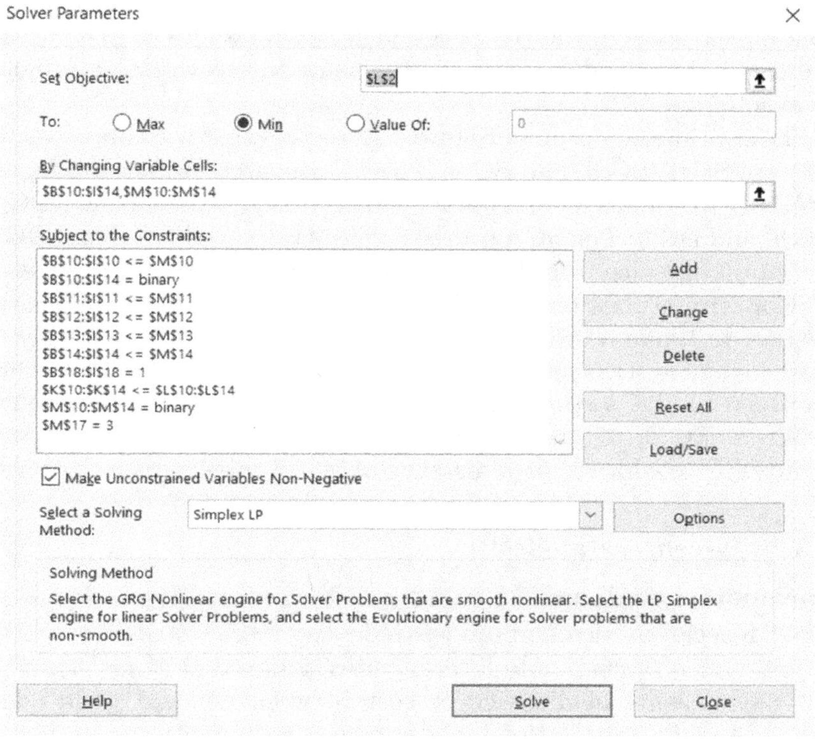

Solver Parameters ✕

Se<u>t</u> Objective: `$L$2|` ⬆

To: ◯ <u>M</u>ax ⦿ Mi<u>n</u> ◯ <u>V</u>alue Of: 0

<u>B</u>y Changing Variable Cells:

`$B$10:$I$14,$M$10:$M$14` ⬆

S<u>u</u>bject to the Constraints:

```
$B$10:$I$10 <= $M$10
$B$10:$I$14 = binary
$B$11:$I$11 <= $M$11
$B$12:$I$12 <= $M$12
$B$13:$I$13 <= $M$13
$B$14:$I$14 <= $M$14
$B$18:$I$18 = 1
$K$10:$K$14 <= $L$10:$L$14
$M$10:$M$14 = binary
$M$17 = 3
```

<u>A</u>dd
<u>C</u>hange
<u>D</u>elete
<u>R</u>eset All
<u>L</u>oad/Save

☑ Ma<u>k</u>e Unconstrained Variables Non-Negative

S<u>e</u>lect a Solving Method: Simplex LP ⌄ O<u>p</u>tions

Solving Method

Select the GRG Nonlinear engine for Solver Problems that are smooth nonlinear. Select the LP Simplex engine for linear Solver Problems, and select the Evolutionary engine for Solver problems that are non-smooth.

<u>H</u>elp <u>S</u>olve Cl<u>o</u>se

Figure 5.13.   Solver dialog box for Example 5.6.

opening of SCs are in cells M10:M14. The total distance between DCs and the SCs to which they are assigned is computed as the sumproduct of B2:I6 and B10:I14 in cell L2, which is the 'Set Objective' cell in the Solver dialog box in Figure 5.13.

Cells K10:K14 contain the sumproducts of the DC to SC assignment variables and demand, representing the amount supplied by each SC. These cells are constrained to be equal to or less than the SC capacities in cells L10:L14. The sum of the binary variables in cells M10:M14 is located in cell M17 and this cell is constrained to equal the number of SCs that should be opened, which is three for this example. Each variable in the matrix defined by cells B10:I14 must be constrained to be equal to or less than the binary variable for its corresponding row in cells M10:M14. For example, the constraint B10:I10 <= M10 in the Solver dialog box ensures that none of the

DCs can be assigned to SC 1 unless SC 1 is opened. Cells B18:I18 contain the sums of the DC assignment variables for each DC. These cells are constrained to equal one to ensure that each DC is assigned to exactly one SC.

The optimal solution in Figure 5.12 reveals that the three selected SCs are SC 1, SC 3, and SC 4. The SC 1 source facility is assigned DC 1, DC 5, and DC 8. The SC 3 source facility is assigned DC 2, DC 4, and DC 6. The SC 4 source facility is assigned DC 3 and DC 7. The total distance from DCs to their assigned SC is 268.22 miles.

Capacity utilization of the SCs is very good. The excess capacities for SCs 1, 3, and 4 are only 200, 400, and 100 units, respectively. In most instances, DCs are assigned to their nearest SC. The exceptions are DC 6 and DC 7. In particular, DC 6 is closer to three other source facilities (SC 2, SC 4, and SC 5) than it is to its assigned source facility SC 3, which is at a distance of 50.57 miles.

5.3.3   p-centers model

Parameters and sets

$I =$ a set of m 'demand' nodes, indexed $(1 \leq i \leq m)$;

$J =$ a set of n candidate location sites, indexed $(1 \leq j \leq n)$;

$d_{ij} =$ a measure of distance or cost between demand node i and candidate site j (for $1 \leq i \leq m$ and $1 \leq j \leq n$)

$g_i =$ a measure of demand at node i, for $(1 \leq i \leq m)$;

$h_j =$ a measure of capacity at candidate site j, for $(1 \leq j \leq n)$;

$p =$ the number of facilities to be located.

Decision variables

$y_j =$ binary decision variables such that $y_j = 1$ if candidate site j is selected as a facility location and $y_j = 0$ otherwise (for $1 \leq j \leq n$);

$x_{ij} =$ binary decision variables such that $x_{ij} = 1$ if demand node i is assigned to candidate site j and $x_{ij} = 0$ otherwise (for $1 \leq i \leq m$ and $1 \leq j \leq n$);

$C =$ a continuous variable corresponding to the maximum (across all demand nodes) distance/cost from the demand node to its assigned facility.

Objective function

$$\text{Minimize: } Z_{PCP} = C. \tag{5.29}$$

Subject to constraints

$$\sum_{j \in J} x_{ij} = 1 \quad \forall i \in I, \tag{5.30}$$

$$x_{ij} \le y_j \quad \forall i \in I, \ j \in J, \tag{5.31}$$

$$\sum_{j \in J} y_j = p, \tag{5.32}$$

$$C \ge \sum_{j \in J} d_{ij} x_{ij} \quad \forall i \in I, \tag{5.33}$$

$$\sum_{i \in I} g_i x_{ij} \le h_j \quad \forall j \in J, \tag{5.34}$$

$$x_{ij} \in \{0, 1\} \quad \forall i \in I, \ j \in J, \tag{5.35}$$

$$y_j \in \{0, 1\} \quad \forall j \in J. \tag{5.36}$$

The objective function of the p-centers model in Equation (5.29) is to minimize the maximum of the distances of the demand nodes to the source facility to which they are assigned. Constraint set (5.30) ensures that each demand node is assigned to exactly one source facility. Constraint set (5.31) guarantees that a demand node can only be assigned to a candidate source facility that has been selected (i.e., opened). Constraint (5.32) requires that exactly p source facilities are selected. Constraint set (5.33) ensures that C equals or exceeds the distances of all demand nodes from their source facility and, therefore, C will correspond to the maximum of the distances across all demand nodes.

Constraint set (5.34) is not part of the p-centers model in its most basic form. However, this constraint set is relevant for the *capacitated* p-centers model where the supply capacities of the source facilities are limited. Specifically, constraint set (5.34) requires that the sum of the requirements of the demand nodes assigned to a particular source facility is less than the capacity of that source facility. Constraint sets (5.35) and (5.36) place binary restrictions on the decision variables.

5.3.4 *Example 5.7 — p-centers example*

Like the p-median example in Section 5.3.2., we will present a long-hand version of the p-centers formulation for Example 5.7.

A firm has five *potential* locations for supply centers (SCs), 1, 2, 3, 4, and 5, that can supply eight existing demand centers (DCs), 1, 2, 3, 4, 5, 6, 7, and 8. The distance (in miles) of assigning each DC to each SC is shown in the table, as are the capacities for the SCs and demand levels for the DCs. [Note: The distances in the table that are written in longhand in the formulation are rounded but the Excel worksheet preserved the precision of the measurements.] The management would like to select three SCs to open and assign each DC to one of the three opened SCs (the demand at each DC will be met entirely by the SC to which it is assigned). The goal is to minimize the maximum distance of a DC to its SC. The optimal plan should ensure that the capacities of the SCs are not exceeded.

	DC 1	DC 2	DC 3	DC 4	DC 5	DC 6	DC 7	DC 8	Capacity
SC 1	30	69	44	32	50	62	60	27	7200
SC 2	46	52	36	35	51	45	43	34	5200
SC 3	68	33	55	31	75	51	49	59	6100
SC 4	83	67	28	90	51	18	19	58	4400
SC 5	97	43	47	85	74	16	17	74	4200
Demand	2000	1700	2500	1300	2800	2700	1800	2200	

Decision variables

$x_{jk} = 1$ if DC k is assigned to SC j and 0 otherwise, for $j = 1$ to 5 and $k = 1$ to 8;

$y_j = 1$ if SC j is opened and 0 otherwise (for $j = 1$ to 5).

Objective function

$$\text{Minimize: } C$$

Subject to constraints

$$y_1 + y_2 + y_3 + y_4 + y_5 = 3$$

$$x_{11} + x_{21} + x_{31} + x_{41} + x_{51} = 1 \quad x_{12} + x_{22} + x_{32} + x_{42} + x_{52} = 1$$

$$x_{13} + x_{23} + x_{33} + x_{43} + x_{53} = 1 \quad x_{14} + x_{24} + x_{34} + x_{44} + x_{54} = 1$$

$$x_{15} + x_{25} + x_{35} + x_{45} + x_{55} = 1 \quad x_{16} + x_{26} + x_{36} + x_{46} + x_{56} = 1$$

$$x_{17} + x_{27} + x_{37} + x_{47} + x_{57} = 1 \quad x_{18} + x_{28} + x_{38} + x_{48} + x_{58} = 1$$

$$x_{jk} \leq y_j \quad \text{for all } j = 1 \text{ to } 5 \text{ and } k = 1 \text{ to } 8.$$

$$C \geq 30x_{11} + 46x_{21} + 68x_{31} + 83x_{41} + 97x_{51}$$

$$C \geq 69x_{12} + 52x_{22} + 33x_{32} + 67x_{42} + 43x_{52}$$

$$C \geq 44x_{13} + 36x_{23} + 55x_{33} + 28x_{43} + 47x_{53}$$

$$C \geq 32x_{14} + 35x_{24} + 31x_{34} + 90x_{44} + 85x_{54}$$

$$C \geq 50x_{15} + 51x_{25} + 75x_{35} + 51x_{45} + 74x_{55}$$

$$C \geq 62x_{16} + 45x_{26} + 51x_{36} + 18x_{46} + 16x_{56}$$

$$C \geq 60x_{17} + 43x_{27} + 49x_{37} + 19x_{47} + 17x_{57}$$

$$C \geq 27x_{18} + 34x_{28} + 59x_{38} + 58x_{48} + 74x_{58}$$

$$2000x_{11} + 1700x_{12} + 2500x_{13} + 1300x_{14} + 2800x_{15} + 2700x_{16}$$
$$+ 1800x_{17} + 2200x_{18} \leq 7200$$

$$2000x_{21} + 1700x_{22} + 2500x_{23} + 1300x_{24} + 2800x_{25} + 2700x_{26}$$
$$+ 1800x_{27} + 2200x_{28} \leq 5200$$

$$2000x_{31} + 1700x_{32} + 2500x_{33} + 1300x_{34} + 2800x_{35} + 2700x_{36}$$
$$+ 1800x_{37} + 2200x_{38} \leq 6100$$

$$2000x_{41} + 1700x_{42} + 2500x_{43} + 1300x_{44} + 2800x_{45} + 2700x_{46}$$
$$+ 1800x_{47} + 2200x_{48} \leq 4400$$

$$2000x_{51} + 1700x_{52} + 2500x_{53} + 1300x_{54} + 2800x_{55} + 2700x_{56}$$
$$+ 1800x_{57} + 2200x_{58} \leq 4200$$

The Excel worksheet for this problem is displayed in Figure 5.14, and the corresponding Solver dialog box is in Figure 5.15.

The layout of Figure 5.14 is very similar to that of Figure 5.12. Figure 5.14 contains the Euclidean distances between SCs and DCs in cells B2:I6, the SC capacity parameters in cells L10:L14, and the demand parameters in cells B17:I17. The DC to SC assignment

	DC 1	DC 2	DC 3	DC 4	DC 5	DC 6	DC 7	DC 8				MAXIMUM	
SC 1	29.73	69.34	44.38	32.25	49.58	61.85	59.91	27.46			49.58	DISTANCE	
SC 2	46.32	52.33	35.61	35.36	51.26	45.18	43.19	34.00					
SC 3	68.35	32.56	55.23	31.05	75.43	50.57	48.75	59.01					
SC 4	83.49	66.65	27.86	89.81	50.91	18.36	19.42	58.31					
SC 5	96.83	43.14	46.87	85.00	74.30	15.81	17.03	74.28					
										Actual	Potential		
	DC 1	DC 2	DC 3	DC 4	DC 5	DC 6	DC 7	DC 8		SC Supply	SC Capacity	OPEN?	
SC 1	1	0	0	0	1	0	0	1		7,000	7,200	1	
SC 2	0	0	1	0	0	1	0	0		5,200	5,200	1	
SC 3	0	1	0	1	0	0	1	0		4,800	6,100	1	
SC 4	0	0	0	0	0	0	0	0		0	4,400	0	
SC 5	0	0	0	0	0	0	0	0		0	4,200	0	
Cust. Demand	2,000	1,700	2,500	1,300	2,800	2,700	1,800	2,200				3	
DC Assigned	1	1	1	1	1	1	1	1					
MAX DISTANCE	29.73	32.56	35.61	31.05	49.58	45.18	48.75	27.46					

Figure 5.14.   Excel worksheet for Example 5.7.

variables are located in cells B10:I14 and the variables controlling the opening of SCs are in cells M10:M14. Unlike cell L2 in Figure 5.12, cell L2 in Figure 5.14 is not the total distance between DCs and the SCs computed as a sumproduct. Instead, this cell is simply the decision variable C, the maximum distance between a DC and the SC to which it is assigned. This cell is the 'Set Objective' cell in the Solver dialog box in Figure 5.15.

Cells K10:K14 contain the sumproducts of the DC to SC assignment variables and demand, representing the amount supplied by each SC. These cells are constrained to be equal to or less than the SC capacities in cells L10:L14. The sum of the binary variables in cells M10:M14 is located in cell M17 and this cell is constrained to equal the number of SCs that should be opened, which is three for this example. Each variable in the matrix defined by cells B10:I14 must be constrained to be equal to or less than the binary variable for its corresponding row in cells M10:M14. For example, the constraint B12:I12 <= M12 in the Solver dialog box ensures that none of the DCs can be assigned to SC 3 unless SC 3 is opened. Cells B18:I18 contain the sums of the DC assignment variables for each DC. These cells are constrained to equal one to ensure that each DC is assigned to exactly one SC.

A key feature of Figure 5.14 is the cell range B20:I20, which contains the distance between each DC and its assigned SC. For example, cell B20 contains the sumproduct of B2:B6 and B10:B14. Cell L2 is

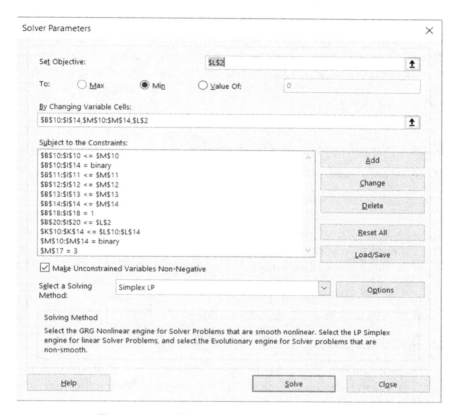

Figure 5.15. Solver dialog box for Example 5.7.

constrained to equal or exceed the values in cells B20:I20 to ensure it represents the maximum distance.

The optimal solution in Figure 5.14 reveals that the three selected SCs for the p-centers example (Example 5.7) are not exactly the same as those in Figure 5.12 for the p-median example (Example 5.6). For the p-centers example, the selected SCs are SC 1, SC 2, and SC 3. The SC 1 source facility is assigned DC 1, DC 5, and DC 8. The SC 2 source facility is assigned DC 3 and DC 6. The SC 3 source facility is assigned DC 2, DC 4, and DC 7. The maximum distance from a DC to its assigned SC is 49.58 miles, which occurs for the assignment of DC 5 to SC 1. Notice that this distance is smaller than the maximum distance of 50.57 miles that was realized from the assignment of DC 6 to SC 3 is the p-median solution in Example 5.6

Capacity utilization of the SCs is not quite as good as it was for the p-median solution in Example 5.6. Although SC 2 is used to full capacity and SC 1 has an excess capacity of only 200 units, the excess capacity for SC 3 is a substantial 1300 units. Once again, most of the DCs are assigned to their nearest SC. The exceptions are DC3, DC 6, and DC 7.

5.3.5   *Simple plant location model*

Parameters and sets

$I =$ a set of m 'demand' nodes, indexed $(1 \leq i \leq m)$;
$J =$ a set of n candidate location sites, indexed $(1 \leq j \leq n)$;
$d_{ij} =$ a measure of distance or cost between demand node i and candidate site j (for $1 \leq i \leq m$ and $1 \leq j \leq n$);
$f_j =$ a measure of fixed cost for opening a facility at site j (for $j \in J$);
$g_i =$ a measure of demand at node i, for $(1 \leq i \leq m)$;
$h_j =$ a measure of capacity at candidate site j, for $(1 \leq j \leq n)$;

Decision variables

$y_j =$ binary decision variables such that $y_j = 1$ if candidate site j is selected as a facility location and $y_j = 0$ otherwise (for $1 \leq j \leq n$);
$x_{ij} =$ binary decision variables such that $x_{ij} = 1$ if demand node i is assigned to candidate site j and $x_{ij} = 0$ otherwise (for $1 \leq i \leq m$ and $1 \leq j \leq n$).

Objective function

$$\text{Minimize: } Z_{SPLP} = \sum_{i \in I} \sum_{j \in J} d_{ij} x_{ij} + \sum_{j \in J} f_j y_j. \qquad (5.37)$$

Subject to constraints

$$\sum_{j \in J} x_{ij} = 1 \quad \forall i \in I, \qquad (5.38)$$

$$x_{ij} \leq y_j \quad \forall i \in I, \; j \in J, \qquad (5.39)$$

$$\sum_{i \in I} g_i x_{ij} \leq h_j \quad \forall j \in J, \qquad (5.40)$$

$$x_{ij} \in \{0, 1\} \quad \forall i \in I, \; j \in J, \qquad (5.41)$$

$$y_j \in \{0, 1\} \quad \forall j \in J. \qquad (5.42)$$

The objective function of the simple plant location model in Equation (5.37) is to minimize the sum of the variable costs of satisfying the demand nodes by the source facility to which they are assigned plus the fixed operating costs of the selected source facilities. Unlike the p-median model, there is no prespecification of the number of source facilities, as the number of source facilities will be determined by the trade-off between fixed and variable costs. However, all of the other constraints of the p-median model are necessary. Constraint set (5.38) ensures that each demand node is assigned to exactly one source facility. Constraint set (5.39) guarantees that a demand node can only be assigned to a candidate source facility that has been selected (i.e., opened). Constraint set (5.40) is not part of the simple plant location model in its most basic form. However, this constraint set is relevant for the *capacitated* simple plant location model where the supply capacities of the source facilities are limited. Constraint set (5.40) requires that the sum of the requirements of the demand nodes assigned to a particular source facility is less than the capacity of that source facility. Constraint sets (5.41) and (5.42) place binary restrictions on the decision variables.

5.3.6 *Example 5.8 — Simple plant location problem example*

A firm has five *potential* locations for supply centers (SCs), 1, 2, 3, 4, and 5, that can supply eight existing demand centers (DCs), 1, 2, 3, 4, 5, 6, 7, and 8. The distances (in miles) from each DC to each SC are shown in the table, as are the capacities for the SCs and demand levels for the DCs. The translation from distance to cost should be based on $1000 per mile. The fixed costs of operating supply centers SC1, SC2, SC3, SC4, and SC5 are $100,000, $50,000, $75,000, $72,000, and $33,000, respectively. The management would like to build an integer linear programming model that, when solved, will select 1, 2, 3, 4, or 5 SCs to open and assign each DC to exactly one selected SC with the goal of minimizing the total variable costs associated with travel distance plus the fixed costs. The optimal plan should ensure that the capacities of the SCs are not exceeded.

	DC 1	DC 2	DC 3	DC 4	DC 5	DC 6	DC 7	DC 8	Capacity
SC 1	30	69	44	32	50	62	60	27	7200
SC 2	46	52	36	35	51	45	43	34	5200
SC 3	68	33	55	31	75	51	49	59	6100
SC 4	83	67	28	90	51	18	19	58	4400
SC 5	97	43	47	85	74	16	17	74	4200
Demand	2000	1700	2500	1300	2800	2700	1800	2200	

Decision variables

$x_{jk} = 1$ if DC k is assigned to SC j and 0 otherwise, for $j = 1$ to 5 and $k = 1$ to 8;

$y_j = 1$ if SC j is opened and 0 otherwise (for $j = 1$ to 5).

Objective function

Minimize: $100,000y_1 + 50000y_2 + 75000y_3 + 72000y_4 + 33000y_5$

$+ 1000[30x_{11} + 69x_{12} + 44x_{13} + 32x_{14} + 50x_{15} + 62x_{16} + 60x_{17}$

$+ 27x_{18} + 46x_{21} + 52x_{22} + 36x_{23} + 35x_{24} + 51x_{25} + 45x_{26}$

$+ 43x_{27} + 34x_{28} + 68x_{31} + 33x_{32} + 55x_{33} + 31x_{34} + 75x_{35}$

$+ 51x_{36} + 49x_{37} + 59x_{38} + 83x_{41} + 67x_{42} + 28x_{43} + 90x_{44}$

$+ 51x_{45} + 18x_{46} + 19x_{47} + 58x_{48} + 97x_{51} + 43x_{52} + 47x_{53}$

$+ 85x_{54} + 74x_{55} + 16x_{56} + 17x_{57} + 74x_{58}]$

Subject to constraints

$x_{11} + x_{21} + x_{31} + x_{41} + x_{51} = 1x_{12} + x_{22} + x_{32} + x_{42} + x_{52} = 1$

$x_{13} + x_{23} + x_{33} + x_{43} + x_{53} = 1x_{14} + x_{24} + x_{34} + x_{44} + x_{54} = 1$

$$x_{15} + x_{25} + x_{35} + x_{45} + x_{55} = 1 \quad x_{16} + x_{26} + x_{36} + x_{46} + x_{56} = 1$$

$$x_{17} + x_{27} + x_{37} + x_{47} + x_{57} = 1 \quad x_{18} + x_{28} + x_{38} + x_{48} + x_{58} = 1$$

$$x_{jk} \leq y_j \quad \text{for all } j = 1 \text{ to } 5 \text{ and } k = 1 \text{ to } 8.$$

$$2000x_{11} + 1700x_{12} + 2500x_{13} + 1300x_{14} + 2800x_{15} + 2700x_{16}$$
$$+ 1800x_{17} + 2200x_{18} \leq 7200$$

$$2000x_{21} + 1700x_{22} + 2500x_{23} + 1300x_{24} + 2800x_{25} + 2700x_{26}$$
$$+ 1800x_{27} + 2200x_{28} \leq 5200$$

$$2000x_{31} + 1700x_{32} + 2500x_{33} + 1300x_{34} + 2800x_{35} + 2700x_{36}$$
$$+ 1800x_{37} + 2200x_{38} \leq 6100$$

$$2000x_{41} + 1700x_{42} + 2500x_{43} + 1300x_{44} + 2800x_{45} + 2700x_{46}$$
$$+ 1800x_{47} + 2200x_{48} \leq 4400$$

$$2000x_{51} + 1700x_{52} + 2500x_{53} + 1300x_{54} + 2800x_{55} + 2700x_{56}$$
$$+ 1800x_{57} + 2200x_{58} \leq 4200$$

The Excel worksheet for this problem is displayed in Figure 5.16, and the corresponding Solver dialog box is in Figure 5.17.

The layout of Figure 5.16 is similar to that of Figures 5.12 and 5.14. Figure 5.16 contains the Euclidean distances between SCs and DCs in cells B2:I6, the SC capacity parameters in cells L10:L14, the SC fixed costs in cells N10:N14, and the demand parameters in cells B17:I17. The DC to SC assignment variables are located in cells B10:I14 and the variables controlling the opening of SCs are in cells M10:M14. Cell L2, which is the 'Set Objective' cell in the Solver dialog box in Figure 5.17, is computed as the sumproduct of cells M10:M14 and N10:N14 (fixed costs), plus $1000 times the sumproduct of cell B2:I6 and B10:I14 (variable costs).

Cells K10:K14 contain the sumproducts of the DC to SC assignment variables and demand, representing the amount supplied by each SC. These cells are constrained to be equal to or less than the SC capacities in cells L10:L14. The sum of the binary variables in cells M10:M14 is located in cell M17; however, unlike the p-median example, this is only included for reporting purposes and is not relevant to the constraint set. Each variable in the matrix defined by cells B10:I14 must be constrained to be equal to or less than the binary

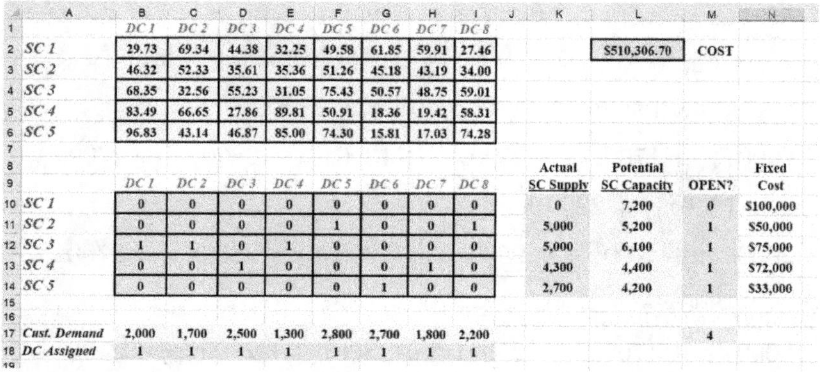

	A	B	C	D	E	F	G	H	I	J	K	L	M	N
1		DC 1	DC 2	DC 3	DC 4	DC 5	DC 6	DC 7	DC 8					
2	SC 1	29.73	69.34	44.38	32.25	49.58	61.85	59.91	27.46			$510,306.70	COST	
3	SC 2	46.32	52.33	35.61	35.36	51.26	45.18	43.19	34.00					
4	SC 3	68.35	32.56	55.23	31.05	75.43	50.57	48.75	59.01					
5	SC 4	83.49	66.65	27.86	89.81	50.91	18.36	19.42	58.31					
6	SC 5	96.83	43.14	46.87	85.00	74.30	15.81	17.03	74.28					
7														
8											Actual	Potential		Fixed
9		DC 1	DC 2	DC 3	DC 4	DC 5	DC 6	DC 7	DC 8		SC Supply	SC Capacity	OPEN?	Cost
10	SC 1	0	0	0	0	0	0	0	0		0	7,200	0	$100,000
11	SC 2	0	0	0	0	1	0	0	1		5,000	5,200	1	$50,000
12	SC 3	1	1	0	1	0	0	0	0		5,000	6,100	1	$75,000
13	SC 4	0	0	1	0	0	0	1	0		4,300	4,400	1	$72,000
14	SC 5	0	0	0	0	0	1	0	0		2,700	4,200	1	$33,000
15														
16														
17	Cust. Demand	2,000	1,700	2,500	1,300	2,800	2,700	1,800	2,200				4	
18	DC Assigned	1	1	1	1	1	1	1	1					

Figure 5.16.   Excel spreadsheet for Example 5.8.

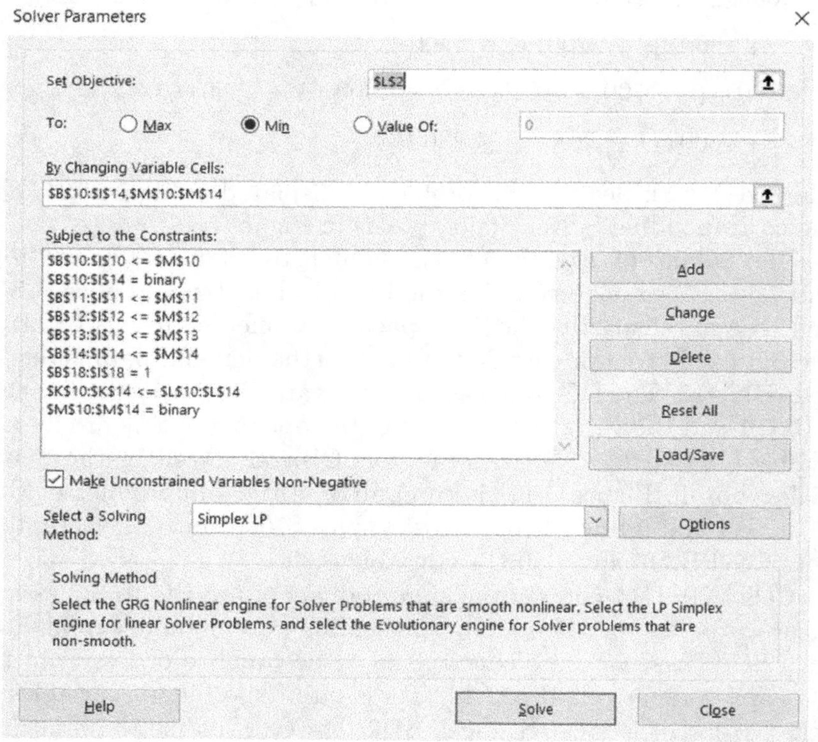

Figure 5.17.   Solver dialog box for Example 5.8.

variable for its corresponding row in cells M10:M14. For example, the constraint B11:I11 $<=$ M11 in the Solver dialog box ensures that none of the DCs can be assigned to SC 2 unless SC 2 is opened. Cells B18:I18 contain the sums of the DC assignment variables for each DC. These cells are constrained to equal one to ensure that each DC is assigned to exactly one SC.

The optimal solution in Figure 5.16 reveals that four SCs are selected for the simple plant location example. Only SC 1 is not selected. The SC 2 source facility is assigned DC 5 and DC 8. The SC 3 source facility is assigned DC 1, DC 2, and DC 4. The SC 4 source facility is assigned DC 3 and DC 7. The SC 5 source facility is assigned only D6. The fixed costs are $50,000 + $75,000 + $72,000 + $33,000 = $230,000. The variable costs are $280,306.70 for a total cost of $510,306.70.

Given the use of four source facilities, the capacity utilization of the SCs is not especially good for the solution in Figure 5.16. Unused capacity is only 200 for SC 2 and 100 for SC 4. However, unused capacity is 1100 for SC3 and 1500 for SC 5. Only half of the DCs (DC 2, DC3, DC4, and DC 6) are assigned to their nearest SC. DC 7 and DC 8 are each assigned to their second-nearest SC. DC 1 and DC 5 are assigned to their third-nearest SC. In particular, DC 1 is 68.35 miles from its source facility SC 3 and this is the maximum distance of any DC from its source facility. We note that this maximum distance of 68.35 miles is appreciably larger than the known smallest maximum distance of 49.58 miles from the p-centers solution in Figure 5.14.

Chapter 6

Routing Problems

Whereas Chapter 5 considered some network problems related to location and distribution, our focus in Chapter 6 is on a couple of important types of routing problems. Section 6.1 examines the shortest route problem, where the goal is to find the shortest route (or path) between two nodes in a network. Although the linear programming approach used to solve this problem is not necessarily the most efficient method, it is conceptually easy to understand and some aspects of the formulation are relevant to other applications.

Section 6.2 considers the classic traveling salesperson problem. In the most basic version of this problem, the goal is to leave a particular node in the network (or point in the plane), visit all other nodes in the network (or in the plane) one time each, and then return to the starting node so as to minimize travel distance. Typically, travel distances are assumed to be symmetric between pairs of nodes or points. However, the problem generalizes easily to the asymmetric case, and we will consider an example where there are asymmetric travel 'costs' between pairs of nodes. Unlike the shortest route problem, the traveling salesperson problem is a combinatorial optimization problem (in fact, a special case of the quadratic assignment problem). It has a very rich history not only in the operations research literature but also as a more general data analysis tool.

6.1   Shortest Route Problem

6.1.1   *A general formulation*

We begin with a very general formulation of the shortest route problem as a linear programming problem.

Sets and parameters

$V =$ a set of n nodes or vertices, $V = \{1, 2, \ldots, n\}$, where, without loss of generality, we assume that node 1 is the starting node and node n is the ending node.

$E =$ the set of all possible node pairs $\{i, j\}$ for which travel can occur.

$d_{ij} =$ the distance from node i to node j for all $\{i, j\}$ pairs in E.

$U_j =$ the set of nodes with incoming edges/arcs to node j (for $1 \leq j \leq n$).

$V_j =$ the set of nodes with outgoing edges/arcs from node j (for $1 \leq j \leq n$).

Decision variables

$x_{ij} =$ 1 if the edge/arc from node i to node j is selected, 0 otherwise, for all $\{i, j\}$ pairs in E.

Objective function

$$\text{Minimize:} \quad \sum_{\{i,j\} \in E} d_{ij} x_{ij}. \qquad (6.1)$$

Subject to constraints

$$\sum_{j \in V_1} x_{1j} = 1, \qquad (6.2)$$

$$\sum_{i \in U_j} x_{ij} = \sum_{k \in V_j} x_{jk} \quad \text{for } 2 \leq j \leq n - 1, \qquad (6.3)$$

$$\sum_{i \in U_n} x_{in} = 1. \qquad (6.4)$$

The objective function in Equation (6.1) is to minimize total travel distance. Constraint (6.2) ensures that the route begins by leaving the starting node (node 1) and traveling to a node for which node 1 has an outgoing edge (node 1 has no incoming edges). Constraint (6.4) guarantees that exactly one of the incoming edges for the ending node (node n) is selected (node n has no outgoing edges). Constraint set (6.3) requires that, for all other edges (2 to $n-1$), if the node has an incoming edge, then it must also have an outgoing edge and vice versa.

To facilitate template construction in Excel, we define an $n \times n$ matrix $\mathbf{A} = [a_{ij}]$, whereby $a_{ij} = d_{ij}$ for node pairs $\{i, j\}$ where travel is permitted to occur and $a_{ij} = M$ for all other pairs. The value of M is chosen to be sufficiently large so as to ensure that the edge would never be selected as part of the route. Accordingly, $U_j = V_j = V$, and the objective function simplifies to the following:

$$\text{Minimize} : \sum_{i=1}^{n}\sum_{j=1}^{n} a_{ij}x_{ij}. \tag{6.5}$$

6.1.2 *Example 6.1 — Shortest route #1*

The network graph for Example 6.1 is displayed in the Excel worksheet in Figure 6.1. There are $n = 7$ nodes and the distances between pairs of nodes are displayed on the edges of the graph. The goal is to find the shortest route from node 1 to node 7.

Cells B2:H8 contain the travel distance parameters (a_{ij}) between pairs of nodes. Values of 9999 in these cells indicate pairs of nodes where direct travel is not possible. Cells B11:H17 contain the x_{ij} decision variables. The distance objective function in cell K3 is the sumproduct of cells B2:H8 and B11:H17, which is commensurate with

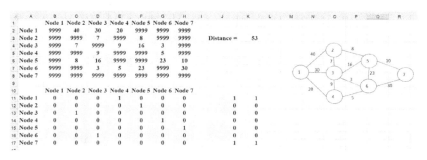

	Node 1	Node 2	Node 3	Node 4	Node 5	Node 6	Node 7			
Node 1	9999	40	30	20	9999	9999	9999			
Node 2	9999	9999	7	9999	8	9999	9999	Distance =	53	
Node 3	9999	7	9999	9	16	3	9999			
Node 4	9999	9999	9	9999	9999	5	9999			
Node 5	9999	8	16	9999	9999	23	10			
Node 6	9999	9999	3	5	23	9999	30			
Node 7	9999	9999	9999	9999	9999	9999	9999			
	Node 1	Node 2	Node 3	Node 4	Node 5	Node 6	Node 7			
Node 1	0	0	0	1	0	0	0		1	1
Node 2	0	0	0	0	1	0	0		0	0
Node 3	0	1	0	0	0	0	0		0	0
Node 4	0	0	0	0	0	1	0		0	0
Node 5	0	0	0	0	0	0	1		0	0
Node 6	0	0	1	0	0	0	0		0	0
Node 7	0	0	0	0	0	0	0		1	1

Figure 6.1.   Excel worksheet for Example 6.1.

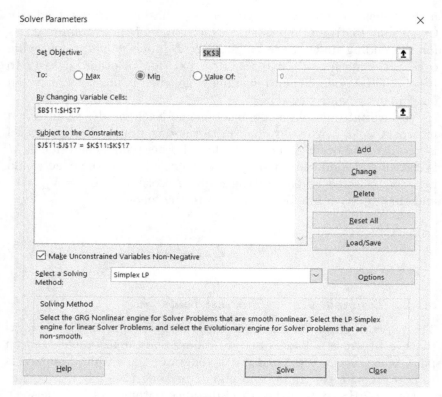

Figure 6.2. Solver dialog box for Example 6.1.

Equation (6.5). Cell K3 is the 'Set Objective' cell in the Solver dialog box in Figure 6.2.

Cell J11 contains the sum of cells B11:H11 and J11 is constrained to equal the value of one in cell K11 to enforce constraint (6.2). Similarly, cell J17 contains the sum of cells H11:H17 and J17 is constrained to equal the value of one in cell K17 to enforce constraint (6.4). Cells J12:J16 contain the sum of the variables in the node row minus the sum of the variables in the node column and these cells are constrained to equal the zeros in cells K12:K16 to enforce constraint set (6.3).

The optimal route for this problem $(1 - 4 - 6 - 3 - 2 - 5 - 7)$ is interesting because it traverses all of the nodes in the network. It has a travel distance of $20 + 5 + 3 + 7 + 8 + 10 = 53$.

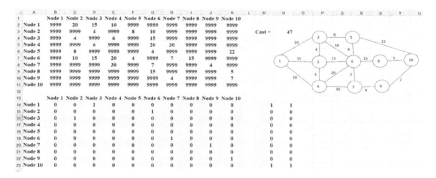

Figure 6.3. Excel worksheet for Example 6.2.

6.1.3 Example 6.2 — Shortest route #2

The network graph for Example 6.2 is displayed in the Excel worksheet in Figure 6.3. There are $n = 10$ nodes and the distances between pairs of nodes are displayed on the edges of the graph. The goal is to find the shortest route from node 1 to node 10.

Example 6.2 is slightly larger than Example 6.1 but the worksheet construction is basically the same. For example, the sumproduct of the distance parameters in cells B2:K11 and the variables in cells B14:23 is the total distance in cell N3, which is to be minimized. The constraints for the starting, ending, and intermediate nodes are also constructed in the same fashion as they were for Example 6.1. The Solver dialog box for Example 6.2 is the same basic structure as the one for Example 6.1 and is, therefore, not displayed.

Travel begins at node 1 and proceeds a distance of 15 miles to node 3. The next step is to travel 4 miles from node 3 to node 2. Travel then proceeds a distance of 10 miles from node 2 to node 6, then a distance of 7 miles from node 6 to node 7, then 4 miles from node 7 to node 9, and finally 7 miles from node 9 to node 10. The optimal route is $1 - 3 - 2 - 6 - 7 - 9 - 10$ for a distance of 47 miles.

6.2 Traveling Salesperson Problem

6.2.1 A general formulation

We begin with a very general formulation of the traveling salesperson problem as an integer (binary) linear programming problem.

V = a set of n nodes or vertices, $V = \{1, 2, \ldots, n\}$,
d_{ij} = the distance from node i to node j (or, possibly, the cost of traveling from node i to node j) for all $1 \leq i \leq n$ and $1 \leq j \leq n$ (we assume here that $d_{ii} = \infty$ for all $1 \leq i \leq n$).

Decision variables

x_{ij} = 1 if the edge/arc from node i to node j is selected, 0 otherwise, for all $1 \leq i \leq n$ and $1 \leq j \leq n$.

Objective function

$$\text{Minimize: } \sum_{i=1}^{n} \sum_{j=1}^{n} d_{ij} x_{ij}. \tag{6.6}$$

Subject to constraints

$$\sum_{i=1}^{n} x_{ij} = 1 \quad \text{for } 1 \leq j \leq n, \tag{6.7}$$

$$\sum_{j=1}^{n} x_{ij} = 1 \quad \text{for } 1 \leq i \leq n, \tag{6.8}$$

$$\sum_{i \in S} \sum_{j \in S, j \neq i} x_{ij} \leq |S| - 1 \quad \text{for } S \subset V, \ |S| \geq 2, \tag{6.9}$$

$$x_{ij} \in \{0, 1\} \quad \text{for } 1 \leq i \leq n \quad \text{and} \quad 1 \leq j \leq n. \tag{6.10}$$

The objective function in Equation (6.6) is to minimize total travel distance (or possibly travel cost). Constraint (6.7) ensures that every node has exactly one incoming edge. Constraint set (6.8) guarantees that every node has exactly one outgoing edge.

Constraint (6.9) prevents *subtours* from occurring in the solution. A *single tour* is defined by leaving a particular node, visiting each of the other $n - 1$ nodes once in some sequence, and then returning to the particular node. For example, if $n = 6$, then the single tour $1 - 2 - 3 - 4 - 5 - 6 - 1$ would correspond to decision variable values $x_{12} = x_{23} = x_{34} = x_{45} = x_{56} = x_{61} = 1$ and all other variables would be zero. Notice that this assignment satisfies constraint sets (6.7) and (6.8).

Now consider the assignment $x_{12} = x_{23} = x_{31} = x_{45} = x_{56} = x_{64} = 1$, and all other variables equal zero. This assignment also satisfies constraint sets (6.7) and (6.8). However, this is not a single tour but rather two subtours: $1-2-3-1$ and $4-5-6-4$. These types of subtours are prevented by constraint set (6.9). Unfortunately, the number of subsets in the constraint set grows exponentially as a function of n. Therefore, a common practice is to iteratively append constraints of type (6.9) as needed. This will be demonstrated in our examples. Finally, binary restrictions must be enforced on the x_{ij} variables via constraint set (6.10).

6.2.2 *Example 6.3: Symmetric distances*

The network graph for Example 6.3 is displayed in the Excel worksheet in Figure 6.4. There are $n = 8$ nodes corresponding to points in the two-dimensional plane. The x and y coordinates are located in cells M2:N9. From these coordinates, rectangular distances were computed between pair of nodes, and these distances are shown in cells C3:J10. Although the distance between any node and itself is obviously zero, we place '999' in the main diagonal cells of the C3:J10 range because we do not want travel from a node to itself. The decision variables are in cells C15:J22. The sumproduct of C3:J10 and C15:J22 is the travel distance in cell Q2. This is the 'Set Objective' cell in the Solver dialog box in Figure 6.5.

Cells C24:J24 contain the column sums of the incoming decision variables for each node. These sums are constrained to equal

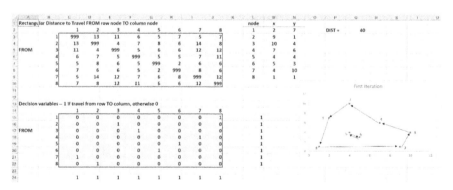

Figure 6.4.   Excel worksheet for Example 6.3.

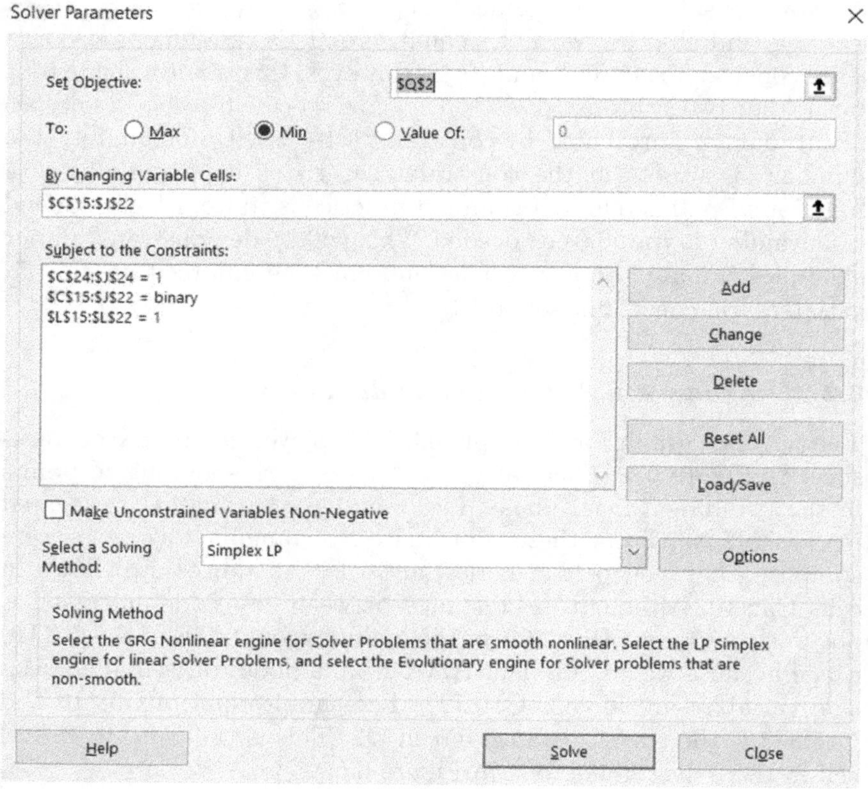

Figure 6.5.   Solver dialog box for Example 6.3.

one to ensure that constraint set (6.7) is satisfied. Likewise, cells L15:L22 contain the row sums of the outgoing decision variables for each node. These values are constrained to equal one to enforce constraint set (6.8). As observed in the Solver dialog box, binary restrictions are placed on the decision variables. Initially, no constraints of the type shown in Equation (6.9) have been incorporated to prevent subtours.

As can be determined from the decision variable values and the map in Figure 6.4, there are two subtours in this first iteration: $1 - 8 - 2 - 3 - 4 - 7 - 1$ and $5 - 6 - 5$. Therefore, this solution is infeasible for the traveling salesperson problem and constraints based on Equation (6.9) should be appended.

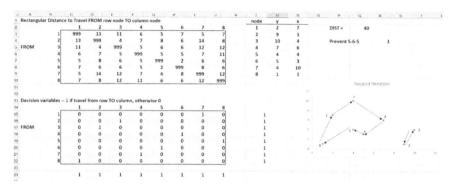

Figure 6.6. Excel worksheet for Example 6.3 (second iteration).

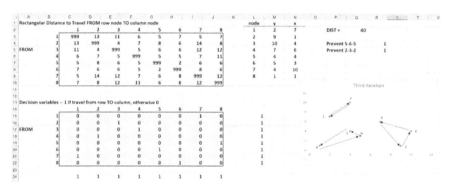

Figure 6.7. Excel worksheet for Example 6.3 (third iteration).

We append the constraint $x_{56} + x_{65} \leq 1$ to the model, which will prevent both of the subtours in the solution in Figure 6.4 from occurring. This is accomplished in Figure 6.6 by placing H19 + G20 in cell R4 and adding the constraint that this cell must be equal to or less than one. When solving the new formulation at this second iteration, we again observed two subtours, $1 - 7 - 4 - 6 - 5 - 8$ and $2 - 3 - 2$, and the distance remained at 40.

We append the constraint $x_{23} + x_{32} \leq 1$ to the model, which will prevent both of the subtours in the solution in Figure 6.6 from occurring. This is accomplished in Figure 6.7 by placing E16 + D17 in cell R5 and adding the constraint that this cell must be equal to or less than one.

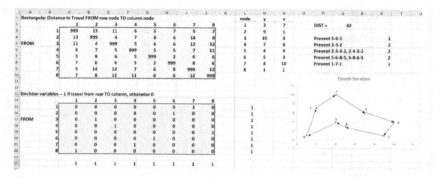

Figure 6.8.   Excel worksheet for Example 6.3 (fourth iteration).

When solving the augmented formulation at the third iteration, the distance remained at 40 and we observed three subtours, $1-7-1$, $5-8-6-5$, and $2-3-4-2$, as shown in Figure 6.7. We append the constraints $x_{17} + x_{71} \leq 1, x_{56} + x_{58} + x_{65} + x_{68} + x_{85} + x_{86} \leq 2$ and $x_{23} + x_{24} + x_{32} + x_{34} + x_{42} + x_{43} \leq 2$ to the model. This will prevent the subtours in the solution in Figure 6.7 from occurring. This is accomplished in Figure 6.8 by the following steps: (i) placing E16 + D17 in cell R8 and adding the constraint that this cell must be equal to or less than one, (ii) placing H19 + J20 + G22 + J19 + H22 + G20 in cell R7 and adding the constraint that this cell must be equal to or less than two, and (iii) placing E16 + F17 + D18 + F16 + E18 + D17 in cell R6 and adding the constraint that this cell must be equal to or less than two. The resulting updated Solver dialog box is shown in Figure 6.9.

The results in Figure 6.8 reveal a single tour, which is an optimal solution to the traveling salesperson problem. The tour is $1-7-4-3-2-6-5-8-1$ with a travel distance of 42. It should be noted that this optimal tour is not unique as there are other single tours that also have a distance of 42.

6.2.3   *Example 6.4: Asymmetric costs*

The network graph for Example 6.4 is displayed in the Excel worksheet in Figure 6.10. Unlike Example 6.3, where the relationships between nodes were symmetric rectangular distances in the two-dimensional plane, in Example 6.4, the relationships are asymmetric travel costs between the $n = 5$ nodes. Specifically, these costs

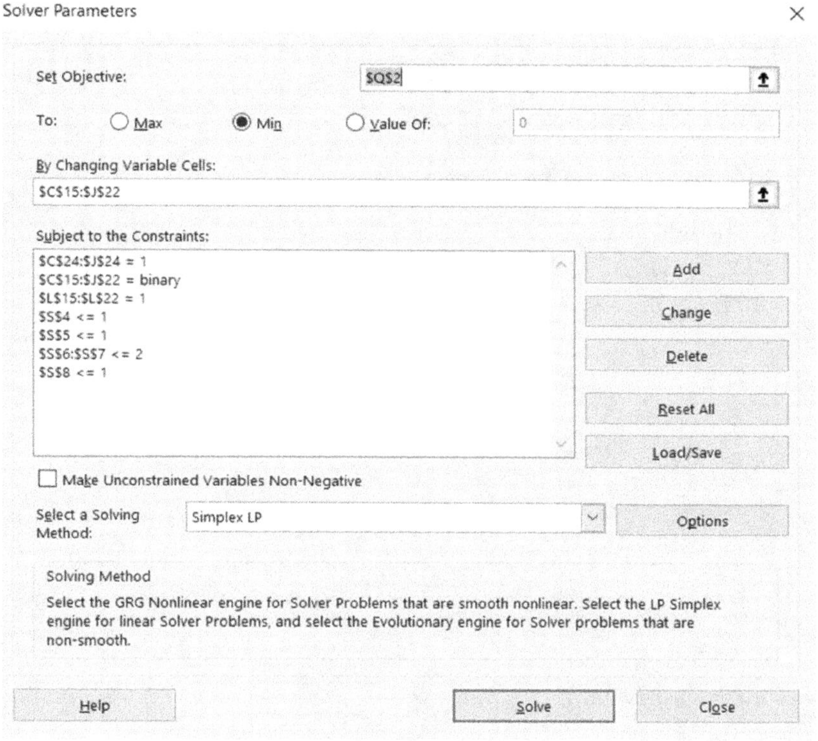

Figure 6.9. Solver dialog box for Example 6.3 (fourth iteration).

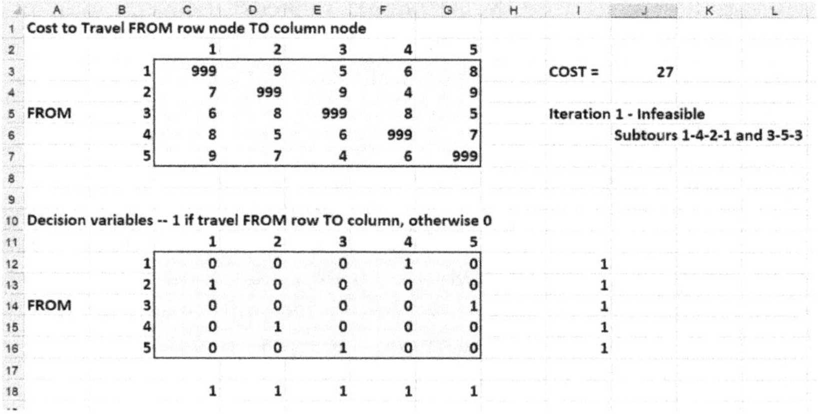

	A	B	C	D	E	F	G	H	I	J	K	L
1	Cost to Travel FROM row node TO column node											
2			1	2	3	4	5					
3		1	999	9	5	6	8		COST =	27		
4		2	7	999	9	4	9					
5	FROM	3	6	8	999	8	5		Iteration 1 - Infeasible			
6		4	8	5	6	999	7			Subtours 1-4-2-1 and 3-5-3		
7		5	9	7	4	6	999					
8												
9												
10	Decision variables -- 1 if travel FROM row TO column, otherwise 0											
11			1	2	3	4	5					
12		1	0	0	0	1	0		1			
13		2	1	0	0	0	0		1			
14	FROM	3	0	0	0	0	1		1			
15		4	0	1	0	0	0		1			
16		5	0	0	1	0	0		1			
17												
18			1	1	1	1	1					

Figure 6.10. Excel worksheet for Example 6.4 (Iteration 1).

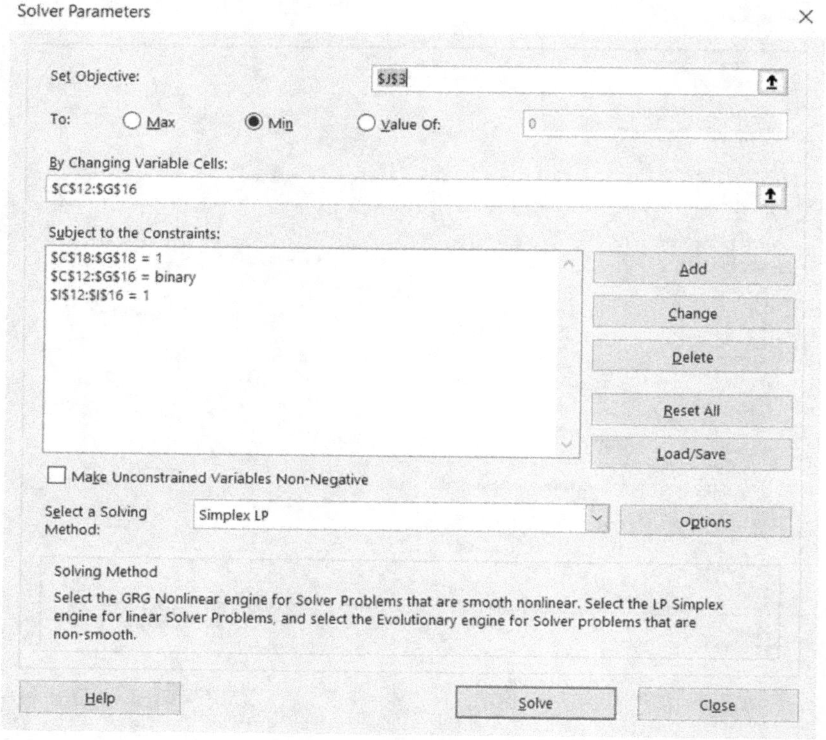

Figure 6.11.   Solver dialog box for Example 6.4.

are located in cells C3:G7. Although the cost between any node and itself is zero, we place '999' in the main diagonal cells of the C3:G7 range because we do not want travel from a node to itself. The decision variables are in cells C12:G16. The sumproduct of C3:G7 and C12:G16 is the travel distance in cell J3. This is the 'Set Objective' cell in the Solver dialog box in Figure 6.11.

Cells C18:G18 contain the column sums of the incoming decision variables for each node. These sums are constrained to equal one to ensure that constraint set (6.7) is satisfied. Cells I12:I16 contain the row sums of the outgoing decision variables for each node. These values are constrained to equal one to enforce constraint set (6.8). The constraints section of the Solver dialog box also reveals that binary restrictions are placed on the decision variables. Initially, no constraints of the type shown in Equation (6.9) have been incorporated to prevent subtours.

As can be determined from the decision variable values in Figure 6.10, there are two subtours in this first iteration, $1-4-2-1$ and $3-5-3$, and the total cost is 27. Therefore, this solution is infeasible for the traveling salesperson problem, and constraints based on Equation (6.9) should be appended.

We append the constraint $x_{35} + x_{53} \leq 1$ to the model, which will prevent both of the subtours in the solution in Figure 6.10 from occurring. This is accomplished in Figure 6.12 by placing E16 + G14 in cell N6 and adding the constraint that this cell must be equal to or less than one. When solving the new formulation at this second iteration, we again observed two subtours, $1-5-3-1$ and $2-4-2$, and the cost remained at 27.

We append the constraint $x_{24} + x_{42} \leq 1$ to the model, which will prevent both of the subtours in the solution in Figure 6.12 from occurring. This is accomplished in Figure 6.13 by placing D15 + F13 in cell N8 and adding the constraint that this cell must be equal to or less than one. Figure 6.14 displays the Solver dialog box with the constraints appended for the second and third iterations.

The results in Figure 6.13 reveal a single tour, which is an optimal solution to the traveling salesperson problem. The tour is $1-3-5-4-2-1$ with a cost of $5+5+6+5+7 = 28$. It should be noted that, unlike the symmetric distance case in Example 6.3, the reverse of the tour is not optimal. That is, $1-2-4-5-3-1$ has a cost of $9+4+7+4+6 = 30$ (not 28).

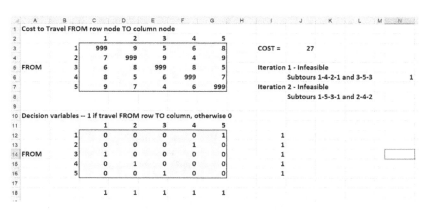

Figure 6.12.   Excel worksheet for Example 6.4 (second iteration).

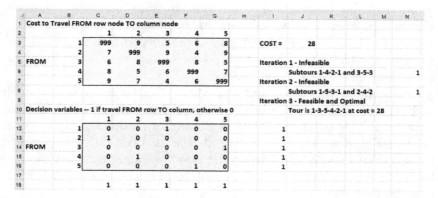

	A	B	C	D	E	F	G	H	I	J	K	L	M	N
1	Cost to Travel FROM row node TO column node													
2			1	2	3	4	5							
3		1	999	9	5	6	8		COST =	28				
4		2	7	999	9	4	9							
5	FROM	3	6	8	999	8	5		Iteration 1 - Infeasible					
6		4	8	5	6	999	7			Subtours 1-4-2-1 and 3-5-3				1
7		5	9	7	4	6	999		Iteration 2 - Infeasible					
8										Subtours 1-5-3-1 and 2-4-2				1
9									Iteration 3 - Feasible and Optimal					
10	Decision variables -- 1 if travel FROM row TO column, otherwise 0									Tour is 1-3-5-4-2-1 at cost = 28				
11			1	2	3	4	5							
12		1	0	0	1	0	0		1					
13		2	1	0	0	0	0		1					
14	FROM	3	0	0	0	0	1		1					
15		4	0	1	0	0	0		1					
16		5	0	0	0	1	0		1					
17														
18			1	1	1	1	1							

Figure 6.13. Excel worksheet for Example 6.4 (third iteration — optimal solution).

Figure 6.14. Solver dialog box for Example 6.4 (third iteration).

Chapter 7

Facility Layout

In this chapter, we focus on examples related to facility layout. Two major classes of layout problems are considered: (i) product layout and (ii) process layout.

In product layout, the layout is based on a sequence of *tasks* that are necessary to produce a highly standardized output. The time to complete each task is a key input to product layout. A *precedence relationship* is also assumed to be available for the tasks, indicating which tasks must be completed before others can begin. More specifically, the *immediate predecessors* for each task are identified. The tasks are to be assigned to a series of workstations along a production *assembly line*. When assigning the tasks, the precedence relationships must be preserved. That is, a task cannot be assigned to an earlier workstation than any of its predecessors. (*Note*: A task can be assigned to the same station as its predecessors, just not to a station before them.)

The goal in assigning tasks to workstations is to maximize efficiency by balancing the workload across the workstations, which is the basis for the expression *assembly line balancing*. We consider two specific versions of assembly line balancing. In the first version, the number of workstations is prespecified and the goal is to assign tasks to stations so as to minimize the *cycle time*. The cycle time is the maximum amount of total task time across all workstations. So, for example, if workstations 1, 2, and 3 had task times of 54, 78, and 63 seconds, respectively, then the cycle time would be 78 seconds and completed units would come off the line every 78 seconds. A perfect balance of 65 seconds for each of the three stations would yield greater efficiency with units coming off the line more frequently based on the cycle time of 65 seconds. In the second version of the assembly line balancing problem, the cycle time is prespecified and the goal is to minimize the number of workstations.

The process layout problem is based on departmental (or work center) arrangement. Here, the output is not standardized and certain pairs of departments might have different degrees of interaction than other pairs. The objectives for departmental layout can vary. For example, it might be desirable to arrange departments to maximize customer travel in a convenience or department store, so as to stimulate impulse buying. However, the particular version that we consider is to minimize travel distance for workers or materials.

The key inputs to our departmental layout problem are (i) the number of trips between pairs of n departments and (ii) the distance between each pair of n locations. The goal is to assign the n departments to the n locations with the goal of minimizing total travel distance. This suggests placement of the departments with a high degree of travel in locations that are close to one another. The underlying mathematical model is the quadratic assignment problem, which can be quite formidable when n exceeds 30.

7.1 Product Layout — Line Balancing

7.1.1 *Example 7.1 — Minimize cycle time*

A general formulation for the problem of assigning tasks to workstations to minimize cycle time subject to assignment and precedence constraints is as follows:

Parameters

$n =$ the number of tasks indexed $1 \leq j \leq n$,
$m =$ the number of workstations indexed $1 \leq i \leq m$,
$t_j =$ the task time for activity j for $1 \leq j \leq n$,
$P_j =$ the set of immediate successors for activity j.

Decision variables

$C =$ cycle time,
$x_{ij} =$ 1 if task j is assigned to workstation i and 0 otherwise for $1 \leq i \leq m$ and $1 \leq j \leq n$.

Objective function

$$\text{Minimize: } C. \tag{7.1}$$

Subject to constraints

$$\sum_{i=1}^{m} x_{ij} = 1 \quad \text{for } 1 \leq j \leq n, \tag{7.2}$$

$$C \geq \sum_{j=1}^{n} t_j x_{ij} \quad \text{for } 1 \leq i \leq m, \tag{7.3}$$

$$\sum_{l=1}^{i} x_{lj} \geq x_{ih} \quad \text{for } 1 \leq j \leq n, \ h \in P_j, \ 1 \leq i \leq m-1. \tag{7.4}$$

The objective function in Equation (7.1) is to minimize cycle time (C). Constraint set (7.2) guarantees that each of the n tasks is assigned to exactly one workstation. Constraint set (7.3) ensures, for each workstation, that C equals or exceeds the sum of the times of the tasks assigned to that station. This is in accordance with the definition of C as the maximum amount of time across all workstations. Constraint set (7.4) requires for a given task h that is an immediate successor to task j that task h can only be assigned to workstation i if task j is assigned to workstation i or an earlier station. It is likely that this constraint set can be reduced by capitalizing on information gleaned from a particular precedence diagram; however, we focus on this general constraint set in our examples.

A 10-task example (Example 7.1) is used to illustrate the formulation. The task times (t_j) and precedence diagram are displayed in Figure 7.1. We desire to assign the $n = 10$ tasks to $m = 3$ stations so as to minimize cycle time. The set of immediate successors for each task can be obtained from the precedence diagram. For example, task 3 is the only immediate successor to task 1, so $P_1 = \{3\}$. Similarly, because task 4 is the only immediate successor to task 2, $P_2 = \{4\}$. $P_3 = \{5\}$ because task 5 is the only immediate successor to task 3. Task 4 has two immediate successors (tasks 6 and 7), so $P_4 = \{6, 7\}$. The immediate successor sets for the remaining tasks are $P_5 = \{8\}$, $P_6 = \{8\}$, $P_7 = \{8, 10\}$, $P_8 = \{9\}$, $P_9 = \emptyset$, and $P_{10} = \emptyset$.

Task #	1	2	3	4	5	6	7	8	9	10
Time in seconds	53	26	28	46	54	40	21	30	15	50

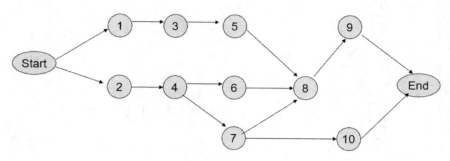

Figure 7.1. Task times and precedence diagram for Example 7.1.

	A	B	C	D	E	F	G	H	I	J	K
1							Task		Cycle Time =		125
2		Station 1	Station 2	Station 3		Asmt	Time				
3	Task 1	1	0	0		1	53		Precedence	Station 1	Station 2
4	Task 2	1	0	0		1	26		1-3	1	0
5	Task 3	0	1	0		1	28		2-4	0	1
6	Task 4	1	0	0		1	46		3-5	0	0
7	Task 5	0	1	0		1	54		4-6	1	0
8	Task 6	0	1	0		1	40		4-7	1	1
9	Task 7	0	0	1		1	21		5-8	0	1
10	Task 8	0	0	1		1	30		6-8	0	1
11	Task 9	0	0	1		1	15		7-8	0	0
12	Task 10	0	0	1		1	50		8-9	0	0
13									7-10	0	0
14	Station										
15	Time	125	122	116							
16											

Figure 7.2. Excel worksheet for Example 7.1.

Figure 7.2 displays the Excel worksheet for Example 7.1.
Figure 7.3 shows the corresponding Solver dialog box.

Cells G3:G12 in Figure 7.2 contain the task time parameters from
Figure 7.1. Cells B3:D12 hold the task assignment decision variables
(i.e., the x_{ij} variables) and cell K1 contains the cycle time decision
variable, C. Cell K1 is the 'Set Objective' cell in the Solver dialog
box in Figure 7.3.

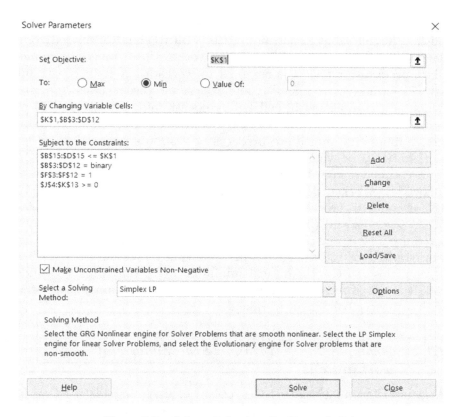

Figure 7.3. Solver dialog box for Example 7.1.

Cells F3:F12 contain the sum of the assignment variables for each task. These cells are constrained to equal one in accordance with constraint set (7.2). Cells B15:D15 are, for each of the three stations, the sumproduct of the assignment variables for the station and the corresponding task times in cells G3:G12. Therefore, these cells contain the total time assigned to the workstation. Cells B15:D15 are constrained to be equal to or less than cell K1 (cycle time) to enforce constraint set (7.3).

Cells J3:K12 are constructed to enforce the precedence constraint set (7.4). For each task and its immediate successors, there is a constraint for the first two stations. For example, consider task $j = 4$ and its set of immediate successors, $P_4 = \{7, 8\}$. In constraint set (7.4) where $j = 4$, $h = 7$, and $i = 1$, we have $x_{14} \geq x_{17}$, which is accomplished in cell J8 by constraining B6 – B9 to equal or exceed zero. That is, task 7 cannot be assigned to workstation 1 unless its immediate predecessor task 4 is assigned to workstation 1. In constraint set (7.4) where $j = 4$, $h = 7$, and $i = 2$, we have $x_{14} + x_{24} \geq x_{27}$, which is accomplished in cell K8 by constraining B6 + C6 – C9 to equal or exceed zero. That is, task 7 cannot be assigned to workstation 2 unless its immediate predecessor task 4 is assigned to either workstation 1 or workstation 2.

As observed from Figure 7.2, the optimal task assignments are as follows: (i) Workstation 1: tasks 1, 2, and 4, which require a total of 125 seconds, (ii) Workstation 2: tasks 3, 5, and 6, which require a total of 122 seconds, and (iii) Workstation 3: tasks 7, 8, 9, and 10, which a require a total of 116 seconds. The maximum time across the three workstations of 125 seconds is the cycle time shown in cell K1. This is very close to the theoretical minimum cycle time of 121 seconds (363 seconds of total task time divided by three stations) associated with a perfect balance of workload.

7.1.2 *Example 7.2 — Minimize the number of workstations*

A general formulation for the problem of assigning tasks to workstations to the number of workstations subject to assignment, precedence, and prespecified cycle time constraints is as follows:

Parameters

$n =$ the number of tasks indexed $1 \leq j \leq n$,
$m =$ an upper bound on the number of workstations indexed $1 \leq i \leq m$,
$t_j =$ the task time for activity j for $1 \leq j \leq n$,
$P_j =$ the set of immediate successors for activity j,
$C =$ cycle time.

Decision variables

$x_{ij} = 1$ if task j is assigned to workstation i and 0 otherwise for $1 \leq i \leq m$ and $1 \leq j \leq n$,
$y_i = 1$ if workstation i is used and 0 otherwise for $1 \leq i \leq m$.

Objective function

$$\text{Minimize: } \sum_{i=1}^{m} y_i. \tag{7.5}$$

Subject to constraints

$$\sum_{i=1}^{m} x_{ij} = 1 \quad \text{for } 1 \le j \le n, \tag{7.6}$$

$$Cy_i \ge \sum_{j=1}^{n} t_j x_{ij} \quad \text{for } 1 \le i \le m, \tag{7.7}$$

$$\sum_{l=1}^{i} x_{lj} \ge x_{ih} \quad \text{for } 1 \le j \le n, \ h \in P_j, \ 1 \le i \le m-1, \tag{7.8}$$

$$y_i \ge y_{i+1} \quad \text{for } 1 \le i \le m-1. \tag{7.9}$$

The objective function in Equation (7.5) is to minimize the number of workstations that are used. Constraint set (7.6) guarantees that each of the n tasks is assigned to exactly one workstation. Constraint set (7.7) ensures, for each possible workstation, that the sum of the times of the tasks assigned to that station is zero if the station is not selected and is less than the cycle time (C) if it is selected. Constraint set (7.8) requires for a given task h that is an immediate successor to task j that task h can only be assigned to workstation i if task j is assigned to workstation i or an earlier station. Finally, constraint set (7.9) ensures that selected workstations are opened in their natural order.

Example 7.2 will use the same $n = 10$ tasks, task times, and precedence relationship shown in Figure 7.1. In addition, the example will make use of an upper bound of $m = 3$ stations and prespecified cycle time of $C = 130$ seconds. Figure 7.4 displays the Excel worksheet for Example 7.2. Figure 7.5 shows the corresponding Solver dialog box.

The layout of Figure 7.4 is very similar to that of Figure 7.2, but there are some important noteworthy differences in Figure 7.4. First, the cycle time of 130 seconds in cell K1 is a parameter, not a decision variable, and it is not the 'Set Objective' cell in the Solver dialog box shown in Figure 7.5. Second, cells B17:D17 have been

	A	B	C	D	E	F	G	H	I	J	K
1							Task		Cycle Time =		130
2		Station 1	Station 2	Station 3		Asmt	Time				
3	Task 1	0	1	0		1	53		Precedence	Station 1	Station 2
4	Task 2	1	0	0		1	26		1-3	0	1
5	Task 3	0	0	1		1	28		2-4	0	1
6	Task 4	1	0	0		1	46		3-5	0	0
7	Task 5	0	0	1		1	54		4-6	0	1
8	Task 6	1	0	0		1	40		4-7	1	0
9	Task 7	0	1	0		1	21		5-8	0	0
10	Task 8	0	0	1		1	30		6-8	1	1
11	Task 9	0	0	1		1	15		7-8	0	1
12	Task 10	0	1	0		1	50		8-9	0	0
13									7-10	0	0
14	Station										
15	Time	112	124	127							
16											
17	Used?	1	1	1					Stations Needed =	3	
18											
19	Cap.	130	130	130							

Figure 7.4.   Excel worksheet for Example 7.2 (cycle time of $C = 130$ seconds).

added to incorporate the binary decision variables (y_i) controlling the opening of the three workstations. Their sum in cell J17 is the objective function. Third, cells B19:D19 contain cells B17:D17 multiplied by the cycle time in cell K1 to represent available capacity at the workstations, and the sum of task times in the workstations in cells B15:D15 is constrained to be equal to or less than the capacities in cells B19:D19. Fourth, constraints have been added to ensure workstations are opened in order: B17 equals or exceeds C17 and C17 equals or exceeds D17.

It should be noted that, based on the results in Figure 7.2 showing that the minimum cycle time for $m = 3$ workstations was 125 seconds, we already knew that three stations were sufficient for a prespecified cycle time of 130 seconds. We also knew that $m = 2$ workstations would be impossible for a cycle time of 130 seconds because $363/130 = 2.79$ and accordingly the theoretical minimum number of workstations is three. The task assignments in Figure 7.4 are rather different from the ones in Figure 7.2; however, the total times at the workstations (112, 124, and 127) are quite comparable. Arguably, the three-workstation layout in Figure 7.2 is preferable to the one in Figure 7.4 because the former has a cycle time of 125, whereas the latter has a cycle time of 127.

We reran the analyses for Example 7.2 after changing the cycle time to C $= 190$ seconds. Because $363/190 = 1.91$, two workstations are at least theoretically possible for the 190-second cycle time.

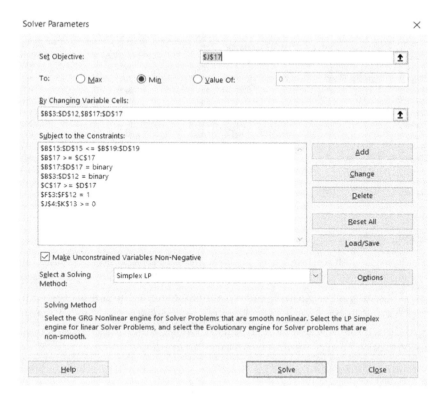

Figure 7.5.   Solver dialog box for Example 7.2.

Our model can be used to determine if this possibility can be realized. The results for the 190-second cycle time are shown in Figure 7.6.

Figure 7.6 reveals that a two-workstation line balance is feasible for a 190-second cycle time. The optimal task assignments are as follows: (i) Workstation 1: tasks $2, 4, 6, 7$, and 10, which require a total of 183 seconds; (ii) Workstation 2: tasks 1, 3, 5, 8, and 9, which require a total of 180 seconds. The maximum time across the two workstations of 183 seconds would be the precise cycle time of the layout, which is even better than the 190 seconds prespecified in the problem. In fact, the only better theoretical two-workstation balance would be 182 seconds in one workstation and 181 seconds in the other and such a solution does not exist.

It is interesting to consider whether or not the two workstations could be interchanged. That is, workstation 1 containing tasks 1, 3, 5,

	A	B	C	D	E	F	G	H	I	J	K
1							Task		Cycle Time =		190
2		Station 1	Station 2	Station 3		Asmt	Time				
3	Task 1	0	1	0		1	53		Precedence	Station 1	Station 2
4	Task 2	1	0	0		1	26		1-3	0	0
5	Task 3	0	1	0		1	28		2-4	0	1
6	Task 4	1	0	0		1	46		3-5	0	0
7	Task 5	0	1	0		1	54		4-6	0	1
8	Task 6	1	0	0		1	40		4-7	0	1
9	Task 7	1	0	0		1	21		5-8	0	0
10	Task 8	0	1	0		1	30		6-8	1	0
11	Task 9	0	1	0		1	15		7-8	1	0
12	Task 10	1	0	0		1	50		8-9	0	0
13									7-10	0	1
14	Station										
15	Time	183	180	0							
16											
17	Used?	1	1	0					Stations Needed =		2
18											
19	Cap.	190	190	0							

Figure 7.6.   Excel worksheet for Example 7.2 (190-second cycle time).

8, and 9 and workstation 2 containing the others. Such an interchange is not possible because of the precedence constraints. Task 8 cannot be assigned to workstation 1 if two of its immediate predecessors (tasks 6 and 7) are assigned to workstation 2.

7.2   Process Layout — Assigning Departments to Locations

7.2.1   *Example 7.3 — Minimize travel distance*

The quadratic assignment problem affords a general formulation to the problem of assigning departments to locations so as to minimize total travel distance.

Parameters

$n =$ the number of departments indexed $1 \leq i, j \leq n$, which is also equal to the number of locations for the departments, and the locations will be indexed $1 \leq k, l \leq n$.

$t_{ij} =$ the number of trips between departments i and j, whereby $t_{ij} = t_{ji}$ for all $1 \leq i < j \leq n$ and $t_{ii} = 0$ for all $1 \leq i \leq n$.

$d_{kl} =$ the distance between locations k and l, whereby $d_{kl} = d_{kl}$ for all $1 \leq k < l \leq n$ and $d_{kk} = 0$ for all $1 \leq k \leq n$.

Decision variables

$x_{ik} = 1$ if department i is assigned to location k and 0 otherwise for $1 \leq i \leq n$ and $1 \leq k \leq n$.

Objective function

$$\text{Minimize:} \quad \frac{1}{2} \sum_{i=1}^{n} \sum_{j=1}^{n} \sum_{k=1}^{n} \sum_{l=1}^{n} t_{ij} d_{kl} x_{ik} x_{jl}. \tag{7.10}$$

Subject to constraints

$$\sum_{i=1}^{n} x_{ik} = 1 \quad \text{for } 1 \leq k \leq n, \tag{7.11}$$

$$\sum_{k=1}^{n} x_{ik} = 1 \quad \text{for } 1 \leq i \leq n, \tag{7.12}$$

$$x_{ik} \in \{0, 1\} \quad \text{for } 1 \leq i \leq n \quad \text{and} \quad 1 \leq k \leq n. \tag{7.13}$$

The objective function in Equation (7.10) is to minimize the total distance traveled. The objective function is quadratic because of the $x_{ik} x_{jl}$ product, and a contribution to the objective function will only be picked up if both these binary variables equal one. In other words, if department i is assigned to location k and department j is assigned to location l, then the contribution to the total distance traveled is the number of trips between departments i and j (f_{ij}) times the distance between locations k and l (d_{kl}). Constraint set (7.11) ensures that each location gets assigned exactly one department and constraint set (7.12) guarantees that each department is assigned to exactly one location. Although superfluous because of the variable definitions, constraint set (7.13) emphasizes the binary nature of the decision variables.

The quadratic assignment problem is notoriously difficult to solve. It can be reformulated as an integer linear program but such an implementation using a spreadsheet would be unwieldy. We have prepared VBA macros for tackling the problem. One of the macros allows the user to input their own assignment of departments to locations and the macro computes the distance. A second, more powerful, macro searches for a good assignment using the computerized relative allocation of facilities technique (CRAFT: Armour & Buffa, 1963).

This is a heuristic approach to the problem that is not guaranteed to find a globally optimal solution. Nevertheless, it often will find the globally optimal solution for small problems and performs well for larger instances too.

Given an initial assignment of departments to locations, the CRAFT procedure examines all possible pairwise interchanges of departments and accepts those interchanges that reduce the total distance. The process continues until no pairwise interchange will further reduce the total distance. Because the algorithm is sensitive to the initial assignment, it is recommended that it be restarted a large number of times to help avoid the potential for a poor local optimum. The user can specify the number of restarts in the worksheet.

We focus on a 16-department layout problem that has been used for a class exercise in an MBA operations management course. The location arrangement is based on a 4×4 grid and letters are used to designate the locations.

A	B	C	D
E	F	G	H
I	J	K	L
M	N	O	P

The distances are measured rectangularly in 'yards' and are shown in the upper triangle of the matrix in cells C6:R21 of Figure 7.7. Rectangular distance measurements are not uncommon in warehouses and other facilities where aisles run horizontally and vertically. The number of daily trips between pairs of departments is shown in the lower triangle of the matrix in cells C6:R21.

The VBA macros read the number of departments/locations from cell H3. Subsequently, the distances between locations and the number of trips between departments are read from the C6:R21 cell range. Although the worksheet is displayed for the 16-department problem of interest, smaller problems can be tackled by specifying a different number of departments in cell H3 and placing the smaller trips/distance matrix beginning in the top left corner in cell C6.

The user can specify a trial assignment of departments to locations in cells C26:R26 and then click the 'Compute Distance' button to get the total distance for that layout. Figure 7.7 shows that the

Process Layout - Minimizing Interdepartmental Travel Distance

		Total Number of Departments				16			TRIPS			DISTANCE			
		A	B	C	D	E	F	G	H	I	J	K	L	M	N	O	P
		1	2	3	4	5	6	7	8	9	10	11	12	13	14	15	16
A	1	0	10	20	30	10	20	30	40	20	30	40	50	30	40	50	60
B	2	47	0	10	20	20	10	20	30	30	20	30	40	40	30	40	50
C	3	54	6	0	10	30	20	10	20	40	30	20	30	50	40	30	40
D	4	75	21	15	0	40	30	20	10	50	40	30	20	60	50	40	30
E	5	51	29	36	33	0	10	20	30	10	20	30	40	20	30	40	50
F	6	28	53	64	14	41	0	10	20	20	10	20	30	30	20	30	40
G	7	41	21	70	62	73	50		10	30	20	10	20	40	30	20	30
H	8	40	13	25	16	73	47	29		40	30	20	10	50	40	30	20
I	9	51	12	14	16	27	55	55	46		10	20	30	10	20	30	40
J	10	27	20	46	51	15	59	58	56	3		10	20	20	10	20	30
K	11	43	19	25	3	7	48	49	46	22	29		10	30	20	10	20
L	12	61	52	27	6	33	6	15	38	3	20	47		40	30	20	10
M	13	55	9	20	25	49	31	60	15	61	47	53	5		10	20	30
N	14	68	54	4	65	70	39	29	77	37	52	31	65	40		10	20
O	15	62	41	75	24	28	74	41	8	3	44	65	49	34	26		10
P	16	80	44	48	3	5	63	26	39	80	22	79	24	12	62	54	

TRY SOLUTIONS ON YOUR OWN

Position	A	B	C	D	E	F	G	H	I	J	K	L	M	N	O	P
Department	1	2	3	4	5	6	7	8	9	10	11	12	13	14	15	16

Total Distance = | 122890 | Compute Distance

LET THE COMPUTER DO THE WORK USING PAIRWISE CRAFT

Number of Craft Iterations | 100

Position	A	B	C	D	E	F	G	H	I	J	K	L	M	N	O	P
Department	13	9	16	11	7	10	6	3	5	8	1	15	4	14	2	12

Total Distance = | 104700 | Run Craft Procedure

Figure 7.7. Excel worksheet for Example 7.3.

default layout of department 1 assigned to location A, department 2 to location B, department 3 to location C, etc., will yield a total travel distance of 122,890 yards.

The pairwise CRAFT method is implemented at the bottom of Figure 7.7. The specified number of iterations (restarts) is 100 in cell E32. This is generally sufficient for problems where $n \leq 16$ but can be increased to 1,000 or even 10,000 for greater confidence in the solution. Clicking the 'Run Craft Procedure' button runs the macro and cells C35:R35 are populated with the assignments. The best

layout found has a distance of 104,700 yards and is displayed in a
4 × 4 grid as follows:

A = 13	B = 9	C = 16	D = 11
E = 7	F = 10	G = 6	H = 3
I = 5	J = 8	K = 1	L = 15
M = 4	N = 14	O = 2	P = 12

When using the VBA macro in a pedagogical setting, it is important to explain to students that, for some data structures, rotations
and mirror images of a given layout can yield the same total distance.
For example, the mirror image of the layout immediately above is as
follows:

A = 11	B = 16	C = 9	D = 13
E = 3	F = 6	G = 10	H = 7
I = 15	J = 1	K = 8	L = 5
M = 12	N = 2	O = 14	P = 4

This layout has the same total distance of 104,700 yards. If we
rotate this layout clockwise so that the rows become the columns
(e.g., the first row becomes the fourth column), then we get the
following layout, which also has a total distance of 104,700 yards:

A = 12	B = 15	C = 3	D = 11
E = 2	F = 1	G = 6	H = 16
I = 14	J = 8	K = 10	L = 9
M = 4	N = 5	O = 7	P = 13

This particular layout was obtained from a second run of the pairwise CRAFT heuristic, as displayed in Figure 7.8. At first glance, cells
C35:R35 in Figure 7.8 look appreciably different from the same cells
in Figure 7.7, despite the fact that the total distance values are the
same. Understanding the issues pertaining to rotations and mirror
images is important for reconciling the two solutions. Of course, the
issue here is created by the definition of the distance measure used
(10-yard increments between locations in the horizontal and verti-
cal directions of the grid) and might be irrelevant for other distance
measures.

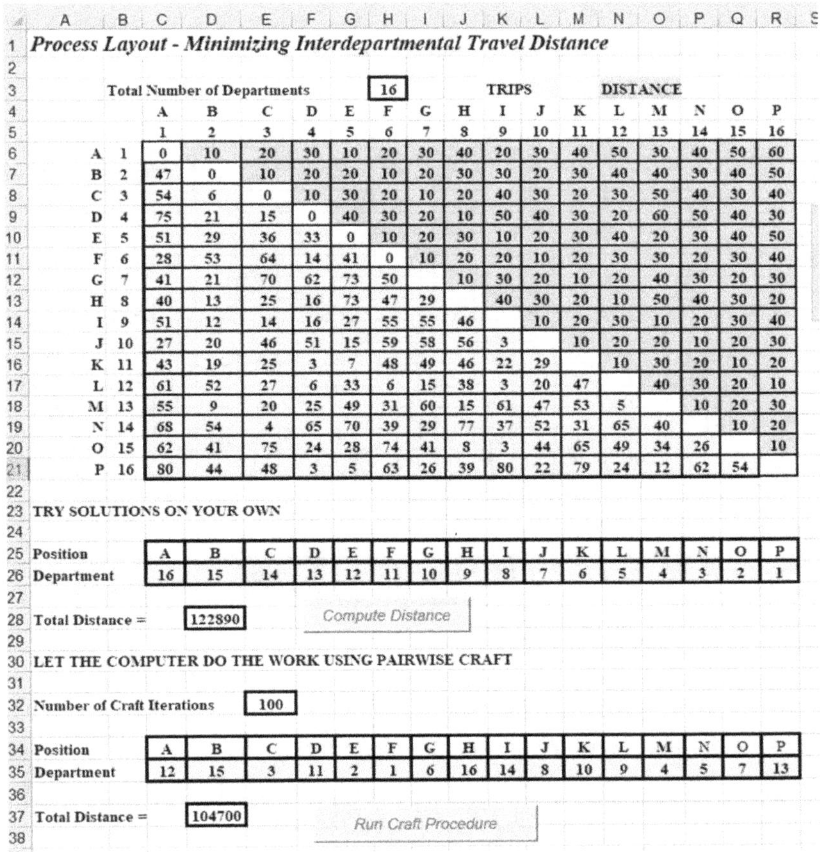

Process Layout - Minimizing Interdepartmental Travel Distance

Total Number of Departments [16]   TRIPS   DISTANCE

		A	B	C	D	E	F	G	H	I	J	K	L	M	N	O	P
		1	2	3	4	5	6	7	8	9	10	11	12	13	14	15	16
A	1	0	10	20	30	10	20	30	40	20	30	40	50	30	40	50	60
B	2	47	0	10	20	20	10	20	30	30	20	30	40	40	30	40	50
C	3	54	6	0	10	30	20	10	20	40	30	20	30	50	40	30	40
D	4	75	21	15	0	40	30	20	10	50	40	30	20	60	50	40	30
E	5	51	29	36	33	0	10	20	30	10	20	30	40	20	30	40	50
F	6	28	53	64	14	41	0	10	20	20	10	20	30	30	20	30	40
G	7	41	21	70	62	73	50		10	30	20	10	20	40	30	20	30
H	8	40	13	25	16	73	47	29		40	30	20	10	50	40	30	20
I	9	51	12	14	16	27	55	55	46		10	20	30	10	20	30	40
J	10	27	20	46	51	15	59	58	56	3		10	20	20	10	20	30
K	11	43	19	25	3	7	48	49	46	22	29		10	30	20	10	20
L	12	61	52	27	6	33	6	15	38	3	20	47		40	30	20	10
M	13	55	9	20	25	49	31	60	15	61	47	53	5		10	20	30
N	14	68	54	4	65	70	39	29	77	37	52	31	65	40		10	20
O	15	62	41	75	24	28	74	41	8	3	44	65	49	34	26		10
P	16	80	44	48	3	5	63	26	39	80	22	79	24	12	62	54	

TRY SOLUTIONS ON YOUR OWN

Position	A	B	C	D	E	F	G	H	I	J	K	L	M	N	O	P
Department	16	15	14	13	12	11	10	9	8	7	6	5	4	3	2	1

Total Distance = [122890]   *Compute Distance*

LET THE COMPUTER DO THE WORK USING PAIRWISE CRAFT

Number of Craft Iterations [100]

Position	A	B	C	D	E	F	G	H	I	J	K	L	M	N	O	P
Department	12	15	3	11	2	1	6	16	14	8	10	9	4	5	7	13

Total Distance = [104700]   *Run Craft Procedure*

Figure 7.8. Excel worksheet for Example 7.3 (alternative run of CRAFT procedure).

Another interesting use of the process layout worksheet is to analyze the impact of various constraints on the layout. For example, we have asked students to analyze the effect on total distance by adding a constraint that departments 5 and 7 must be at least 30 yards apart (they are only 10 yards apart in the earlier layouts).

Typically, students will tackle this problem by using the 'Compute Distance' macro to evaluate various layouts in the neighborhood of the unconstrained solution after manually adjusting the assignments to separate departments 5 and 7. This approach, however, has proven rather suboptimal. The primary problem is that students often limit

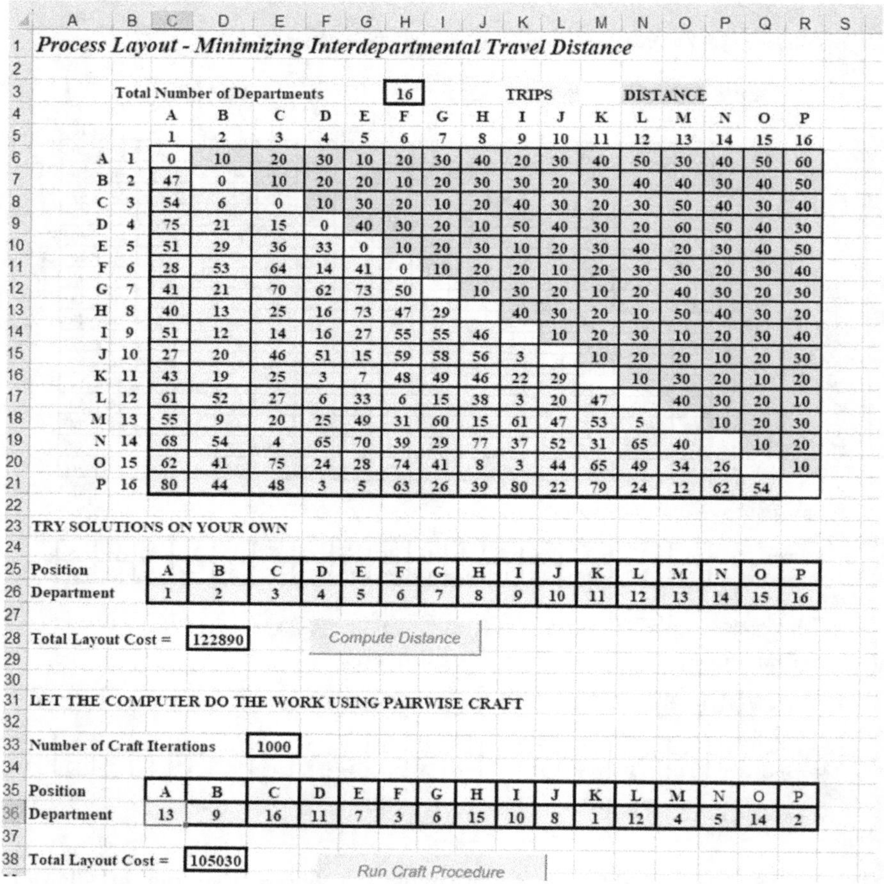

Process Layout - Minimizing Interdepartmental Travel Distance

Total Number of Departments | 16 | TRIPS   DISTANCE

		A	B	C	D	E	F	G	H	I	J	K	L	M	N	O	P
		1	2	3	4	5	6	7	8	9	10	11	12	13	14	15	16
A	1	0	10	20	30	10	20	30	40	20	30	40	50	30	40	50	60
B	2	47	0	10	20	20	10	20	30	30	20	30	40	40	30	40	50
C	3	54	6	0	10	30	20	10	20	40	30	20	30	50	40	30	40
D	4	75	21	15	0	40	30	20	10	50	40	30	20	60	50	40	30
E	5	51	29	36	33	0	10	20	30	10	20	30	40	20	30	40	50
F	6	28	53	64	14	41	0	10	20	20	10	20	30	30	20	30	40
G	7	41	21	70	62	73	50		10	30	20	10	20	40	30	20	30
H	8	40	13	25	16	73	47	29		40	30	20	10	50	40	30	20
I	9	51	12	14	16	27	55	55	46		10	20	30	10	20	30	40
J	10	27	20	46	51	15	59	58	56	3		10	20	20	10	20	30
K	11	43	19	25	3	7	48	49	46	22	29		10	30	20	10	20
L	12	61	52	27	6	33	6	15	38	3	20	47		40	30	20	10
M	13	55	9	20	25	49	31	60	15	61	47	53	5		10	20	30
N	14	68	54	4	65	70	39	29	77	37	52	31	65	40		10	20
O	15	62	41	75	24	28	74	41	8	3	44	65	49	34	26		10
P	16	80	44	48	3	5	63	26	39	80	22	79	24	12	62	54	

TRY SOLUTIONS ON YOUR OWN

Position	A	B	C	D	E	F	G	H	I	J	K	L	M	N	O	P
Department	1	2	3	4	5	6	7	8	9	10	11	12	13	14	15	16

Total Layout Cost = | 122890 | *Compute Distance*

LET THE COMPUTER DO THE WORK USING PAIRWISE CRAFT

Number of Craft Iterations | 1000 |

Position	A	B	C	D	E	F	G	H	I	J	K	L	M	N	O	P
Department	13	9	16	11	7	3	6	15	10	8	1	12	4	5	14	2

Total Layout Cost = | 105030 | *Run Craft Procedure*

Figure 7.9.   Excel spreadsheet for Example 7.3 with separation constraint for 5 and 7.

their focus to department location changes of departments 5, 7, and a couple of other departments that are moved to accomplish the separation. The broader impacts on neighboring (unadjusted) departments are not adequately considered. A better method is to modify the VBA macro to forbid solutions where departments 5 and 7 are not adequately separated. This results in the solution in Figure 7.9.

The constrained layout in Figure 7.9 when expressed in the 4×4 grid is as follows:

A = 13	B = 9	C = 16	D = 11
E = 7	F = 3	G = 6	H = 15
I = 10	J = 8	K = 1	L = 12
M = 4	N = 5	O = 14	P = 2

This layout has a total distance of 105,030 yards, which is only 330 yards (0.3%) more than the layout in Figure 7.7. However, there are some substantial differences between the two layouts.

The separation of departments 5 and 7 to their locations E and N could have been accomplished simply by taking the solution in Figure 7.7 and moving department 5 from I to N and department 14 from N to I, but this is suboptimal. Much more rearrangement of other departments needs to be accomplished when department 5 is moved. A total of seven departments (2, 3, 5, 10, 12, 14, and 15) have different locations in the solution in Figure 7.9 compared to the layout in Figure 7.7.

Chapter 8

Project Scheduling

Time/cost trade-offs in project networks have been studied for nearly six decades (Fulkerson, 1961; Kelley, 1961). Although there are different versions of time/cost trade-off problems, the general principle is to shorten the completion time of some (or all) of the project activities so as to either (1) complete the project at minimum cost subject to a project completion time constraint or (2) complete the project in the minimum amount of time subject to a budget constraint. Shortening the completion time of an activity is known as *crashing*. For each activity, there is a *normal* completion time that represents the time for the activity under 'normal' conditions as well as a *crash* time that represents the minimum possible completion time for the activity if additional resources are committed. It is commonly assumed that the time/cost trade-off for each activity is a continuous linear function over the interval from crash to normal time; however, discrete (nonlinear) time/cost trade-offs can also be modeled (De *et al.*, 1995). In this chapter, we will examine models for both linear and nonlinear cost structures.

8.1   Linear Crashing Costs

We present two formulations for the project network crashing problem under the assumption of a linear cost structure. The first is a traditional formulation that is found in many management science textbooks. The second is an alternative formulation proposed by Huse and Brusco (2021) that commonly has fewer variables and constraints and is conceptually easier to understand. The two formulations share much of the same notation and our presentation is largely taken from Huse and Brusco (2021). We begin with precise definitions of the relevant parameters, sets, and decision variables. This is followed by the presentation of the traditional and alternative formulations. We conclude the section with two numerical examples to help compare the formulations.

8.1.1   *Notation*

Parameters and sets

$n =$ the number of activities, indexed $j = 1, 2, \ldots, n$;

$\tau_{Nj} =$ the normal time for activity j (for $1 \leq j \leq n$);

$\tau_{Cj} =$ the crash time (i.e., minimum possible completion time) for activity j (for $1 \leq j \leq n$);

$\omega_{Nj} =$ the cost for activity j at the normal time (for $1 \leq j \leq n$);

$\omega_{Cj} =$ the cost for activity j at the crash time, such that $\omega_{Cj} \geq \omega_{Nj}$ (for $1 \leq j \leq n$);

$m_j =$ the maximum number of time periods to which activity j can be shortened;

$$m_j = (\tau_{Nj} - \tau_{Cj}) \quad \forall 1 \leq j \leq n; \tag{8.1}$$

$c_j =$ the per-time period crashing cost for activity j (for $1 \leq j \leq n$), which is

$$c_j = \begin{cases} \frac{(\omega C_j - \omega N_j)}{m_j} & \text{if } m_j > 0 \\ 0 & \text{if } m_j = 0 \end{cases} \quad \forall 1 \leq j \leq n; \tag{8.2}$$

$T =$ the target project completion time;

$Q_j =$ the set of activities that are immediate predecessors for activity j (for $1 \leq j \leq n$);

$G =$ the set of activities with no successors.

Decision variables

$x_j =$ the early finish time for activity j (for $1 \leq j \leq n$);

$y_j =$ the number of time periods that activity j is crashed/shortened by (for $1 \leq j \leq n$).

8.1.2 *Traditional formulation (F1)*

With the parameters, sets, and decision variables in place, the traditional formulation for project network crashing (F1) can be stated as follows:

$$\text{min: } Z_1 = \sum_{j=1}^{n} c_j y_j \tag{8.3}$$

subject to

$$y_j \leq m_j \quad \text{for } 1 \leq j \leq n, \tag{8.4}$$

$$x_j \leq T \quad \text{for } j \in G, \tag{8.5}$$

$$x_j \geq (\tau_{Nj} - y_j) + x_k \quad \text{for } 1 \leq j \leq n, \ k \in Q_j, \tag{8.6}$$

$$x_j \geq 0 \quad \text{for } 1 \leq j \leq n, \tag{8.7}$$

$$y_j \geq 0 \quad \text{for } 1 \leq j \leq n. \tag{8.8}$$

The objective function, Z_1, of F1 in Equation (8.3) is the total crashing cost, which is to be minimized. Constraint set (8.4) ensures that the number of time periods to which each activity is crashed does not exceed its maximum possible value. Constraint set (8.5) requires that all terminal activities (i.e., those with no successors) are finished within the target project completion time (T). Constraint set (8.6) guarantees, for each activity j and each of its immediate predecessors ($k \in Q_j$), that the early completion time for activity j equals or exceeds the early completion time for its predecessor plus its own normal completion time minus the number of time periods by which it is crashed. Constraint sets (8.7) and (8.8) are nonnegativity restrictions on the decision variables.

8.1.3 *Alternative formulation (F2)*

When transitioning from the traditional formulation (F1) to the alternative formulation (F2), the x_j variables and constraint sets (8.5) and (8.6) are eliminated. The equivalence of the two formulations is demonstrated by Huse and Brusco (2021). We define p as the number of paths through the network, L_i as the length (total completion time) for path i (for all $1 \leq i \leq p$), and S_i as the set of all activities on path i (for all $1 \leq i \leq p$). Formulation F2 seeks to minimize (8.3) subject to constraint sets (8.4), (8.8), and the following constraint set:

$$\sum_{j \in S_i} y_j \geq L_i - T \quad \text{for } 1 \leq i \leq p : L_i > T. \tag{8.9}$$

Formulation F2 has one-half the variables of F1. It also replaces constraint set (8.6), which is arguably the most conceptually challenging to grasp, with constraint set (8.9). The interpretation of constraint set (8.9) is as follows: For each path (i), such that the length of path i exceeds T (i.e., $L_i > T$), the activities on path i must be shortened by enough time to reduce the length of the path from L_i to T. For most project networks in management science textbooks, the number of constraints in (8.9) is fewer than the number of constraints in (8.6). Although it is easy to synthetically construct high-density (i.e., high interdependence among activities) networks where the number of constraints in (8.9) greatly exceeds the number of constraints in (8.6), such networks are not particularly relevant for pedagogical purposes. Moreover, many real-world networks are of a low to moderate density and, accordingly, F2 is applicable for such networks.

For completeness, as is the case for F1, F2 can be modified to minimize completion time subject to a budget constraint, as opposed to minimizing cost subject to a time constraint. This is accomplished by changing T to a decision variable, changing the objective function to min: T, and limiting the amount spent on crashing to a specified budget value (B) via the constraint

$$\sum_{j=1}^{n} c_j y_j \leq B. \tag{8.10}$$

8.1.4   *Example 8.1 — Minimize cost*

To compare formulations F1 and F2, we provide an example of a small network comparable to the size of those commonly encountered in introductory management science textbooks. The precedence diagram for the example network is displayed in Figure 8.1. The time and cost parameters, sets, and paths are provided in Figure 8.2. The LP formulations for Example 8.1 are provided in Figure 8.3.

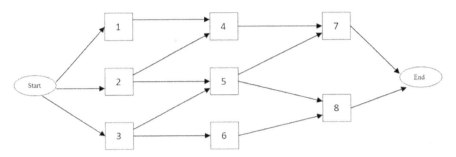

Figure 8.1. Precedence diagram for Example 8.1.

Activity (j)	1	2	3	4	5	6	7	8
Normal cost (ω_{Nj})	3000	1200	1800	4200	1600	1800	1400	3000
Crash cost (ω_{Cj})	5800	2000	3000	6000	2500	2400	3000	5000
Normal time (τ_{Nj})	10	7	8	9	7	5	5	9
Crash time (τ_{Cj})	6	5	4	6	6	3	3	5
Maximum crash time (m_j)	4	2	4	3	1	2	2	4
Per week crash cost (c_j)	700	400	300	600	900	300	800	500

$n = 8, p = 7, T = 20$

$G = \{7, 8\}, Q_1 = Q_2 = Q_3 = \varnothing$

$Q_4 = \{1, 2\}, Q_5 = \{2, 3\}, Q_6 = \{3\}$

$Q_7 = \{4, 5\}, Q_8 = \{5, 6\}$

Path $i = 1$: $S_1 = \{1, 4, 7\}, L_1 = 24$

Path $i = 2$: $S_2 = \{2, 4, 7\}, L_2 = 21$

Path $i = 3$: $S_3 = \{2, 5, 7\}, L_3 = 19$

Path $i = 4$: $S_4 = \{2, 5, 8\}, L_4 = 23$

Path $i = 5$: $S_5 = \{3, 5, 7\}, L_5 = 20$

Path $i = 6$: $S_6 = \{3, 5, 8\}, L_6 = 24$

Path $i = 7$: $S_7 = \{3, 6, 8\}, L_7 = 22$

Figure 8.2. Cost, crashing, and path data for Example 8.1.

The F2 formulation in the bottom panel of Figure 8.3 is clearly more compact than the F1 formulation in the top panel. Although the objective functions (8.3) and upper bound constraints (8.4) for crashing activities are the same for the two formulations, the F1 formulation model requires 14 additional constraints (12 of these are to link the early finish times to the crashing decisions), whereas the F2 formulation requires only five (one for each path of length greater

Min: $700y_1 + 400y_2 + 300y_3 + 600y_4 + 900y_5 + 300y_6 + 800y_7 + 500y_8$

Subject to:

$y_1 \le 4$	$y_2 \le 2$	$y_3 \le 4$	$y_4 \le 3$	$y_5 \le 1$	$y_6 \le 2$	$y_7 \le 2$	$y_8 \le 4$

$x_7 \le 20$ $x_8 \le 20$

$x_1 \ge (10 - y_1)$ $x_4 \ge (9 - y_4) + x_1$ $x_5 \ge (7 - y_5) + x_3$ $x_7 \ge (5 - y_7) + x_4$ $x_8 \ge (9 - y_8) + x_5$

$x_2 \ge (7 - y_2)$ $x_4 \ge (9 - y_4) + x_2$ $x_6 \ge (5 - y_6) + x_3$ $x_7 \ge (5 - y_7) + x_5$ $x_8 \ge (9 - y_8) + x_6$

$x_3 \ge (8 - y_3)$ $x_5 \ge (7 - y_5) + x_2$

Min: $700y_1 + 400y_2 + 300y_3 + 600y_4 + 900y_5 + 300y_6 + 800y_7 + 500y_8$

Subject to:

$y_1 \le 4$	$y_2 \le 2$	$y_3 \le 4$	$y_4 \le 3$	$y_5 \le 1$	$y_6 \le 2$	$y_7 \le 2$	$y_8 \le 4$

$y_1 + y_4 + y_7 \ge 4$ $y_2 + y_4 + y_7 \ge 1$ $y_2 + y_5 + y_8 \ge 3$ $y_3 + y_5 + y_8 \ge 4$ $y_3 + y_6 + y_8 \ge 2$

Figure 8.3.   F1 and F2 formulations for Example 8.1.

	A	B	C	D	E	F	G	H	I	J	K	L	M	N
1	TARGET TIME =	20		Y_1	Y_2	Y_3	Y_4	Y_5	Y_6	Y_7	Y_8		COST =	4300
2														
3	Decision Variables			1	0	1	3	0	0	0	3			
4														
5	Path	Length		Constraint Matrix									LHS	RHS
6	1-4-7	24		1			1			1			4	4
7	2-4-7	21			1		1			1			3	1
8	2-5-7	19			1			1		1			0	-1
9	2-5-8	23			1			1			1		3	3
10	3-5-7	20				1		1		1			1	0
11	3-5-8	24				1		1			1		4	4
12	3-6-8	22				1			1		1		4	2
13														
14	CRASH LIMIT			4	2	4	3	1	2	2	4			
15	PER WEEK COST			700	400	300	600	900	300	800	500			

Figure 8.4.   Excel worksheet for Example 8.1.

than $T = 20$). Although there are seven paths in the network, two are excluded in Figure 8.3 (i.e., paths 3 and 5) because they are already 20 weeks or less. Moreover, it is evident that the F2 formulation does not require the x_j variables.

Figure 8.4 displays the Excel spreadsheet for the F2 formulation, and the corresponding Solver dialog box is shown in Figure 8.5. The target completion time is in cell B1 of Figure 8.4 and the normal time path length parameters are in cells B6:B12. The target completion times minus the normal time path lengths are in cells N6:N12. The maximum crash times for each activity and the per-week crashing

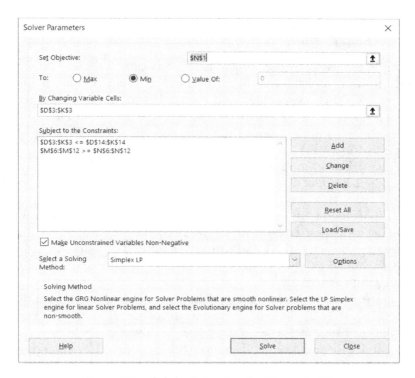

Figure 8.5. Solver dialog box for Example 8.1.

costs are in cells D14:K15. The constraint coefficients for the crashing variables are in cells D6:K12.

The crashing decision variables are in cells D3:K3. The sumproduct of these variables and the per-week crashing costs in cells D15:K15 is entered in cell N1 and this is the objective function. The decision variables in row 3 are constrained to be less than the maximum crashing limits in D14:K14. The decision variables in row 3 are multiplied by their constraint coefficients to produce the left-hand side of constraint set (8.3) in cells M6:M12 and these are constrained to equal or exceed the values in N6:N12.

Figure 8.4 reveals that the optimal solution to the F2 formulation is $y_1^* = y_3^* = 1$, $y_4^* = y_8^* = 3$, and all other $y_j^* = 0$. The total crashing cost is $Z_1^* = \$4,300$. The template for the LP formulation contains all seven paths, even though two of the paths already satisfied the target completion time of $T = 20$.

8.1.5   *Example 8.2 — Minimize time subject to a budget limit*

Example 8.2 uses the same network, cost parameters, and maximum crashing data as Example 8.1. However, in this example, a budget limit of $5,500 is prespecified and the goal is to minimize the project completion time, T. Figure 8.6 displays the Excel spreadsheet for the modified formulation and the corresponding Solver dialog box is shown in Figure 8.7.

The worksheet modifications are modest. Cell N2 is added to enter the budget limit of $5,500 and cell N1 is constrained to be equal to or less than this limit. Cell B1 is now a decision variable rather than a parameter. In cells M6:M12, the sumproducts of the crashing variables and their constraint set (8.3) coefficients are added to the difference between cell B1 and the path length. The values in M6:M12 are constrained to equal or exceed zero, so as to enforce the conditions in constraint set (8.9).

Figure 8.6 reveals that the optimal solution to the formulation is $y_1^* = 2$, $y_3^* = 1$, $y_4^* = 3$, $y_8^* = 4$, and all other $y_j^* = 0$. The project completion time for the $5,500 budget is 19 weeks.

If the F2 formulation for Example 8.1 is modified by changing the completion time from 20 weeks to 19 weeks, then it will be discovered that the minimum cost necessary to achieve the 19-week completion time is $5,500. Thus, an additional cost of $1,200 must be incurred to get the completion time down from 20 to 19 weeks.

	A	B	C	D	E	F	G	H	I	J	K	L	M	N
1	COMP TIME	19		Y_1	Y_2	Y_3	Y_4	Y_5	Y_6	Y_7	Y_8		COST =	5500
2													BUDGET	5500
3	Decision Variables			2	0	1	3	0	0	0	4			
4														
5	Path	Length		Constraint Matrix									LHS	
6	1-4-7	24		1				1		1			0	
7	2-4-7	21			1		1			1			1	
8	2-5-7	19			1			1		1			0	
9	2-5-8	23			1			1			1		0	
10	3-5-7	20				1		1		1			0	
11	3-5-8	24				1		1			1		0	
12	3-6-8	22				1			1		1		2	
13														
14	CRASH LIMIT			4	2	4	3	1	2	2	4			
15	PER WEEK COST			700	400	300	600	900	300	800	500			

Figure 8.6.   Excel spreadsheet for Example 8.2.

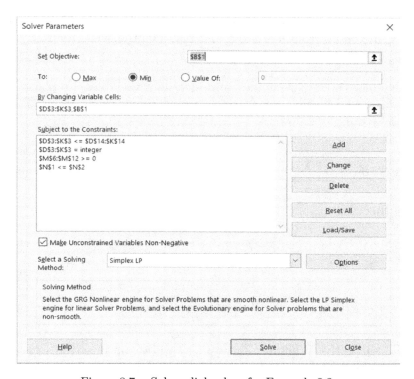

Figure 8.7.   Solver dialog box for Example 8.2.

8.2   Nonlinear Crashing Costs

8.2.1   *Modified F2 formulation for discrete crashing costs (F2-D)*

Next, we describe how the F2 formulation can be adapted to accommodate discrete (nonlinear) crashing costs. The extended formulation (F2-D) requires the definition of the following cost parameters, which will replace the c_j parameters:

b_{jh} = the additional cost that is incurred when increasing the number of time periods that activity j is crashed from $h - 1$ to h (for $1 \leq j \leq n$ and $1 \leq h \leq m_j$).

In addition, it is necessary to replace the y_j variables with the following binary crashing variables:

$v_{jh} = 1$ if the number of time periods that activity j is crashed is increased from $h-1$ to h and $v_{jh} = 0$ otherwise (for $1 \leq j \leq n$ and $1 \leq h \leq m_j$).

With these new parameters and decision variables in place, the extended formulation for discrete project network crashing (F2-D) can be stated as follows:

$$\text{min: } Z_2 = \sum_{j=1}^{n} \sum_{h=1}^{m_j} b_{jh} v_{jh} \tag{8.11}$$

subject to

$$\sum_{j \in S_i} \sum_{h=1}^{m_j} v_{jh} \geq L_i - T \quad \text{for } 1 \leq i \leq p : L_i > T, \tag{8.12}$$

$$v_{jh} \geq v_{j(h+1)} \quad \text{for } 1 \leq j \leq n \text{ and } 1 \leq h \leq m_j - 1, \tag{8.13}$$

$$v_{jh} \in \{0,1\} \quad \text{for } 1 \leq j \leq n, \ 1 \leq h \leq m_j. \tag{8.14}$$

The objective function, Z_2, of F2-D to be minimized in Equation (8.11) is the total crashing cost, which is computed as the product of the binary crashing variables and their corresponding incremental costs for shortening an activity by one additional time period. Constraint set (8.12) is analogous to (8.9) but now captures the crashing of each activity as the sum of binary variables. Constraint set (8.13) ensures that binary crashing variables are selected 'in order'. For example, suppose that the cost of crashing activity 1 is $b_{11} = \$700$ for the first week and the cost is $b_{12} = \$600$ (additional) to shorten the activity by a second week. Without the constraint $v_{11} \geq v_{12}$, the solution to the model could shorten the activity by only one week yet do so by selecting the variable corresponding to the second week (instead of the first) because it is cheaper. Constraint set (8.14) places binary restrictions on the decision variables.

8.2.2 *Example 8.3 — Nonlinear crashing example*

The data for the discrete crashing problem are identical to the data provided in Figure 8.2 for Example 8.1 with one exception: The c_j

	A	B	C	D	E	F	G	H	I	J	K	L	M	N
1	TARGET TIME =	20		Y_1	Y_2	Y_3	Y_4	Y_5	Y_6	Y_7	Y_8		COST =	5100
2														
3	Total Crashing			2	1	2	0	0	0	2	2			
4														
5	Path	Length		Constraint Matrix									LHS	RHS
6	1-4-7	24		1				1		1			4	4
7	2-4-7	21			1			1		1			3	1
8	2-5-7	19			1		1			1			3	-1
9	2-5-8	23			1		1				1		3	3
10	3-5-7	20				1	1			1			4	0
11	3-5-8	24				1	1				1		4	4
12	3-6-8	22				1			1		1		4	2
13														
14	CRASH LIMIT			4	2	4	3	1	2	2	4			
15	PER WEEK COST	Week 1		700	400	300	600	900	300	800	500			
16		Week 2		600	300	400	900	99999	400	600	800			
17		Week 3		1200	99999	500	800	99999	99999	99999	1000			
18		Week 4		500	99999	600	99999	99999	99999	99999	800			
19														
20	Decision Variables	Week 1		1	1	1	0	0	0	1	1			
21		Week 2		1	0	1	0	0	0	1	1			
22		Week 3		0	0	0	0	0	0	0	0			
23		Week 4		0	0	0	0	0	0	0	0			

Figure 8.8. Excel worksheet for Example 8.3.

cost parameters are replaced with b_{jh} parameters, which are displayed in rows 15–18 of the Excel template in Figure 8.8. To illustrate, in Example 8.1, the cost of crashing activity 1 was \$700 per week and it was possible to shorten activity 1 by up to four weeks. In Example 8.3, rows 15–18 of column D in Figure 8.8 show that the cost of crashing activity 1 is \$700 for the first week. However, to shorten the activity by a second week costs only an additional \$600. To shorten the activity by a third week will cost an additional \$1,200, and a fourth week will add another \$500. Accordingly, formulation F2-D allows for modeling of nonlinear time/cost trade-offs using a piecewise linear function. The Solver dialog box for the template shown in Figure 8.8 is displayed in Figure 8.9.

The maximum that any activity can be crashed is four weeks and, therefore, we defined four weeks of v_{jh} variables for each activity in rows 20–23 even though most activities allow fewer weeks of crashing. For example, activity 2 can be crashed by up to only two weeks. To ensure that three or four weeks of crashing were not selected, we imposed prohibitively large cost penalties of 99999 on these variables (see cells E17 and E18 in Figure 8.8). We also computed, in row 3 'Total Crashing', the y_j variables as the sum of the v_{jh} variables. For example, cell D3 in Figure 8.8 is the sum of cells D20:D23.

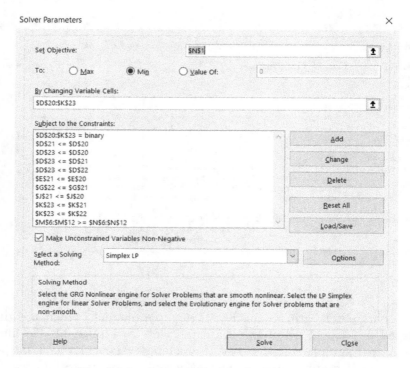

Figure 8.9.   Solver dialog box for Example 8.3.

The optimal solution to the F2-D formulation is $y_1^* = y_3^* = y_7^* = y_8^* = 2$, $y_2^* = 1$, and all other variables equal zero. For example, the binary variables for activity 1 are $v_{11}^* = v_{12}^* = 1$ and $v_{13}^* = v_{14}^* = 0$; therefore, $y_1^* = v_{11}^* + v_{12}^* = 2$. The cost incurred for crashing activity 1 by one week is \$700 and an additional \$600 is incurred for the second week. Likewise, the binary variables for activity 2 are $v_{21}^* = 1$ and $v_{22}^* = v_{23}^* = v_{24}^* = 0$. Therefore, activity 2 is crashed by only one week $(y_2^* = v_{21}^* = 1)$ at a cost of \$400. The total cost of crashing required to shorten the project to 20 weeks is $Z_2^* = \$5,100$.

It is interesting to compare the F2 and F2-D results in Figures 8.4 and 8.8, respectively. The solution in Figure 8.4 shows that activity 4 is crashed by three weeks; however, activity 4 is not crashed at all in the solution in Figure 8.8. Contrastingly, activity 7 is not crashed at all in the Figure 8.4 solution but is crashed by two weeks in the Figure 8.8 solution. Part of the explanation for this result can be deduced from examination of the 1–4–7 path. For the F2 model in

Figure 8.4, the per-week cost of crashing activity 4 ($600) is less expensive than the corresponding costs for activities 1 ($700) and 7 ($800), making it the preferable option for shortening the 1–4–7 path. However, for the nonlinear cost structure for the F2-D model in Figure 8.8, although activity 4 is still less expensive than activities 1 and 7 for the first week of crashing, it is appreciably more expensive for the second week of crashing.

8.2.3 *Example 8.4 — Another nonlinear crashing example*

A project requires the completion of five activities. There are four paths in the network: (i) 1–3–5, (ii) 1–4–5, (iii) 2–3–5, and (iv) 2–4–5. Table 8.1 provides the normal time (in weeks) for each activity, the maximum number of weeks each activity can be crashed, and the cost of crashing each activity by each additional week.

For example, if activity 1 is crashed by one week, then the cost is $850. If activity 1 is crashed by two weeks, then the cost is $1,425 ($850 for the first week plus $575 for the second). If activity 1 is crashed by three weeks, then the cost is $2,325 ($850 for the first week, plus $575 for the second, plus $900 for the third). The following is an integer linear programming formulation that, when solved, will

Table 8.1. Data for nonlinear crashing example.

	Activity 1	Activity 2	Activity 3	Activity 4	Activity 5
Normal completion time	10	9	9	8	10
Maximum crash time	4	3	3	2	3
Cost of crashing by one week	850	375	725	550	975
Additional cost if crashed a second week	575	500	725	925	750
Additional cost if crashed a third week	900	725	950	XXX	600
Additional cost if crashed a fourth week	425	XXX	XXX	XXX	XXX

provide a minimum cost crashing plan that will ensure a project completion time of no more than 24 weeks.

Parameters

m_j = the maximum number of weeks activity j can be crashed to (for $j = 1, 2, 3, 4, 5$).

b_{jh} = the additional cost that is incurred when increasing the number of weeks that activity j is crashed to from $h - 1$ to h (for $1 \leq j \leq 5$ and $1 \leq h \leq 4$), where $b_{jh} = 9{,}999$ for $h > m_j$ (for $j = 1, 2, 3, 4, 5$ and $h = 1, 2, 3, 4$).

Decision variables

v_{jh} = 1 if the number of weeks that activity j is crashed by is increased from $h - 1$ to h and $v_{jh} = 0$ otherwise (for $1 \leq j \leq 5$ and $1 \leq h \leq 4$).

Objective function: Minimize: $\sum_{j=1}^{5} \sum_{h=1}^{4} b_{jh} v_{jh}$

Subject to constraints

$$\sum_{h=1}^{4} (v_{1h} + v_{3h} + v_{5h}) \geq 29 - 24$$

$$\sum_{h=1}^{4} (v_{1h} + v_{4h} + v_{5h}) \geq 28 - 24$$

$$\sum_{h=1}^{4} (v_{2h} + v_{3h} + v_{5h}) \geq 28 - 24$$

$$\sum_{h=1}^{4} (v_{2h} + v_{4h} + v_{5h}) \geq 27 - 24$$

$$v_{jh} \geq v_{j(h+1)} \quad \text{for } j = 1, 2, 3, 4, 5 \ \& \ h = 1, 2, 3.$$

The Excel worksheet for Example 8.4 is displayed in Figure 8.10 and the corresponding Solver dialog box is in Figure 8.11. Like Example 8.3, we prohibited the use of crashing of an activity (j) for more than the permitted duration (m_j) by imposing large penalties (of 9999) on these disallowed variables. The cost parameters are in cells D12:H15. The lengths of each of the four paths under normal

	A	B	C	D	E	F	G	H	I	J	K
1	Target =	24		Activity 1	Activity 2	Activity 3	Activity 4	Activity 5		COST =	$3,900
2											
3	Total crashing			1	0	1	0	3			
4											
5	Path	Length		Constraint Matrix							
6	1-3-5	29		1		1		1		0	
7	1-4-5	28		1			1	1		0	
8	2-3-5	28			1	1		1		0	
9	2-4-5	27			1		1	1		0	
10											
11	CRASH LIMIT			4	3	3	2	3			
12	Per Week Cost	Week 1		850	375	725	550	975			
13		Week 2		575	500	725	925	750			
14		Week 3		900	725	950	9999	600			
15		Week 4		425	9999	9999	9999	9999			
16											
17	DECISION VARIABLES										
18		Week 1		1	0	1	0	1			
19		Week 2		0	0	0	0	1			
20		Week 3		0	0	0	0	1			
21		Week 4		0	0	0	0	0			

Figure 8.10.   Worksheet for Example 8.4.

conditions are in cells B6:B9 and the target completion time of 24 weeks is in cell B1. The constraint coefficient matrix in cells D6:H9 enables the collection of crashing for the activities on each path.

The v_{jh} decision variables are in cells D18:H21. The sums of the columns in this matrix, which represent the total crashing for the activities corresponding to the columns, are located in cells D3:H3. The objective function in cell K1, which is the 'Set Objective' cell in the Solver dialog box in Figure 8.11, is the sumproduct of D18:H21 and D12:H15. Cells J6:J9 collect, for each path, the path length based on normal completion times for the activities on the path minus the total number of weeks the path is shortened from the crashing decisions, minus the target completion time in cell B1. Cells J6:J9 are constrained to be equal to or less than zero to ensure that no path is longer than 24 weeks after the crashing decisions have been implemented. The Solver dialog box also contains the constraints to ensure that crashing order is preserved. For example, if activity 1 is to be crashed by just one week, then we must ensure that the variable v_{11} is used at a cost of 850, not the cheaper alternative of v_{12} at a cost of 575.

The optimal solution in Figure 8.10 indicates that activities 1 and 3 should be crashed by one week and activity 5 should be crashed by three weeks. There is a lot of value to shortening activity 5 because

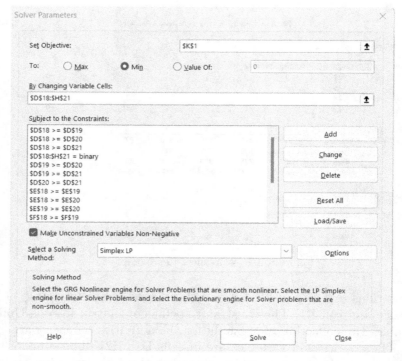

Figure 8.11.   Worksheet for Example 8.4.

it is part of all four paths. Thus, shortening activity 5 by three weeks shortens each of the four paths by three weeks. Path 1–3–5 would then be $29 - 3 = 26$ weeks, path 1–4–5 would then be $28 - 3 = 25$ weeks, path 2–3–5 would be $28 - 3 = 25$ weeks, and path 2–4–5 would be $27 - 3 = 24$ weeks. So, path 2–4–5 now meets the 24-week completion requirement. Of the remaining three paths that need to be shortened, we notice that activity three is on two of them. So, shortening activity 3 by one week will reduce path 1–3–5 to $26 - 1 = 25$ weeks and path 2–3–5 to $25 - 1 = 24$ weeks. Path 2–3–5 now meets the completion time requirement. Paths 1–3–5 and 1–4–5 are both at 25 weeks and, noticing that activity 1 is on both these paths, shortening activity 1 by one week will get both these paths down to the 24-week completion requirement. In conclusion, Excel spreadsheets afford a powerful tool for project scheduling (see, for example, Ragsdale, 2003). In addition to the worksheets provided for the examples in this chapter, we also include the Excel workbook for crashing project networks developed by Huse and Brusco (2021) in our online materials.

Chapter 9

Marketing

In this chapter, we consider examples related to the field of marketing. Several of the examples pertain to the interface between marketing and operations, yet we frame the chapter in terms of marketing. The first two examples pertain to advertising. The first of these two examples is a general example that focuses on advertisement planning where effectiveness varies by days of the week and time of the day. The second example is specific to a real-world application pertaining to the placement of advertisements on web pages and is based on the work of Chickering and Heckerman (2003).

The third and fourth examples focus on revenue management. The third example is based on an airline revenue management problem described by Anderson *et al.* (2019, Chapter 5). The fourth example is based on a revenue management application for Harrahs's Casinos (Metters *et al.*, 2008).

The fifth example considers the problem of product design using conjoint analysis. The first stage of this method requires the estimation of part-worths for each customer on various levels of production attributes. The output of this stage is assumed to be given. We focus on the second stage, where the goal is to choose levels for each attribute so as to maximize the number of customers who have greater utility for the designed product than for the product they are currently using.

The sixth and final example in the chapter corresponds to the prediction of product life cycle. This is accomplished using the Bass (1969) model, which is a nonlinear optimization problem. As noted by

Anderson *et al.* (2019, Chapter 8), it might seem somewhat absurd at first glance to construct a forecast for a product that has already completed its life cycle. However, the information gleaned from the analysis can be used to make predictions for launches of similar products. Moreover, it is possible to use the model on a rolling horizon basis over the course of a product's life cycle.

9.1   Advertising

9.1.1   *Example 9.1 — TV advertising*

A firm would like to allocate a television advertising budget for an upcoming event. A television advertising blitz is planned for the weekend, and the budgetary limit is $500,000. The cost of an advertisement and the audience reached vary by the day and time of the advertisements, as shown in the following table:

	Cost per ad		Contact per ad	
	5–9 pm	9 pm–1 am	5–9 pm	9 pm–1 am
Friday	$15,000	$12,500	126,000	92,000
Saturday	$10,000	$8,500	84,000	72,000
Sunday	$16,500	$11,500	143,000	89,000

The following are some conditions that must be satisfied by the advertising plan:

(i) Each day must contribute a minimum of 25% to the total audience contact across the three days.
(ii) The difference in the number of ads in the two time slots must not exceed 5 for any given day.
(iii) For any given time slot, the difference in the number of ads across the three days must not exceed 5 for any given pair of days.

A linear programming formulation that will maximize total audience contact and yet satisfy all of the constraints is as follows:

Decision variables

X_{ij} = the number of ads scheduled for day i in time slot j (where i = F = Friday, S = Saturday and U = Sunday, and j = E (early, 5–9) and L (late, 9–1).

Objective function

Maximize: $126X_{FE} + 92X_{FL} + 84X_{SE} + 72X_{SL} + 143X_{UE} + 89X_{UL}$.

Subject to constraints

$15X_{FE} + 12.5X_{FL} + 10X_{SE} + 8.5X_{SL} + 16.5X_{UE} + 11.5X_{UL} \le 500$

$126X_{FE} + 92X_{FL} \ge 0.25[126X_{FE} + 92X_{FL} + 84X_{SE} + 72X_{SL} + 143X_{UE} + 89X_{UL}]$

$84X_{SE} + 72X_{SL} \ge 0.25[126X_{FE} + 92X_{FL} + 84X_{SE} + 72X_{SL} + 143X_{UE} + 89X_{UL}]$

$143PX_{UE} + 89X_{UL} \ge 0.25[126X_{FE} + 92X_{FL} + 84X_{SE} + 72X_{SL} + 143X_{UE} + 89X_{UL}]$

$X_{FE} - X_{FL} \le 5, X_{FL} - X_{FE} \le 5$

$X_{SE} - X_{SL} \le 5, X_{SL} - X_{SE} \le 5$

$X_{UE} - X_{UL} \le 5, X_{UL} - X_{UE} \le 5$

$X_{FE} - X_{SE} \le 5, X_{SE} - X_{FE} \le 5$

$X_{FL} - X_{SL} \le 5, X_{SL} - X_{FL} \le 5$

$X_{FE} - X_{UE} \le 5, X_{UE} - X_{FE} \le 5$

$X_{FL} - X_{UL} \le 5, X_{UL} - X_{FL} \le 5$

$X_{SE} - X_{UE} \le 5, X_{UE} - X_{SE} \le 5$

$X_{SL} - X_{UL} \le 5, X_{UL} - X_{SL} \le 5$

The worksheet for this formulation is displayed in Figure 9.1 and the corresponding Solver dialog box is shown in Figure 9.2. The cost parameters are located in cells B2:C4 of Figure 9.1 and the contact parameters are in cells B7:C9. The budget limit of $500,000 is in cell G7.

The decision variables are located in cells B12:C14 in Figure 9.1. The sum of these cells is the total number of ads in cell G1. The sumproduct of B2:C4 and B12:C14 is the total cost in cell G3. This cell is constrained to be equal to or less than the budget limit in cell G7 in the Solver dialog box. Total audience contact is computed in cell G5 as the sumproduct of cells B7:C9 and B12:C14 and this is the 'Set Objective' cell in the Solver dialog box. Cell I5 is set equal

	A	B	C	D	E	F	G	H	I
1		5:00-9:00	9:00-1:00		Total Number of Ads		40		
2	Friday	$15,000	$12,500						
3	Saturday	$10,000	$8,500		Total Advertising Cost		$500,000		
4	Sunday	$16,500	$11,500						
5					Total Audience Contact		4175000		1043750
6		5:00-9:00	9:00-1:00						
7	Friday	126,000	92,000		Budget Limit =		$500,000		
8	Saturday	84,000	72,000						
9	Sunday	143,000	89,000						
10									Daily
11		5:00-9:00	9:00-1:00			5:00-9:00	9:00-1:00		Contact
12	Friday	8	3		Friday	1,008,000	276,000		1,284,000
13	Saturday	8	8		Saturday	672,000	576,000		1,248,000
14	Sunday	9	4		Sunday	1,287,000	356,000		1,643,000
15									
16		slot1 - slot2	slot2 - slot1			5:00-9:00	9:00-1:00		
17	Friday	5	-5		Fri - Sat	0	-5		
18	Saturday	0	0		Fri - Sun	-1	-1		
19	Sunday	5	-5		Sat - Sun	-1	4		
20					Sat -Fri	0	5		
21					Sun - Fri	1	1		
22					Sun - Sat	1	-4		
23									

Figure 9.1.   Excel worksheet for Example 9.1.

to 0.25*G5, which is the threshold that must be met for audience contact on each day. Audience contact figures are computed for each day and each time slot in cells F12:G14 and the total contact values for each day are found in cells I12:I14. Cells I12:I14 are constrained to equal or exceed the value in cell I5 in the Solver dialog box.

Cells B17:B19 contain the number of ads in the 5:00–9:00 pm time slot minus the number of ads in the 9:00–1:00 am time slot for each day of the week. Cells C17:C19 contain the number of ads in the 9:00–1:00 am time slot minus the number of ads in the 5:00–9:00 pm time slot for each day of the week. The cell range of B17:C:19 is constrained to be equal to or less than five to ensure that the difference in the number of ads in the two time slots does not exceed five for any day of the week. For example, constraining cells B17 and C17 to be equal to or less than five enforces, respectively, the constraints $X_{FE} - X_{FL} \le 5$ and $X_{FL} - X_{FE} \le 5$.

Cells F17:G22 contain differences between the number of ads run in a particular time slot for a certain pair of days. For example, cell F19 contains the number of ads in the 5:00–9:00 pm time slot on Saturday minus the number of ads in the same time slot on Sunday. Likewise, cell G19 contains the number of ads in the 5:00–9:00 pm

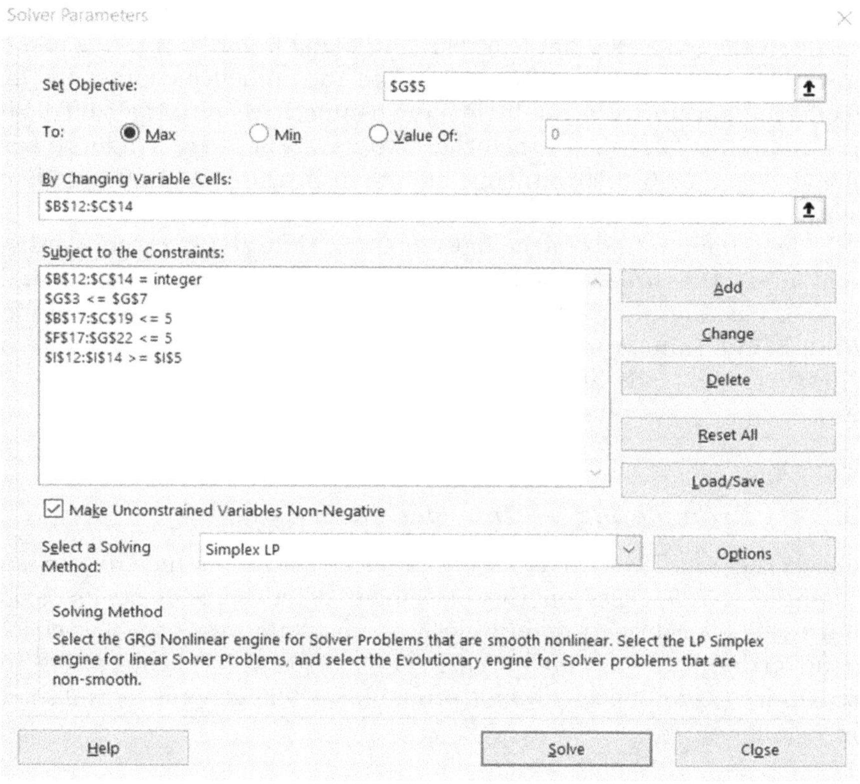

Figure 9.2. Solver dialog box for Example 9.1.

time slot on Sunday minus the number of ads in the same time slot on Saturday. By constraining cells F19 and G19 to be equal to or less than five, we enforce the constraints $X_{SE} - X_{UE} \leq 5$ and $X_{UE} - X_{SE} \leq 5$; that is, the difference between the number of ads in the 5:00–9:00 pm time slot on Saturday vs. Sunday must be five or less. Constraining the range F17:G22 to be equal to or less than five accomplishes this goal for all pairs of days and both time slots.

The optimal solution in Figure 9.1 consists of 40 ads that exhaust the $500,000 budget exactly. The total audience contact reached is 4,175,000. The audience contact for each day easily eclipses the 25% requirement of 1,043,750; hence, these constraints are not binding at the optimal solution.

Twenty-five ads are made in the 5:00–9:00 pm time slot and 15 are made in the 9:00–1:00 am time slot. An equal number of ads is made in the two time slots on Saturday when the audience contact in the 9:00–1:00 am time slot is a little more competitive with the 5:00–9:00 pm time slot. On each of the other two days, there are five more ads made in the 5:00–9:00 pm time slot than in the 9:00–1:00 am time slot.

The number of ads in the 5:00–9:00 pm time slot is rather consistent across the three weekend days (eight on Friday and Saturday and nine on Sunday). In the 9:00–1:00 am time slot, the numbers of ads on Friday and Sunday are comparable at three and four, respectively. However, eight ads are made on Saturday and thus the constraint $X_{SL} - X_{FL} \leq 5$ is binding at the optimal solution.

9.1.2   *Example 9.2 — Internet advertising*

Chickering and Heckerman (2003) describe an interesting advertising example that focuses on the placement of advertisements on web pages. The goal is to determine how many impressions of each of n advertisements to run on each of m website segments so as to maximize overall click-through probability. The general formulation is as follows:

Parameters

$n =$ the number of advertisements indexed $1 \leq i \leq n$;

$m =$ the number of segments on the site indexed $1 \leq j \leq m$;

$q_i =$ a quota for advertisement i that corresponds to the minimum number of impressions of the advertisement that must be run within some prescribed time interval (e.g., a week or a month), for $1 \leq i \leq n$;

$s_j =$ the capacity of segment j, which corresponds to the maximum number of impressions that could be shown on the segment in the prescribed time interval, for $1 \leq j \leq m$;

$p_{ij} =$ the estimated probability that a website visitor will click on advertisement i in segment j (for $1 \leq i \leq n$ and $1 \leq j \leq m$);

$N =$ the total number of impressions to be delivered in the prescribed time interval.

Decision variables

x_{ij} = the number of impressions of advertisement i to run on segment j in the prescribed time interval (for $1 \leq i \leq n$ and $1 \leq j \leq m$).

Objective function

$$\text{Maximize:} \quad \frac{1}{N} \sum_{i=1}^{n} \sum_{j=1}^{m} p_{ij} x_{ij}. \tag{9.1}$$

Subject to constraints

$$\sum_{j=1}^{m} x_{ij} \geq q_i \quad \text{for } 1 \leq i \leq n; \tag{9.2}$$

$$\sum_{i=1}^{n} x_{ij} \leq s_j \quad \text{for } 1 \leq j \leq m; \tag{9.3}$$

$$\sum_{i=1}^{n} \sum_{j=1}^{m} x_{ij} = N. \tag{9.4}$$

The formulation is very similar to the transportation problem covered in Chapter 5. The objective function in Equation (9.1) is to maximize click-through probability. Constraint set (9.2) guarantees that the quota is met for each advertisement. Constraint set (9.3) ensures that segment capacity is not exceeded for any segment. Constraint (9.4) requires the exact number of impressions desired to be accomplished.

We consider an example where $n = 10$ advertisements are to be run on $m = 4$ segments. The problem is displayed in the worksheet in Figure 9.3 and the corresponding Solver dialog box in Figure 9.4. The click-through probability parameters (p_{ij}) are located in cells B3:E12 of Figure 9.3. The ad quota parameters (q_i) are located in cells H15:H24 and the segment capacity parameters (s_j) are in cells B27:E27. The total number of impressions to run is $N = 60,000$, which is in cell H3.

The decision variables (x_{ij}) are located in cells B15:E24. Their row sums are located in cells G15:G24 and these cells are constrained

	A	B	C	D	E	F	G	H
1								Total
2	Advertisement	Seg 1	Seg 2	Seg 3	Seg 4			Impressions
3	1	0.05	0.03	0.01	0.01		Target	60000
4	2	0.04	0.07	0.03	0.02		Actual	60000
5	3	0.03	0.05	0.04	0.02			
6	4	0.06	0.02	0.03	0.04		Clickthrough	
7	5	0.02	0.01	0.05	0.07		Sum	3400.25
8	6	0.04	0.03	0.02	0.05		Probability	0.0567
9	7	0.02	0.03	0.06	0.04			
10	8	0.01	0.03	0.05	0.02			
11	9	0.08	0.05	0.06	0.07			
12	10	0.01	0.04	0.02	0.03			
13							Impressions	Impressions
14	Advertisement	Seg 1	Seg 2	Seg 3	Seg 4		Run	Quota
15	1	6500	0	0	0		6500	6500
16	2	0	4600	0	0		4600	4000
17	3	0	3750	0	0		3750	3750
18	4	3000	0	0	0		3000	3000
19	5	0	0	0	6450		6450	4500
20	6	0	0	0	8750		8750	8750
21	7	0	0	4675	0		4675	2500
22	8	0	0	9725	0		9725	9725
23	9	5700	0	0	0		5700	3750
24	10	0	6850	0	0		6850	6850
25								
26	Impression in Segment	15200	15200	14400	15200			
27	Segment capacity	15200	15200	15200	15200			

Figure 9.3. Excel worksheet for Example 9.2.

to equal or exceed the quotas in H15:H24 to satisfy constraint set (9.2). The column sums of the decision variables are located in cells B26:E26 and these cells are constrained to be equal to or less than the segment capacities in cells B27:E27 to satisfy constraint set (9.3). The total sum of the decision variables is in cell H4 and this cell is constrained to equal cell H3 to satisfy constraint (9.4).

Cell H7 contains the sumproduct of B3:E12 and B15:E24. This cell is the 'Set Objective' cell in the Solver dialog box in Figure 9.4. To convert this figure to a probability, it is divided by $N = 60{,}000$ in cell H8.

The optimal solution in Figure 9.3 reveals that 60,000 impressions yield an overall click-through probability of 0.0567. Segments 1, 2, and 4 are used to capacity, and segment 3 has an unused capacity of 800 impressions.

The quotas are met exactly for advertisements 1, 3, 4, 6, 8, and 10. The quotas are exceeded for other advertisements because of their high click-through probabilities on certain segments. For example, 5700 impressions of advertisement 9 were made in segment 1 (far

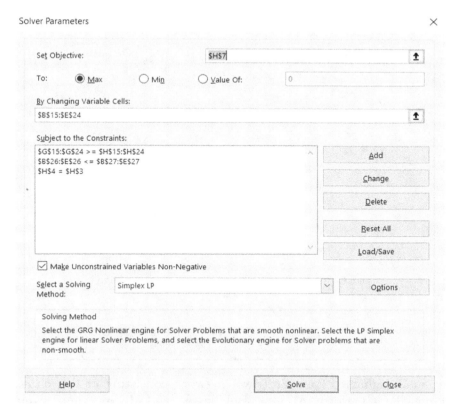

Figure 9.4.   Solver dialog box for Example 6.2.

exceeding the quota of 3700 for advertisement 9) because of the high click-through probability of 0.08. Similar observations can be made for advertisement 2 in segment 2 (click-through probability of 0.07), advertisement 5 in segment 4 (click-through probability 0.07), and advertisement 7 in segment 3 (click-through probability of 0.06).

9.2   Revenue Management

9.2.1   *Example 9.3. — Airplane seats*

Example 9.3. draws heavily from the airline revenue management example described by Anderson *et al.* (2019, Chapter 5). We consider

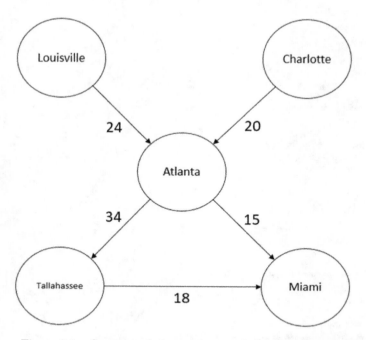

Figure 9.5.   Segment of air travel network for Example 9.3.

a small segment of an air travel network consisting of five cities: Louisville, Charlotte, Atlanta, Tallahassee, and Miami. There is one aircraft in Louisville and one in Charlotte. Each of these aircraft will first fly to the hub in Atlanta. The aircraft from Louisville will then proceed to Tallahassee and subsequently to Miami. The aircraft from Charlotte will proceed directly to Miami after its layover in Atlanta. A graphical view of the network is provided in Figure 9.5 and each edge of the network is labeled with the remaining capacity of the aircraft for that flight leg.

Itineraries are represented by origin, destination, and layovers. For example, LAM originates in Louisville and its destination is Miami with a layover in Atlanta. Similarly, CATM originates in Charlotte and its destination is Miami with layovers in both Atlanta and Tallahassee.

The parameters for the linear programming formulation of the revenue maximization problem are the per-passenger fares for each itinerary (cells E2:E13 of the worksheet in Figure 9.6), the demand forecasts for each itinerary (F2:F13), and the capacities for each flight leg (cells D20:D24). The decision variables for the revenue maximization problem are the number of seats to allocate to each itinerary (cells G2:G13).

Denoting the decision variables using the itinerary labels as subscripts, we have

X_{LA} = the number of seats to allocate for Louisville–Atlanta,

X_{LAT} = the number of seats to allocate for Louisville–Atlanta–Tallahassee,

X_{LAM} = the number of seats to allocate for Louisville–Atlanta–Miami,

X_{LATM} = the number of seats to allocate for Louisville–Atlanta–Tallahassee–Miami,

X_{CA} = the number of seats to allocate for Charlotte–Atlanta,

X_{CAT} = the number of seats to allocate for Charlotte–Atlanta–Tallahassee,

X_{CAM} = the number of seats to allocate for Charlotte–Atlanta–Miami,

X_{CATM} = the number of seats to allocate for Charlotte–Atlanta–Tallahassee–Miami,

X_{AT} = the number of seats to allocate for Atlanta–Tallahassee,

X_{AM} = the number of seats to allocate for Atlanta–Miami,

X_{ATM} = the number of seats to allocate for Atlanta–Tallahassee–Miami,

X_{TM} = the number of seats to allocate for Tallahassee–Miami.

The objective function, which is the sumproduct of E2:E13 and G2:G13 contained in cell C15 of Figure 9.6, is as follows:

$$\text{Maximize: } 339X_{LA} + 418X_{LAT} + 455X_{LAM} + 306X_{LATM} + 314X_{CA}$$
$$+ 424X_{CAT} + 399X_{CAM} + 461X_{CATM} + 244X_{AT}$$
$$+ 345X_{AM} + 286X_{ATM} + 497X_{TM}.$$

The demand constraints are accomplished by constraining cells G2:G13 to be equal to or less than cells F2:F13 and are explicitly written as follows:

$X_{LA} \leq 9$, $X_{LAT} \leq 7$, $X_{LAM} \leq 5$, $X_{LATM} \leq 11$, $X_{CA} \leq 10$, $X_{CAT} \leq 8$, $X_{CAM} \leq 7$, $X_{CATM} \leq 10$, $X_{AT} \leq 6$, $X_{AM} \leq 7$, $X_{ATM} \leq 4$, $X_{TM} \leq 8$.

Cells C20:C24 contain the number of seats allocated for each flight leg and these are constrained to be equal to or less than cells D20:D24.

	A	B	C	D	E	F	G
1	Origin	Layover 1	Layover 2	Destination	Fare	Forecast	Allocation
2	Louisville	none	none	Atanta	$339	9	9
3	Louisville	Atlanta	none	Tallahassee	$418	7	7
4	Louisville	Atlanta	none	Miami	$455	5	5
5	Louisville	Atlanta	Tallahassee	Miami	$306	11	3
6	Charlotte	none	none	Atanta	$314	10	6
7	Charlotte	Atlanta	none	Tallahassee	$424	8	8
8	Charlotte	Atlanta	none	Miami	$399	7	3
9	Charlotte	Atlanta	Tallahassee	Miami	$461	10	3
10	Atlanta	none	none	Tallahassee	$244	6	6
11	Atlanta	none	none	Miami	$345	7	7
12	Atlanta	Tallahassee	none	Miami	$286	4	4
13	Tallahassee	none	none	Miami	$497	8	8
14							
15	Total Revenue =		26025				
16							
17							
18				SEATS	SEATS		
19	FLIGHT LEGS		ALLOCATED	REMAINING			
20	1. Louisville to Atlanta		24	24			
21	2. Charlotte to Atlanta		20	20			
22	3. Atlanta to Tallahasee		31	34			
23	4. Atlanta to Miami		15	15			
24	5. Tallahassee to Miami		18	18			

Figure 9.6.   Excel worksheet for Example 9.3.

The constraints are explicitly written as follows:

Louisville to Atlanta $\quad X_{LA} + X_{LAT} + X_{LAM} + X_{LATM} \le 24,$

Charlotte to Atlanta $\quad X_{CA} + X_{CAT} + X_{CAM} + X_{CATM} \le 20,$

Atlanta to Tallahassee $\quad X_{LAT} + X_{LATM} + X_{CAT} + X_{CATM}$
$$+ X_{AT} + X_{ATM} \le 34,$$

Atlanta to Miami $\quad X_{LAM} + X_{CAM} + X_{AM} \le 15,$

Tallahassee to Miami $\quad X_{LATM} + X_{CATM} + X_{ATM} + X_{TM} \le 18.$

The solution is displayed in Figure 9.6 and the Solver dialog box is in Figure 9.7.

The results in Figure 9.6 reveal that the maximum revenue is \$26,025. Four of the five flight legs are used to capacity. The exception is the Atlanta to Tallahassee flight leg, as only 31 of the 34 remaining seats have been allocated.

Eight of the 12 itineraries have seat allocations that match their demand forecasts. Three of the four itineraries that do not have

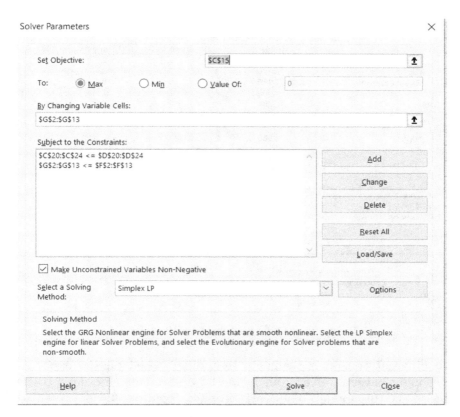

Figure 9.7. Solver dialog box for Example 9.3.

allocations matching the forecast demand have a Charlotte origin. This seems a bit surprising because some of the underutilized Charlotte origins have high fares (e.g., CATM fare is \$461 but only 3 of 10 possible seats are allocated). However, this can be at least partially explained by the fact that itineraries with more layovers consume seats on more flight legs than those with fewer layovers.

9.2.2 *Example 9.4 — Hotel revenue management*

Metters *et al.* (2008) offer a comprehensive coverage of a revenue management application for Harrah's Cherokee Casino and Hotel. Part of the application is a bid-price control system that uses linear programming to determine the number of rooms to allocate

	A	B	C	D	E	F	G	H
1		Customer	Average	Demand	Rooms			
2		Segment	Revenue	Forecast	Reserved		Fridays	Saturdays
3	Friday	1	$1,000	2	2		2	
4	Night	2	$500	3	3		3	
5	Reservations	3	$300	5	5		5	
6		4	$100	7	7		7	
7		5	$50	12	2		2	
8		6	$25	17	0		0	
9	Saturday	1	$1,000	1	1			1
10	Night	2	$500	2	2			2
11	Reservations	3	$300	4	4			4
12		4	$100	5	4			4
13		5	$50	11	0			0
14		6	$25	14	0			0
15	Two-Night	1	$1,775	2	2		2	2
16	Night	2	$875	3	3		3	3
17	Reservations	3	$440	3	3		3	3
18		4	$145	5	0		0	0
19		5	$85	8	0		0	0
20		6	$40	9	0		0	0
21								
22		Total Rooms Reserved					27	19
23		Total Capacity (Rooms Left)					27	19
24								
25		Total Revenue			$16,895			

Figure 9.8.   Excel worksheet for Example 9.4.

to different consumer segments that differ based on the expected per-night revenue of customers in those segments. Example 9.4 is based on this model. The worksheet for Example 9.4 is presented in Figure 9.8 and the corresponding Solver dialog box is shown in Figure 9.9.

The example assumes six segments of customers. For each segment, expected revenues are available for Friday night stays, Saturday night stays, and two night stays (see cells C3:C20 in Figure 9.8). There are also forecast levels of demand in each of these revenue categories (see cells D3:D20 in Figure 9.8). In addition, there are currently 27 rooms available for Friday night and 19 available for Saturday night (see cells G23:H23). The goal is to develop a linear programming formulation that will determine how many reservations to take for each segment stay combination to maximize revenue.

Indices

Segment index goes from $i = 1, 2, 3, 4, 5, 6$ and stay type index j is $1 =$ Fri, $2 =$ Sat, and $3 =$ two-night.

Parameters

r_{ij} = the average revenue for segment i and stay type j;
d_{ij} = the demand forecast for segment i and stay type j;

Decision variables

x_{ij} = the number of reservations to take for the combination of segment i and stay type j.

Objective function

$$\text{Maximize: } \sum_{i=1}^{6} \sum_{j=1}^{3} r_{ij} x_{ij}. \tag{9.5}$$

Subject to constraints

$$X_{ij} \le d_{ij} \quad \text{for all } 1 \le i \le 6 \quad \text{and} \quad 1 \le j \le 3, \tag{9.6}$$

$$\sum_{i=1}^{6} (x_{i1} + x_{i3}) \le 27, \tag{9.7}$$

$$\sum_{i=1}^{6} (x_{i2} + x_{i3}) \le 19. \tag{9.8}$$

The objective function in Equation (9.5) is to maximize total expected revenue, which is computed in cell E25 in Figure 9.8 as the sumproduct of C3:C20 and E3:E20 and is the 'Set Objective' cell in the Solver dialog box in Figure 9.9.

Constraint set (9.6) ensures that the number of reservations taken for a given segment/stay type combination does not exceed the forecast demand for that combination. This is accomplished in the worksheet by constraining cells E3:E20 to be equal to or less than cells D3:D20. Constraints (9.7) and (9.8), ensure that the number of reservations taken does not exceed the remaining room capacities for Friday and Saturday, respectively. Cells G3:G20 and H3:H20 collect the number of guests on Friday and Saturday, respectively, and the sums of these columns are in cells G22:H22. The constraints are accomplished by requiring G22:H22 to be equal to or less than H23:G23.

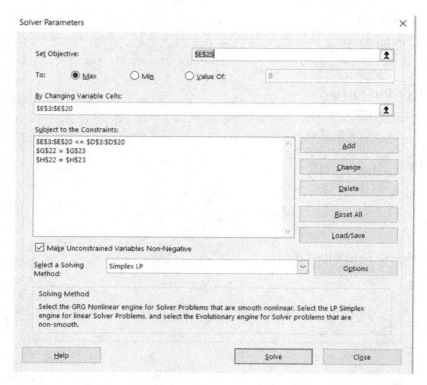

Figure 9.9.   Solver dialog box for Example 9.4.

The results in Figure 9.8 reveal an optimal reservation plan with an expected total revenue of \$16,895. Notice that, based on this plan, room reservations would not be taken for two night stays for segments 4, 5, and 6, nor would they be taken for segments 5 and 6 for Saturday night only. Only segment 6 is excluded on Friday night only.

9.3   Product Design

9.3.1   *Conjoint analysis overview*

Example 9.5 is an integer programming model that could be used for the second stage in conjoint analysis for product design. In the first stage, which we do not address here, the market research analysts would survey a set of customers regarding their attitudes toward different levels of various attributes of a product. Statistical analyses would be completed to establish 'part-worths' for each customer on each level of each attribute. These are measures of how much the attribute levels would contribute to the customers' overall utility. In the second stage that we consider, the part-worths are taken as inputs and the goal is to construct a model that will select the level for each attribute so as to design a product such that as many customers as possible will have a greater utility for the designed product than they do for the product they currently use (or for some other competitor's product). The following is a generic formulation for the problem:

Sets and parameters

$K =$ the number of consumers/customers (indexed $1 \leq k \leq K$),

$J =$ the number of product attributes (indexed $1 \leq j \leq J$),

$I_j =$ the set of levels for attribute j (indexed $i \in I_j$ for all $1 \leq j \leq J$),

$P_{ijk} =$ the part-worth measure (expressed as an integer) for consumer k on level i of attribute j (for $1 \leq k \leq K$, $1 \leq j \leq J$, $i \in I_j$),

$U_k =$ the utility threshold for consumer k (for $1 \leq k \leq K$).

Decision variables

$Y_k = 1$ if consumer k chooses the designed product, 0 otherwise (for $1 \leq k \leq K$),

$X_{ij} = 1$ if the designed product selection is level i for attribute j, and 0 otherwise (for all $1 \leq j \leq J$ and $i \in I_j$).

Objective function

$$\text{Maximize: } \sum_{k=1}^{K} Y_k. \tag{9.9}$$

Subject to

$$\sum_{i \in I_j} X_{ij} = 1 \quad \text{for all } 1 \leq j \leq J, \tag{9.10}$$

$$\sum_{j=1}^{J} \sum_{i \in I_j} P_{ijk} X_{ij} \geq (U_k + 1) Y_k \quad \text{for all } 1 \leq k \leq K. \tag{9.11}$$

The objective function in Equation (9.9) is to maximize the number of customers (or consumers) who 'choose' the designed product. Here, 'choose' means that the customer has a greater utility for the designed product than for the utility threshold, which might come from the product they currently use or the product of some competitor. Constraint set (9.10) ensures, for each attribute, that exactly one level of the attribute is selected. Constraint set (9.11) will ensure that, for each customer, the binary variable for choosing the designed product can only be one if the utility of the designed product exceeds the threshold.

9.3.2   *Example 9.5 — Salad design*

A local restaurant (Salem) is designing a new premium salad to compete with the premium salad of their primary competitor (Winston). Some attributes have been predetermined, but four remain to be chosen. The four attributes and their levels (in parentheses) are as follows: (1) Greens (iceberg lettuce, romaine lettuce, spinach), (2) Meat (chicken, turkey, salmon), (3) Fruit (grapes, cranberries), and (4) Dressing (ranch, blue cheese). Winston's premium salad consists of romaine lettuce, salmon, cranberries, and ranch dressing. The part-worth values for each level of each attribute are provided for eight customers in cells B2:I11 of the worksheet in Figure 9.10. This worksheet provides the information for the integer programming formulation, and the corresponding Solver dialog box is in Figure 9.11.

Cells K2:K11 contain the binary selection variables (X_{ij}) for levels of each attribute. Cells B18:B21 contain the sums of these binary variables for each of the four attributes. These cells are constrained to equal one to enforce constraint set (9.10). Cells B15:I15 contain the binary customer choice variables (Y_k). The sum of these cells corresponds to the objective function in Equation (9.9). The sum of B15:I15 is contained in cell K18, which is the 'Set Objective' cell of the Solver dialog box in Figure 9.11.

The binary variables in row 15 are also used to establish constraint set (9.11). The first step is to compute the utility for each customer for the Winston product in cells B13:I13 (using romaine, salmon, cranberries, and ranch). So, for customer 1, the Winston utility is $15 + 19 + 20 + 14 = 68$. The utility for Salem is computed in cells B14:I14 as the sumproduct of K2:K11 and each customer's column of part-worths. Cells B16:I16 contain the formula: Salem's utility

	A	B	C	D	E	F	G	H	I	J	K
1		Cust 1	Cust 2	Cust 3	Cust 4	Cust 5	Cust 6	Cust 7	Cust 8		Chosen?
2	Iceberg lettuce	18	11	5	14	12	21	13	10		0
3	Romaine lettuce	15	14	12	18	9	17	22	16		1
4	Spinach	23	4	19	6	5	16	15	23		0
5	Chicken	12	13	20	11	6	12	21	19		0
6	Turkey	15	18	10	19	11	20	15	14		0
7	Salmon	19	16	8	17	23	13	13	18		1
8	Grapes	17	23	18	15	19	23	9	14		1
9	Cranberries	20	6	16	21	17	22	24	15		0
10	Ranch	14	17	11	7	19	18	20	14		0
11	Blue Cheese	18	16	18	16	21	15	13	19		1
12											
13	Winston Utility	68	53	47	63	68	70	79	63		
14	Salem Utility	69	69	56	66	72	68	57	67		
15	Choose Salem?	1	1	1	1	1	0	0	1		
16	Choice Constraint	0	15	8	2	3	67	56	3		
17											
18	Greens	1		Number of customers preferring Salem							6
19	Meat	1									
20	Fruit	1									
21	Dressing	1									

Figure 9.10. Excel worksheet for Example 9.5.

minus (Winston's utility times the binary choice variable plus one), and these cells are constrained to equal or exceed zero. So, when Salem's utility exceeds Winston's, the Y_k variable is one and when it does not the variable value is zero.

Like Winston's salad, the optimal salad design for Salem uses Romaine lettuce and salmon. However, the Salem design uses grapes and blue cheese, as opposed to cranberries and ranch. Salem's design leads to six of the eight customers having greater utility for the Salem salad relative to the Winston salad.

9.4   Product Life Cycle Forecasting

9.4.1   *Overview of the Bass model*

The problem of forecasting the life cycle of a product is of considerable relevance. The Bass model (Bass, 1969) is a classic formulation of this particular problem. The Bass model distinguished between two motivations for adopting a product: imitation and innovation. The innovation component of the model is based on word of mouth (or word of social media) and is a function of the number of adoptees

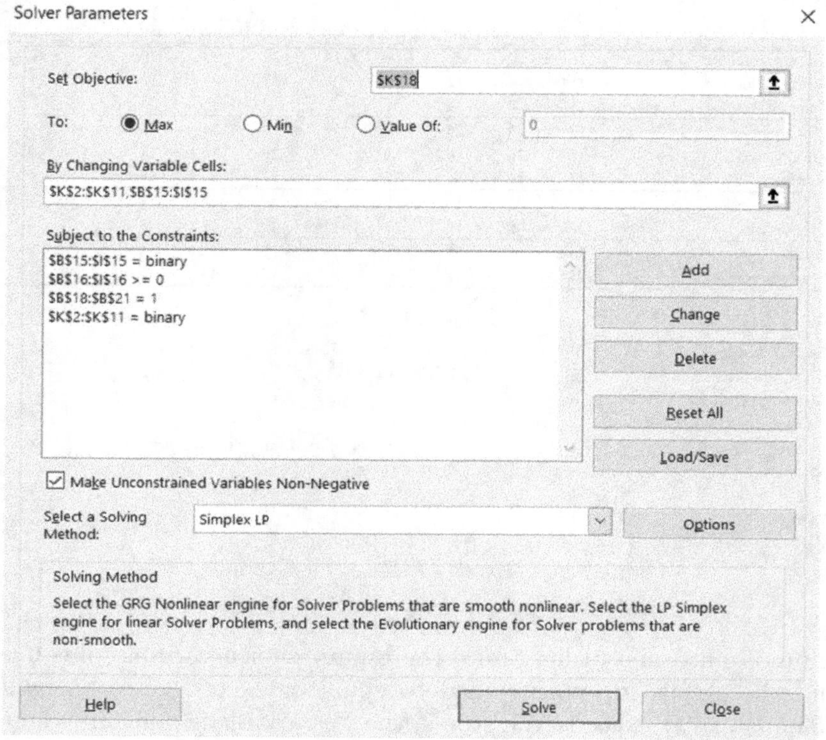

Figure 9.11.   Solver dialog box for Example 9.5.

at the particular juncture of the life cycle. By contrast, the innovation component is independent of the current level of adoption. A general nonlinear programming formulation of the Bass model is as follows:

Parameters

$T =$ the number of time periods (indexed $1 \leq t \leq T$),

$S_t =$ the 'demand' in time period t (for all $1 \leq t \leq T$). We use the term 'demand' here in a generic sense (it could be 'number of adoptees', 'revenues', or some other measure of demand).

$C_t = \sum_{j=1}^{t} S_j$, is the cumulative demand through time period t (for $1 \leq t \leq T$) and $C_0 = 0$.

Decision variables

m = estimated total demand over the life cycle of the product,
q = imitation coefficient – adoption based on word of mouth,
p = innovation coefficient – adoption not based on another's input,
F_t = forecast demand in time period t (for $1 \le t \le T$).

Objective function

$$\text{Minimize: } \sum_{t=1}^{T}(F_t - S_t)^2. \tag{9.12}$$

Subject to

$$F_t = [p + q(C_{t-1}/m)][m - C_{t-1}] \quad \text{for all } 1 \le t \le T. \tag{9.13}$$

The key parameters for the Bass model are demand values (S_t) for some number of time periods, T. The cumulative demand values (C_t) are simply computed from the S_t values. The three key decision variables are m, q, and p. The forecasts (i.e., the F_t values) are computed from these variables based on the identity relationship in Equation (9.13), and so the F_t variables are primarily for convenience of expression. The objective function in Equation (9.12) is the sum (across all time periods) of squared forecast errors.

9.4.2 *Example 9.6 — Pharmaceutical drug*

We consider the following data pertaining to the number of new patients (in thousands) prescribed a particular drug since its introduction to the market in the Winter of 2015 (i.e., the 'Win 15' quarter). The goal is to prepare a nonlinear program to fit the Bass model to the data.

Quarter	Win 15	Sp 15	Su 15	Fa 15	Win 16	Sp 16	Su 16	Fa 16	Win 17	Sp 17	Su 17
Patients	11	23	48	76	90	74	65	49	38	24	10

The worksheet for this problem is displayed in Figure 9.12 and the corresponding Solver dialog box is shown in Figure 9.13. The parameters corresponding to the number of new patients adopting the drug in each quarter are contained in cells B3:B13 of Figure 9.12.

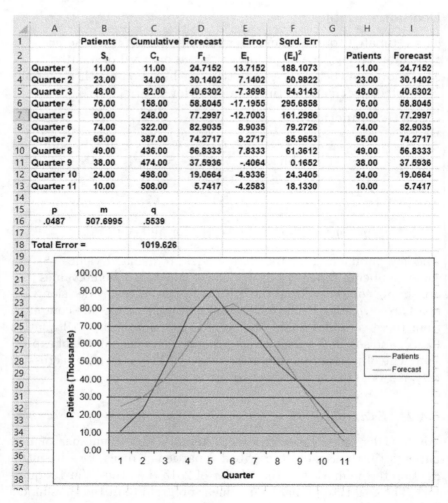

	A	B	C	D	E	F	G	H	I
1		Patients	Cumulative	Forecast	Error	Sqrd. Err			
2		S_t	C_t	F_t	E_t	$(E_t)^2$		Patients	Forecast
3	Quarter 1	11.00	11.00	24.7152	13.7152	188.1073		11.00	24.7152
4	Quarter 2	23.00	34.00	30.1402	7.1402	50.9822		23.00	30.1402
5	Quarter 3	48.00	82.00	40.6302	-7.3698	54.3143		48.00	40.6302
6	Quarter 4	76.00	158.00	58.8045	-17.1955	295.6858		76.00	58.8045
7	Quarter 5	90.00	248.00	77.2997	-12.7003	161.2986		90.00	77.2997
8	Quarter 6	74.00	322.00	82.9035	8.9035	79.2726		74.00	82.9035
9	Quarter 7	65.00	387.00	74.2717	9.2717	85.9653		65.00	74.2717
10	Quarter 8	49.00	436.00	56.8333	7.8333	61.3612		49.00	56.8333
11	Quarter 9	38.00	474.00	37.5936	-.4064	0.1652		38.00	37.5936
12	Quarter 10	24.00	498.00	19.0664	-4.9336	24.3405		24.00	19.0664
13	Quarter 11	10.00	508.00	5.7417	-4.2583	18.1330		10.00	5.7417
14									
15	p	m	q						
16	.0487	507.6995	.5539						
17									
18	Total Error =		1019.626						

Figure 9.12.   Excel worksheet for Example 9.6.

From these values, the cumulative number of patients is computed in cells C3:C13. The p, m, and q decision variables are located in cells A16, B16, and C16, respectively. From these variables, the forecasts are computed in cells D3:D13 using Equation (9.13). The forecast errors in cells E3:E13 are computed from B3:B13 minus D3:D13. Cells F3:F13 contain the squared forecast errors and the sum of these cells is found in cell C18. Cell C18 is the objective function corresponding

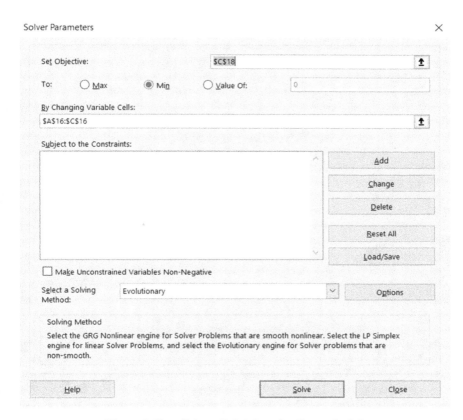

Figure 9.13. Solver dialog box for Example 9.6.

to Equation (9.12) and the 'Set Objective' cell in the Solver dialog box in Figure 9.13.

Because this is a nonlinear program, the 'Simplex LP' engine is not an option. As shown in the Solver dialog box in Figure 9.13, we used the 'Evolutionary' engine to obtain a solution. The total squared error for the predictive model is 1019.626 and the graph at the bottom of Figure 9.12 shows that the fit is reasonably good for most quarters.

Chapter 10

Finance

In this chapter, we focus on financial examples of linear and nonlinear optimization. Three categories of problems are considered: (i) payout models, (ii) portfolio optimization, and (iii) productivity measurement.

Payout models pertain to situations where payments must be made over several years to satisfy requirements. For example, payouts might be necessary to satisfy a retirement program or to meet the conditions of a lawsuit. An initial payout value is made at the beginning of the planning horizon and the funds are invested in various securities, such that the proceeds from the investments can help make necessary payments in subsequent years.

Portfolio optimization problems typically focus on the trade-offs between risk and return. We use four examples, based on those found in the textbook by Anderson *et al.* (2019, Chapters 5 & 8), to demonstrate how these trade-offs can be made. The first example uses linear programming to find a portfolio investment strategy that maximizes the minimum return across several scenarios. This conservative approach effectively seeks to keep the worst-case scenario as good as possible and, therefore, emphasizes the minimization of risk. The second example allows the decision-maker to control the degree of risk by specifying a threshold for the minimum return that will be tolerated. The goal is to maximize the expected return subject to the threshold constraint. The third example is the classic Markowitz (1952, 1956) model, which is a nonlinear optimization problem that seeks to minimize the variance of the return subject to a constraint on the expected or average return. A fourth example considers the impact of side constraints on portfolio selection.

The productivity measurement example focuses on the determination of which business units are making efficient use of their inputs when delivering outputs and which ones are not. The specific model is data envelopment analysis (Charnes *et al.*, 1978). This model has been applied to measure the relative efficiency of sets of various types of business units, such as branch banks, county schools, hospitals, restaurant chains, and academic programs. We will consider both the primal and dual formulations of the data envelopment analysis model in our analyses.

10.1   Example 10.1 — Payout Model

A donor would like to develop a funding plan for university scholarships for the next nine years. The funding requirements (in thousands of $) for payments at the beginning of each of the nine years are displayed in cells C20:K20 of Figure 10.1. To help meet these payments, the donor may invest in each of three bonds. The par value of each bond is $1000 but the prices of the bonds per $1000 purchased differ. Specifically, the bond prices per $1000 purchased are $1025, $1035, and $1045 for bonds 1, 2, and 3, respectively. The annual return rates for bonds 1, 2, and 3 are 3.75%, 5.25%, and 5.75%, respectively. The years to maturity for bonds 1, 2, and 3

	A	B	C	D	E	F	G	H	I	J	K
1		Decision	Year 1	Year 2	Year 3	Year 4	Year 5	Year 6	Year 7	Year 8	Year 9
2		Variables									
3	Payout	$1,702.92	1								
4	Bond 1	$226.20	-1.025	0.0375	0.0375	0.0375	1.0375				
5	Bond 2	$309.69	-1.035	0.0525	0.0525	0.0525	0.0525	1.0525			
6	Bond 3	$679.22	-1.045	0.0575	0.0575	0.0575	0.0575	0.0575	1.0575		
7	Savings - yr 1	$355.75	-1	1.0275							
8	Savings - yr 2	$289.33		-1	1.0275						
9	Savings - yr 3	$186.09			-1	1.0275					
10	Savings - yr 4	$0.00				-1	1.0275				
11	Savings - yr 5	$0.00					-1	1.0275			
12	Savings - yr 6	$0.00						-1	1.0275		
13	Savings - yr 7	$423.27							-1	1.0275	
14	Savings - yr 8	$184.91								-1	1.0275
15	Savings - yr 9	$0.00									-1
16											
17											
18											
19	Payments		$85.00	$140.00	$175.00	$255.00	$290.00	$365.00	$295.00	$250.00	$190.00
20	Requirement		$85.00	$140.00	$175.00	$255.00	$290.00	$365.00	$295.00	$250.00	$190.00

Figure 10.1.   Excel worksheet for Example 10.1.

are 4 years, 5 years, and 6 years, respectively. Each year, the remaining funds after the payment is made are placed in savings where they earn a rate of 2.75% annually. The donor would like to determine the initial payout and investment plan that will meet the requirements. The objective is to minimize the initial payout, and the linear programming formulation is as follows:

Decision variables

P = the initial payout (in thousands of $);
B_k = the investment in bond k (in thousands of $) for $1 \leq k \leq 3$;
S_t = savings (in thousands of $) at the beginning of year t, for $1 \leq t \leq 9$.

Objective function

Minimize: P.

Subject to constraints

Year 1: $P - 1.025B_1 - 1.035B_2 - 1.045B_3 - S_1 = 85$,
Year 2: $0.0375B_1 + 0.0525B_2 + 0.0575B_3 + 1.0275S_1 - S_2 = 140$,
Year 3: $0.0375B_1 + 0.0525B_2 + 0.0575B_3 + 1.0275S_2 - S_3 = 175$,
Year 4: $0.0375B_1 + 0.0525B_2 + 0.0575B_3 + 1.0275S_3 - S_4 = 255$,
Year 5: $1.0375B_1 + 0.0525B_2 + 0.0575B_3 + 1.0275S_4 - S_5 = 290$,
Year 6: $1.0525B_2 + 0.0575B_3 + 1.0275S_5 - S_6 = 365$,
Year 7: $1.0575B_3 + 1.0275S_6 - S_7 = 295$,
Year 8: $1.0275S_7 - S_8 = 250$,
Year 9: $1.0275S_8 - S_9 = 190$.

The objective function is to minimize the payout variable, P. At the beginning of year 1, this payout is broken into parts. Part of the payout is invested in the purchase of the bonds at the appropriate purchase price (e.g., $1000 in bond 1 costs $1025, hence the $1.025B_1$ in the constraint for year 1). A portion of the payout is used to meet the $85,000 requirement in year 1 and the rest goes into savings at the beginning of year 1 (i.e., S_1). At the beginning of year 2, the funds available are the annual returns from the bonds (i.e., $0.0375B_1 + 0.0525B_2 + 0.0575B_3$) and the savings in year 1, which earned 2.75% interest (i.e., $1.0275S_1$). These funds are uses to meet the $140,000 requirement in year 2 and the rest goes into savings at the beginning of year 2 (i.e., S_2).

At the beginning of year 5, bond 1 matures. Therefore, the contribution of bond 1 in year 5 is $1.0375B_1$ because both the return and the principal investment are realized at this time. After year 5, bond 1 no longer appears in the constraints. In a similar fashion, bond 2 matures at the beginning of year 6 and bond 3 matures at the beginning of year 7. At the beginning of years 8 and 9, only contributions from savings are available.

Figure 10.1 contains the worksheet for the formulation. As noted previously, the requirements for each year are located in cells C20:K20. The constraint parameters are in cells C3:K15. The decision variables are located in cells B3:B15. Cell B3 is the objective function and is the 'Set Objective' cell in the Solver dialog box in Figure 10.2.

Cells C19:K19 contain the sumproducts of B3:B15 and each column of the constraint matrix in cells C3:K15. Cells

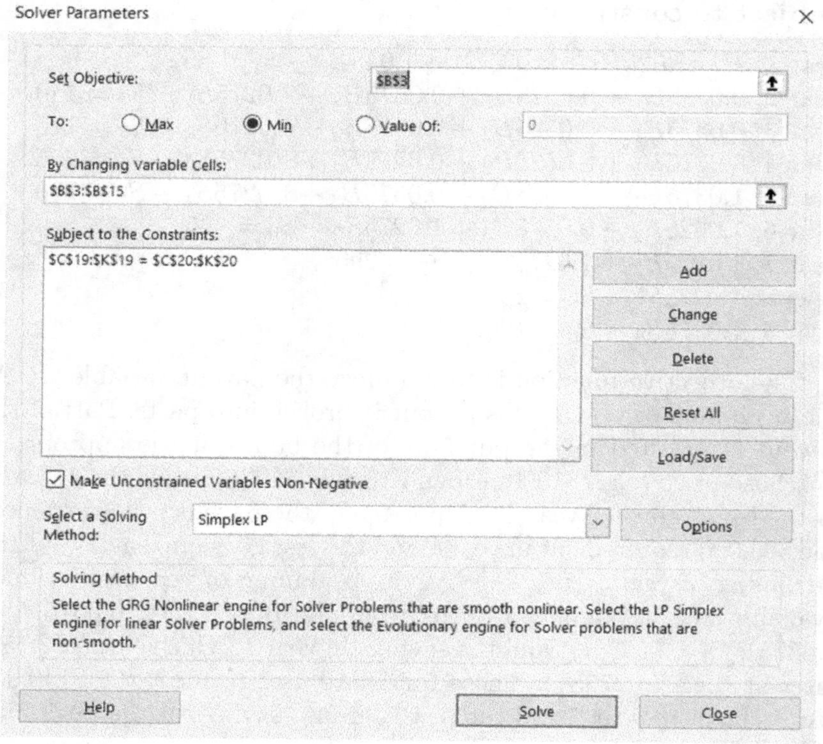

Figure 10.2. Solver dialog box for Example 10.1.

C19:K19 are constrained to equal the yearly requirements in cells C20:K20.

As shown in cell B3, the optimal initial payout value is approximately \$1.703 million. There is investment in all three bonds but the investment in bond 3 is more than the investment in the other two bonds combined. About \$355,000 goes into savings at the beginning of year 1 because some liquidity is needed to meet requirements in years 2 through 4. There are no savings at the beginning of years 4, 5, and 6. However, savings are required in years 7 and 8 because only savings are available to meet requirements in years 8 and 9 after all three bonds have matured.

10.2   Portfolio Optimization

10.2.1   *Example 10.2 — Maximize the minimum return*

The first portfolio optimization problem we consider is the maximization of the minimum return. The formulation is as follows:

Parameters

J = the number of assets (typically stocks or mutual funds), which are indexed $1 \leq j \leq J$;

I = the number of return scenarios, which are typically framed as the returns for different years and are indexed $1 \leq i \leq I$;

p_{ij} = the return of asset j in scenario i (for $1 \leq i \leq I$ and $1 \leq j \leq J$).

Decision variables

M = the minimum return;

x_j = the proportion of the portfolio invested in asset j (for $1 \leq j \leq J$).

Objective function

$$\text{Maximize:} \ M \qquad\qquad (10.1)$$

Subject to constraints

$$\sum_{j=1}^{J} p_{ij} x_j \geq M \quad \text{for all } 1 \leq i \leq I, \qquad (10.2)$$

$$\sum_{j=1}^{J} x_j = 1. \tag{10.3}$$

The objective function in Equation (10.1) is the maximization of the minimum return. Constraint set (10.2) ensures that M is not larger than the return for each of the I scenarios. The return for a scenario is a weighted (by the proportion of the investment in the asset) average of the returns for the assets in that scenario. Because M must not be larger than the return for all scenarios, while at the same time the objective function seeks to maximize M, the value of M will be equal to the minimum of the returns across all scenarios. Constraint (10.3) requires the investment proportions to sum to one.

Cells B3:I7 of Figure 10.3 provide the data (i.e., the p_{ij} parameters) for Example 10.2, which corresponds to the returns for $J = 8$ assets for each of $I = 5$ years (scenarios). Cells B9:I9 contain the investment proportion decision variables. Cell J9 contains the decision variable M (the minimum return), which is the objective function, and the 'Set Objective' cell in the Solver dialog box in Figure 10.4.

Cell L9 contains the sum of the proportions in B9:I9 and is constrained to equal one in accordance with constraint (10.3). Cells B12:I16 are the products of the returns in cells B3:I7 and the proportions in cells B9:I9. The value of M is subtracted from the sum of the returns for each year and these quantities are placed in cells L12:L16. Effectively, this is subtracting M from both sides of

	A	B	C	D	E	F	G	H	I	J	K	L
1												
2		Asset 1	Asset 2	Asset 3	Asset 4	Asset 5	Asset 6	Asset 7	Asset 8	M		
3	Year 1	4.25	9.79	34.75	16.22	17.74	-4.45	20.02	-3.35			
4	Year 2	11.79	18.63	23.89	5.74	41.35	11.38	-7.59	12.51			
5	Year 3	15.23	31.18	-14.45	9.33	-8.45	32.56	-4.82	9.49			
6	Year 4	-1.74	-12.44	11.36	2.19	-2.68	24.25	-1.68	26.82			
7	Year 5	3.86	5.19	17.29	11.41	27.22	-8.82	38.44	8.43			
8												
9	Proportion	.00000	.05393	.18447	.00000	.07099	.43551	.25511	.00000	11.36676		1.00000
10	Invested											
11												
12	Year 1	.00000	.52802	6.41019	.00000	1.25930	-1.93800	5.10725	.00000	-11.36676		.00000
13	Year 2	.00000	1.00481	4.40689	.00000	2.93528	4.95605	-1.93627	.00000	-11.36676		.00000
14	Year 3	.00000	1.68169	-2.66553	.00000	-.59983	14.18005	-1.22962	.00000	-11.36676		.00000
15	Year 4	.00000	-.67095	2.09553	.00000	-.19024	10.56101	-.42858	.00000	-11.36676		.00000
16	Year 5	.00000	.27992	3.18941	.00000	1.93225	-3.84116	9.80634	.00000	-11.36676		.00000

Figure 10.3.   Excel worksheet for Example 10.2.

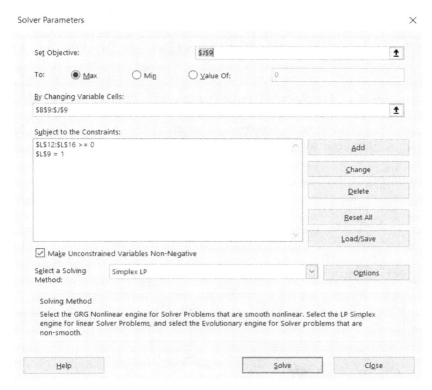

Figure 10.4.   Solver dialog box for Example 10.2.

constraint set (10.2) and leaving zero on the right. By constraining the values in cells L12:L16 to equal or exceed zero, constraint set (10.2) is accomplished.

The results in Figure 10.3 reveal that the largest minimum return is 11.36676%, which seems rather good for a worst-case scenario. It is accomplished by investing large proportions in assets 6 (0.43551), 7 (0.25511), and 3 (0.18847). There is also some modest investment in assets 5 (0.07099) and 2 (0.05393).

10.2.2   *Example 10.3 — Maximize the expected return*

The second portfolio optimization problem we consider is the maximization of the expected return, subject to a threshold constraint on the minimum return. The formulation is as follows:

Parameters

J = the number of assets (typically stocks or mutual funds), which are indexed $1 \leq j \leq J$;

I = the number of return scenarios, which are typically framed as the returns for different years and are indexed $1 \leq i \leq I$;

p_{ij} = the return of asset j in scenario i (for $1 \leq i \leq I$ and $1 \leq j \leq J$);

q_i = the probability of scenario i occurring (for $1 \leq i \leq I$);

t = a threshold for a minimum level of return.

Decision variables

x_j = the proportion of the portfolio invested in asset j (for $1 \leq j \leq J$);

R_i = the return associated with scenario/year i (for $1 \leq i \leq I$).

Objective function

$$\text{Maximize:} \sum_{i=1}^{I} q_i R_i. \tag{10.4}$$

Subject to constraints

$$\sum_{j=1}^{J} p_{ij} x_j = R_i \quad \text{for all } 1 \leq i \leq I, \tag{10.5}$$

$$\sum_{j=1}^{J} x_j = 1, \tag{10.6}$$

$$R_i \geq t \quad \text{for all } 1 \leq i \leq I. \tag{10.7}$$

The objective function in Equation (10.4) is the maximization of expected return. Our example will assume equal probabilities for each scenario (i.e., $q_i = 1/I$ for all $1 \leq i \leq I$) but the potential is there to evaluate other distributions. Constraint set (10.5) establishes the returns for each scenario. As the R_i are linked to the x_j via an equality constraint, the definitions of the R_i variables are more for convenience of presentation than necessity. Constraint (10.6) ensures the proportions sum to one. Constraint set (10.7) guarantees that the minimum level of return is achieved for each scenario.

	A	B	C	D	E	F	G	H	I	J	K	L
1												
2		Asset 1	Asset 2	Asset 3	Asset 4	Asset 5	Asset 6	Asset 7	Asset 8			
3	Year 1	4.25	9.79	34.75	16.22	17.74	-4.45	20.02	-3.35			
4	Year 2	11.79	18.63	23.89	5.74	41.35	11.38	-7.59	12.51			
5	Year 3	15.23	31.18	-14.45	9.33	-8.45	32.56	-4.82	9.49			
6	Year 4	-1.74	-12.44	11.36	2.19	-2.68	24.25	-1.68	26.82			
7	Year 5	3.86	5.19	17.29	11.41	27.22	-8.82	38.44	8.43			
8												
9	Proportion	.00000	.00000	.00000	.00000	.79395	.20605	.00000	.00000		1.00000	
10	Invested											Scenario
11												Probability
12	Year 1	.00000	.00000	.00000	.00000	14.08472	-.91691	.00000	.00000		13.16781	0.2
13	Year 2	.00000	.00000	.00000	.00000	32.82994	2.34482	.00000	.00000		35.17476	0.2
14	Year 3	.00000	.00000	.00000	.00000	-6.70890	6.70890	.00000	.00000		.00000	0.2
15	Year 4	.00000	.00000	.00000	.00000	-2.12779	4.99665	.00000	.00000		2.86885	0.2
16	Year 5	.00000	.00000	.00000	.00000	21.61139	-1.81734	.00000	.00000		19.79406	0.2
17												
18									Expected value		14.2011	

Figure 10.5. Excel worksheet for Example 10.3.

Cells B3:I7 of Figure 10.5 provide the data (i.e., the p_{ij} parameters) for Example 10.3, which corresponds to the same returns for the $J = 8$ assets for each of $I = 5$ years in Example 10.2. Cells B9:I9 contain the investment proportion decision variables. Cells B12:I16 are the products of the returns in cells B3:I7 and the proportions in cells B9:I9. The row sum of cells B12:I16 is contained in cells K12:K16 and these are the returns for each year. The sumproduct of cells K12:K16 and the scenario probabilities in cells L12:L16 are contained in cell K18. Cell K18 is, therefore, the expected return objective function and is the 'Set Objective' cell in the Solver dialog in Figure 10.6.

Cell K9 contains the sum of the proportions in B9:I9 and is constrained to equal one. In addition, cells K12:K16 must be constrained to equal or exceed the selected return threshold to enforce constraint set (10.7). As can be seen from the Solver dialog box in Figure 10.6, we constrained these cells to equal or exceed a threshold of $t = 0$. This is a somewhat high level of risk as the results for Example 10.2 show that a minimum return of 11.36676% could be achieved.

The results in Figure 10.5 reveal a substantially different investment portfolio relative to the one in Figure 10.3. The solution in Figure 10.5 requires investment in only two assets. The bulk of the portfolio (0.79395) is invested in asset 5 and the remaining proportion (0.20605) is invested in asset 6. The expected return is 14.20110%, which is greater than the corresponding expected value of 11.36676% for the portfolio in Figure 10.3. However, the portfolio in Figure 10.3

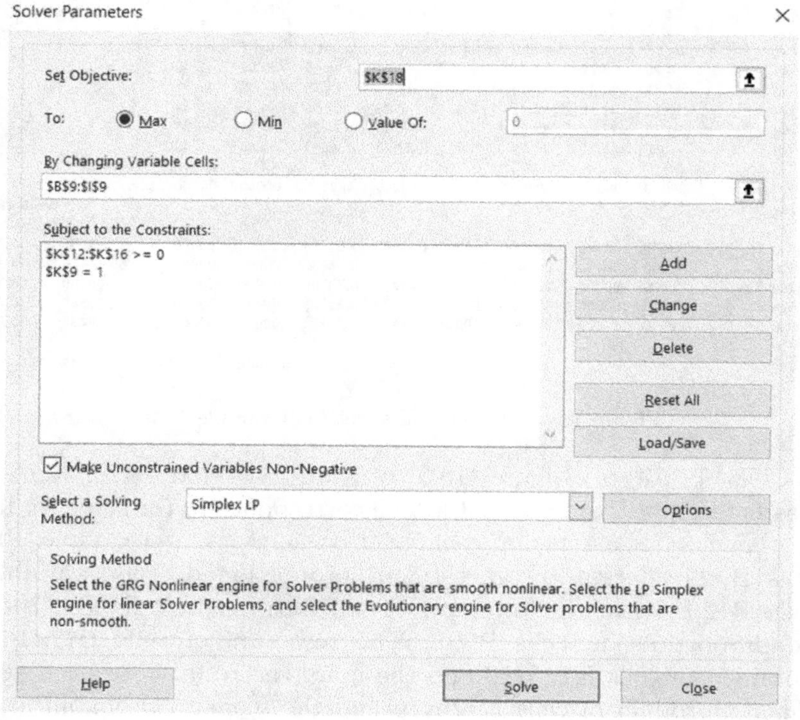

Figure 10.6.   Solver dialog box for Example 10.3.

had zero variance (all years had a return of 11.36676%), but we can see from Figure 10.5 that there is considerable variance for this portfolio. For example, year 3 has a return of 0.00% and year 4 has a return of only 2.87%. By contrast, year 2 has a return of 35.17%. In the next section, we will examine the use of the Markowitz model to examine return variance within the rather narrow expected return range from 11.36676% to 14.20110%.

10.2.3   *Example 10.4 — Minimize return variance (Markowitz model)*

The third portfolio optimization problem we consider is the Markowitz model, which seeks the minimization of the variance of the return subject to a constraint on the expected (or average) return. The formulation is as follows:

Parameters

$J =$ the number of assets (typically stocks or mutual funds), which are indexed $1 \leq j \leq J$;

$I =$ the number of return scenarios, which are typically framed as the returns for different years and are indexed $1 \leq i \leq I$;

$p_{ij} =$ the return of asset j in scenario i (for $1 \leq i \leq I$ and $1 \leq j \leq J$);

$u =$ a threshold for a minimum level of expected or average return.

Decision variables

$x_j =$ the proportion of the portfolio invested in asset j (for $1 \leq j \leq J$);

$R_i =$ the return associated with scenario/year i (for $1 \leq i \leq I$);

$\bar{R} =$ the mean of the returns.

Objective function

$$\text{Minimize:} \quad \frac{1}{I} \sum_{i=1}^{I} (R_i - \bar{R})^2. \tag{10.8}$$

Subject to constraints

$$\sum_{j=1}^{J} p_{ij} x_j = R_i \quad \text{for all } 1 \leq i \leq I, \tag{10.9}$$

$$\sum_{j=1}^{J} x_j = 1, \tag{10.10}$$

$$\bar{R} = \frac{1}{I} \sum_{i=1}^{I} R_i, \tag{10.11}$$

$$\bar{R} \geq u. \tag{10.12}$$

The objective function in Equation (10.8) is the minimization of the variance of the return. Variance is computed as the sum of the squared deviations of the returns for each scenario from the mean return, divided by the number of scenarios. Constraint set (10.9) establishes the returns for each scenario. Constraint (10.10)

	A	B	C	D	E	F	G	H	I	J	K
1											
2		Asset 1	Asset 2	Asset 3	Asset 4	Asset 5	Asset 6	Asset 7	Asset 8		
3	Year 1	4.25	9.79	34.75	16.22	17.74	-4.45	20.02	-3.35		
4	Year 2	11.79	18.63	23.89	5.74	41.35	11.38	-7.59	12.51		
5	Year 3	15.23	31.18	-14.45	9.33	-8.45	32.56	-4.82	9.49		
6	Year 4	-1.74	-12.44	11.36	2.19	-2.68	24.25	-1.68	26.82		
7	Year 5	3.86	5.19	17.29	11.41	27.22	-8.82	38.44	8.43		
8											
9	Proportion	.00000	.00000	.34576	.00000	.20150	.43393	.01881	.00000		1.00000
10	Invested										
11											
12	Year 1	.00000	.00000	12.01504	.00000	3.57468	-1.93100	.37654	.00000		14.03526
13	Year 2	.00000	.00000	8.26012	.00000	8.33219	4.93814	-.14275	.00000		21.38770
14	Year 3	.00000	.00000	-4.99618	.00000	-1.70271	14.12881	-.09066	.00000		7.33926
15	Year 4	.00000	.00000	3.92779	.00000	-.54003	10.52284	-.03160	.00000		13.87900
16	Year 5	.00000	.00000	5.97813	.00000	5.48494	-3.82728	.72298	.00000		8.35878
17											
18								Expected value			13.0000
19											
20								Variance			25.1566

Figure 10.7.   Excel worksheet for Example 10.4.

ensures the proportions sum to one. Constraint (10.11) computes the expected (average) return from the returns for the individual scenarios. Constraint (10.12) guarantees that the expected return exceeds the prescribed threshold.

Cells B3:I7 of Figure 10.7 provide the data (i.e., the p_{ij} parameters) for Example 10.4, which corresponds to the same returns for the $J = 8$ assets for each of $I = 5$ years in Examples 10.2 and 10.3. Cells B9:I9 contain the investment proportion decision variables. Cells B12:I16 are the products of the returns in cells B3:I7 and the proportions in cells B9:I9. The row sum of cells B12:I16 is contained in cells K12:K16 and these are the returns for each year. The average of cells K12:K16 is contained in cell K18. Cell K20 contains the variance of the returns, which is the sum of squared deviations of cells K12:K16 from their average in cell K18, divided by the number of years (i.e., 5). Cell K20 is the 'Set Objective' cell in the Solver dialog box in Figure 10.8. Because this objective function is quadratic, the GRG Nonlinear engine of the Solver is used.

Cell K9 contains the sum of the proportions in B9:I9 and is constrained to equal one. In addition, cell K18 must be constrained to equal or exceed the selected expected return threshold to enforce constraint set (10.12). As can be seen from the Solver dialog box in Figure 10.8, we constrained this cell to equal or exceed a threshold of $u = 13$. This threshold falls between the expected returns for the solutions for Examples 10.2 (11.36676% expected return, zero variance)

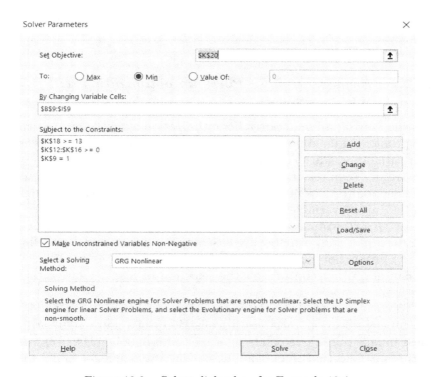

Figure 10.8. Solver dialog box for Example 10.4.

and 10.3 (14.20110% expected return, high variance of more than 160).

The results in Figure 10.7 reveal a portfolio that is different from the ones in Figures 10.3 and 10.5, yet more similar to the one in 10.3 than the one in 10.5. The solution in Figure 10.7 requires investment in four assets: assets 6 (0.43393), 3 (0.34576), 5 (0.20150), and .7 (0.01881). Thus, relative to the results for Example 10.2 in Figure 10.3, the big difference in the results for Example 10.4 in Figure 10.7 is far less investment in asset 7 and much greater investment in assets 3 and 5. The benefit realized from these changes is a greater expected return of 13% but a greater variance of 25.1566.

Although not shown, we also obtained results for the Markowitz model using thresholds of $u = 12$ and $u = 14$. The variance for the $u = 12$ threshold is only 3.7549, but there were two scenarios where the return was less than 11%, which is lower than the minimum return realized in Figure 10.3. The variance for the $u = 14$ threshold is 111.4856, which is dramatically worse than the variance for $u = 13$.

10.2.4 Example 10.5 — Portfolio optimization with side constraints

For the fourth and final portfolio optimization example, we return to the problem of maximization of the minimum return but now assume there are other side constraints to consider. The incorporation of side constraints requires the introduction of binary variables to control the selection (or lack of selection) of each asset. Specifically, we define a set of binary variables: $y_j = 1$ if asset j is selected and 0 otherwise, for $1 \leq j \leq J$. We can then use these variables in a variety of ways. For example, they can be used to impose upper and lower limits on the proportion of the portfolio invested in each asset. Moreover, we can add other controls on the selection of assets relative to one another.

To construct Example 10.5, we return to Example 10.2, adding the binary selection variables in cells B9:I9 of Figure 10.9. We use these binary variables to place an upper limit of 0.4 on the proportion of the portfolio that might be invested in any one asset and a lower proportion limit of 0.1 on any selected asset. The constraints are of

	A	B	C	D	E	F	G	H	I	J	K	L
1												
2		Asset 1	Asset 2	Asset 3	Asset 4	Asset 5	Asset 6	Asset 7	Asset 8	M		
3	Year 1	4.25	9.79	34.75	16.22	17.74	-4.45	20.02	-3.35			
4	Year 2	11.79	18.63	23.89	5.74	41.35	11.38	-7.59	12.51			
5	Year 3	15.23	31.18	-14.45	9.33	-8.45	32.56	-4.82	9.49			
6	Year 4	-1.74	-12.44	11.36	2.19	-2.68	24.25	-1.68	26.82			
7	Year 5	3.86	5.19	17.29	11.41	27.22	-8.82	38.44	8.43			
8												
9	Invested (Y/N)?	0	1	1	0	0	1	1	1			
10												
11	Proportion	.00000	.14210	.21095	.00000	.00000	.29052	.21203	.14439	11.19018		1.00000
12	Invested											
13												
14	Lower bound	0.0	0.1	0.1	0.0	0.0	0.1	0.1	0.1			
15	Upper bound	0.0	0.4	0.4	0.0	0.0	0.4	0.4	0.4			
16												
17	Year 1	.00000	1.39117	7.33065	.00000	.00000	-1.29284	4.24489	-.48370	-11.19018		.00000
18	Year 2	.00000	2.64735	5.03969	.00000	.00000	3.30617	-1.60932	1.80629	-11.19018		.00000
19	Year 3	.00000	4.43073	-3.04828	.00000	.00000	9.45949	-1.02200	1.37024	-11.19018		.00000
20	Year 4	.00000	-1.76774	2.39644	.00000	.00000	7.04523	-.35621	3.87247	-11.19018		.00000
21	Year 5	.00000	.73751	3.64739	.00000	.00000	-2.56243	8.15052	1.21718	-11.19018		.00000
22												
23												
24	No more than two of Assets 3, 4, 5, and 6 can be chosen:					2						
25												
26	If Assets 6 and 7 are chosen, then 8 must be chosen also:				1							

Figure 10.9. Excel worksheet for Example 10.5.

the form $0.1y_j \le x_j \le 0.4y_j$ for $1 \le j \le 8$. They are accomplished in the worksheet in Figure 10.9 by adding rows 14 and 15. Cells B14:I14 are constructed by multiplying the binary variables in cells B9:I9 by 0.1. Similarly, cells B15:I15 are constructed by multiplying the binary variables in cells B9:I9 by 0.4. Cells B11:I11, which contain the proportion decision variables, are then constrained to be equal to or greater than the cells in B14:I14 and equal to or less than the cells in B15:I15, as shown in the Solver dialog box in Figure 10.10.

We also impose two additional side constraints using the binary variables. First, we add the condition that no more than two of four assets (assets 3, 4, 5, and 6) can be selected. This is accomplished by summing cells D9:G9 in cell G24 and then constraining G24 to be equal to or less than 2. The second side constraint is that if assets 6 and 7 are selected, then asset 8 must also be selected. The constraint is $y_8 + 1 \ge y_6 + y_7$. In other words, if both assets 6 and 7 are selected,

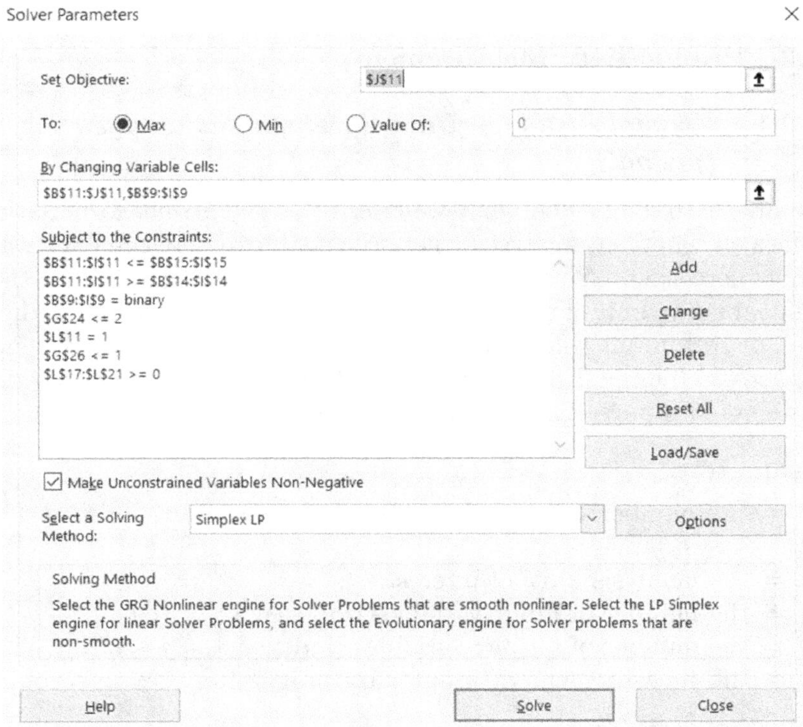

Figure 10.10.   Solver dialog box for Example 10.5.

then y_8 must be one to satisfy the constraint; otherwise, if 6 and 7 are not both selected, then the constraint is satisfied regardless of the value of y_8. We operationalize this constraint in cell G26 with the formula G9 + H9 − I9 and then constrain this cell to be equal to or less than one.

The optimal solution in Figure 10.9 reveals that, despite the somewhat restrictive constraints that have been added, a new portfolio has been identified with only a slightly lower minimum return (11.19018%) than the one in Example 10.2 (11.36676%). Moreover, the portfolio in Figure 10.9 is much more balanced, with proportions ranging from 0.14210 for asset 2 to 0.29052 for asset 6.

Example 10.5 reveals that balanced portfolios can be accomplished in different ways. One approach is to place bounds on the proportions of the portfolio that can be invested in different assets. A second approach is to enforce restrictions on the selection of assets from different categories by using binary variables.

10.3 Productivity Measurement

10.3.1 *Example 10.6 — Data envelopment analysis (primal)*

Example 10.6 focuses on what we refer to as the *primal* formulation of data envelopment analysis, where the goal is to determine whether or not there exists a set of weights on input and output measures that will enable the unit of analysis (i.e., the particular branch bank, hospital, or restaurant that is of interest) to have the greatest efficiency (E) across all units. Efficiency is defined as weighted outputs divided by weighted inputs. The formulation is as follows:

Parameters

K = the number of units (e.g., banks or hospitals, indexed $1 \leq k \leq K$),

$\bar{k}$ = the 'analysis unit' of interest,

I = the number of input variables (indexed $1 \leq i \leq I$),

J = the number of output variables (indexed $1 \leq j \leq J$),

A_{ik} = the measurement of input i on unit k (for $1 \leq i \leq I$ and $1 \leq k \leq K$),

B_{jk} = the measurement of output j on unit k (for $1 \leq j \leq J$ and $1 \leq k \leq K$).

Decision variables

$U_i =$ the weight on input variable i (for $1 \leq i \leq I$),
$V_j =$ the weight on output variable j (for $1 \leq j \leq J$).

Objective function (fractional)

$$\text{Maximize:} \ E = \frac{\sum_{j=1}^{J} B_{j\bar{k}} V_j}{\sum_{i=1}^{I} A_{i\bar{k}} U_i}. \tag{10.13}$$

Subject to constraints (fractional)

$$\frac{\sum_{j=1}^{J} B_{jk} V_j}{\sum_{i=1}^{I} A_{ik} U_i} \leq 1 \quad \text{for } 1 \leq k \leq K. \tag{10.14}$$

The objective function in Equation (10.13) is to maximize the efficiency of the unit of analysis, $\bar{k}$, which is the ratio of the unit's weighted outputs to weighted inputs. Constraint set (10.14) ensures that none of the K units will have an efficiency greater than one. Therefore, if a set of weights can be found such that $E = 1$ for the unit of analysis, then that unit will have an efficiency that equals or exceeds the efficiency of all other units, and we can judge the unit of analysis as 'relatively efficient'.

Because the objective function is fractional, we adopt the common convention of normalizing weighted input for the unit of analysis to sum to one. The objective function becomes Equation (10.15) and the normalizing constraint is shown in Equation (10.17). We also rewrite constraint set (10.14) by multiplying through by weighted inputs and then subtracting the weighted input from both sides as shown in Equation (10.16). The objective function and constraints for the operational model are as follows:

Objective function

$$\text{Maximize:} \ E = \sum_{j=1}^{J} B_{j\bar{k}} V_j. \tag{10.15}$$

Subject to constraints

$$\sum_{j=1}^{J} B_{jk} V_j - \sum_{i=1}^{I} A_{ik} U_i \leq 0 \quad \text{for } 1 \leq k \leq K, \tag{10.16}$$

$$\sum_{i=1}^{I} A_{i\bar{k}} U_i = 1. \tag{10.17}$$

Example 10.6 is a small illustration of how data envelopment analysis might be used to examine the relative efficiency of $K = 5$ schools that offer MBA programs. The example assumes $I = 2$ input variables: number of MBA faculty and advertising expenditure on the program. There are also $J = 2$ output variables: number of Fortune 500 recruiters and average starting salary. Figure 10.11 displays the worksheet for the example, and the corresponding Solver dialog box is shown in Figure 10.12. We also present the sensitivity report for the solution in Figure 10.13.

Cells B3:E7 in Figure 10.11 contain the raw data for the analysis, that is, the input and output measures for each of the five schools. Cells B9:E9 are the decision variables, namely, the weights on the input and output variables. Cells B11:B15 contain weighted outputs minus weighted inputs for the five schools. As shown in the Solver dialog box, these cells are constrained to be equal to or less than zero in accordance with constraint set (10.16).

	A	B	C	D	E	F
1		Number of MBA	Advertising	Number of F500	Avg. Starting	
2		Faculty	Budget	Recruiters	Salary	
3	School 1	22	$20,400	34	$76,100	
4	School 2	13	$23,200	28	$84,200	
5	School 3	18	$19,500	31	$68,500	
6	School 4	24	$16,900	39	$85,300	
7	School 5	20	$17,500	35	$81,000	
8						
9	Weights	0.0349014	0.0000114	0.0256341	0.0000000	
10						
11	Bound School 1	-0.1284414	Stand. School 1	1.000	Eff School 1	0.8715586
12	Bound School 2	0.0000000	Stand. School 2	0.718	Eff School 2	0.7177542
13	Bound School 3	-0.0554953	Stand. School 3	0.850	Eff School 3	0.7946564
14	Bound School 4	-0.0302406	Stand. School 4	1.030	Eff School 4	0.9997290
15	Bound School 5	0.0000000	Stand. School 5	0.897	Eff School 5	0.8971927
16						
17		Maximum	Number of MBA	Advertising	Number of F500	Avg. Starting
18		Efficiency	Faculty	Budget	Recruiters	Salary
19	School 1	0.8716	0.03490137	0.00001138	0.02563408	0.00000000
20	School 2	1.0000	0.02037332	0.00003169	0.00000000	0.00001188
21	School 3	0.9347	0.04105310	0.00001339	0.03015236	0.00000000
22	School 4	1.0000	0.00000000	0.00005917	0.00000000	0.00001172
23	School 5	1.0000	0.01864649	0.00003583	0.00000000	0.00001235

Figure 10.11.   Excel worksheet for Example 10.6.

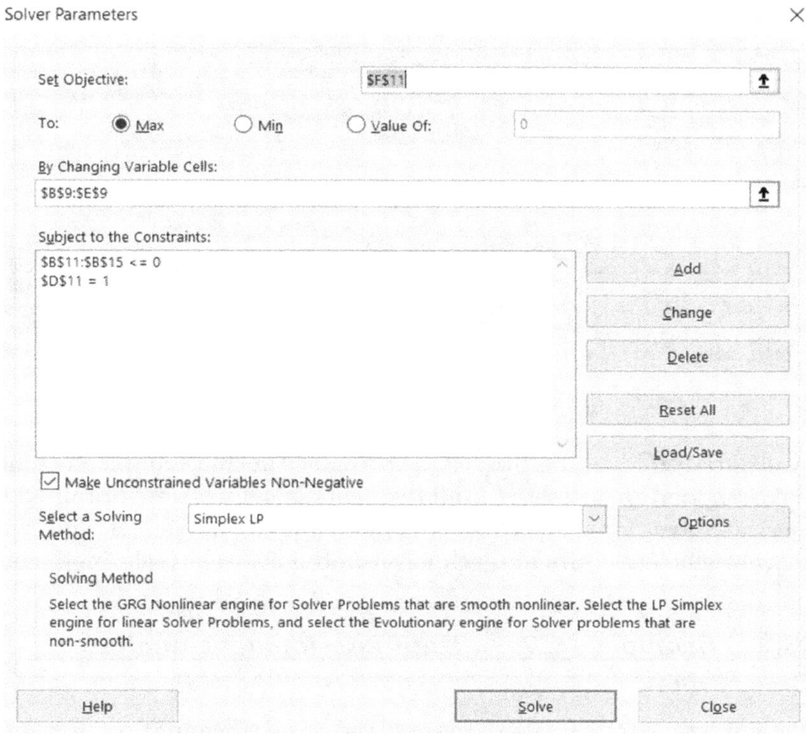

Figure 10.12. Solver dialog box for Example 10.6.

It should the clarified that the Solver will be run five times, once for each school. The Solver dialog box in Figure 10.12 is configured for school 1. The 'Set Objective' cell is F11, which contains weighted output for school 1: $D3*D9+E3*E9$. Likewise, weighted input for school 1, which is $B3*B9+C3*C9$ is in cell D11 and this cell is constrained to equal one in the Solver dialog box.

The optimal objective function value for school 1 in cell F11 is 0.8716, which reveals that this school is not relatively efficient. The sensitivity report in Figure 10.13 indicates that the 'reference set' of schools that shape the inefficiency of school 1 includes schools 2 and 5, which have non-zero shadow prices of 0.0848 and 0.9036, respectively. In fact, a closer examination of the input data reveals that school 5 dominates school 1, as it uses less of both inputs and has more of both outputs.

6	Variable Cells						
7			Final	Reduced	Objective	Allowable	Allowable
8	Cell	Name	Value	Cost	Coefficient	Increase	Decrease
9	$B$9	Weights Faculty	0.0349014	0.0000000	0.0000000	14.5952381	1.0952381
10	$C$9	Weights Budget	0.0000114	0.0000000	0.0000000	1015.5844156	13533.7662338
11	$D$9	Weights Recruiters	0.0256341	0.0000000	34.0000000	INF	1.7903917
12	$E$9	Weights Salary	0	-4230.067201	76100	4230.067201	1E+30
13							
14	Constraints						
15			Final	Shadow	Constraint	Allowable	Allowable
16	Cell	Name	Value	Price	R.H. Side	Increase	Decrease
17	$B$11	Bound School 1 Faculty	-0.1284414	0.0000000	0	1E+30	0.1284414
18	$B$12	Bound School 2 Faculty	0.0000000	0.0847605	0	0.136363636	0.0864848
19	$B$13	Bound School 3 Faculty	-0.0554953	0.0000000	0	1E+30	0.0554953
20	$B$14	Bound School 4 Faculty	-0.0302406	0.0000000	0	1E+30	0.0302406
21	$B$15	Bound School 5 Faculty	0.0000000	0.9036202	0	0.021693155	0.1704545
22	$E$11	Stand. School 1 Salary	1.0000000	0.8715586	1	1E+30	1.0000000

Figure 10.13.   Sensitivity report for Example 10.6.

Cells B19:F23 of Figure 10.11 provide the maximum efficiency and corresponding weights for all five schools, which were obtained by using the Solver for each school independently. Schools 2, 4, and 5 are relatively efficient units and schools 1 and 3 are relatively inefficient.

10.3.2   *Example 10.7 — Data envelopment analysis (dual)*

For completeness, we also present the *dual* formulation for data envelopment analysis. Here, the goal is to determine a set of weights for the units so as to create a 'composite' unit with its own outputs and inputs. If a set of weights can be found such that the composite unit can deliver the same level of outputs as the unit of analysis, while using less input than the unit of analysis, then the unit of analysis can be judged as relatively inefficient. The formulation is as follows:

Parameters

K = the number of units (e.g., banks or hospitals, indexed $1 \le k \le K$),

$\bar{k}$ = the 'analysis unit' of interest,

I = the number of input variables (indexed $1 \le i \le I$),

J = the number of output variables (indexed $1 \le j \le J$),

A_{ik} = the measurement of input i on unit k (for $1 \le i \le I$ and $1 \le k \le K$),

B_{jk} = the measurement of output j on unit k (for $1 \leq j \leq J$ and $1 \leq k \leq K$).

Decision variables

E = the percentage of the analysis unit's inputs needed by the composite unit to deliver the same level of outputs,

X_k = the weight on inputs and outputs for unit k (for $1 \leq k \leq K$).

Objective function

$$\text{Minimize: } E. \tag{10.18}$$

Subject to constraints

$$\sum_{k=1}^{K} B_{jk} X_k \geq B_{j\bar{k}} \quad \text{for all } 1 \leq j \leq J, \tag{10.19}$$

$$\sum_{k=1}^{K} A_{ik} X_k \leq A_{i\bar{k}} E \quad \text{for all } 1 \leq i \leq I. \tag{10.20}$$

The objective function (10.18) is to minimize E, the percentage of the analysis unit's inputs needed to deliver the outputs. Constraint set (10.19) ensures, for each output measure, that the output for the composite unit equals or exceeds the output for the unit of analysis. Constraint set (10.20) guarantees, for each input measure, that the input for the composite unit is equal to or less than the input for the unit of analysis multiplied by E. If E can be less than one, then that means that the composite unit can use less of the inputs to deliver the outputs.

Example 10.7 uses the same data as Example 10.6. Figure 10.14 displays the worksheet for the example, and the corresponding Solver dialog box is shown in Figure 10.15. We also present the sensitivity report for the solution in Figure 10.16, which will allow us to map results from Example 10.7 back to Example 10.6.

Cells B3:E7 in Figure 10.14 contain the raw data for the analysis, that is, the input and output measures for each of the five schools. Cells G3:G7 are the X_j decision variables, namely, the weights on the five schools. The sum of the weights in cells G3:G7 is contained in cell G9. This cell is not used in our analyses. However,

	A	B	C	D	E	F	G
1		Number of MBA	Advertising	Number of F500	Avg. Starting		School
2		Faculty	Budget	Recruiters	Salary		Weights
3	School 1	22	$20,400	34	$76,100		0.0000000
4	School 2	13	$23,200	28	$84,200		0.0847605
5	School 3	18	$19,500	31	$68,500		0.0000000
6	School 4	24	$16,900	39	$85,300		0.0000000
7	School 5	20	$17,500	35	$81,000		0.9036202
8							
9					Weights sum =		0.9883807
10					E =		0.8715586
11	Weigthed						
12	Sums	0.0000000	0.0000000	0.0000000	4230.0672014		

Figure 10.14.   Excel worksheet for Example 10.7.

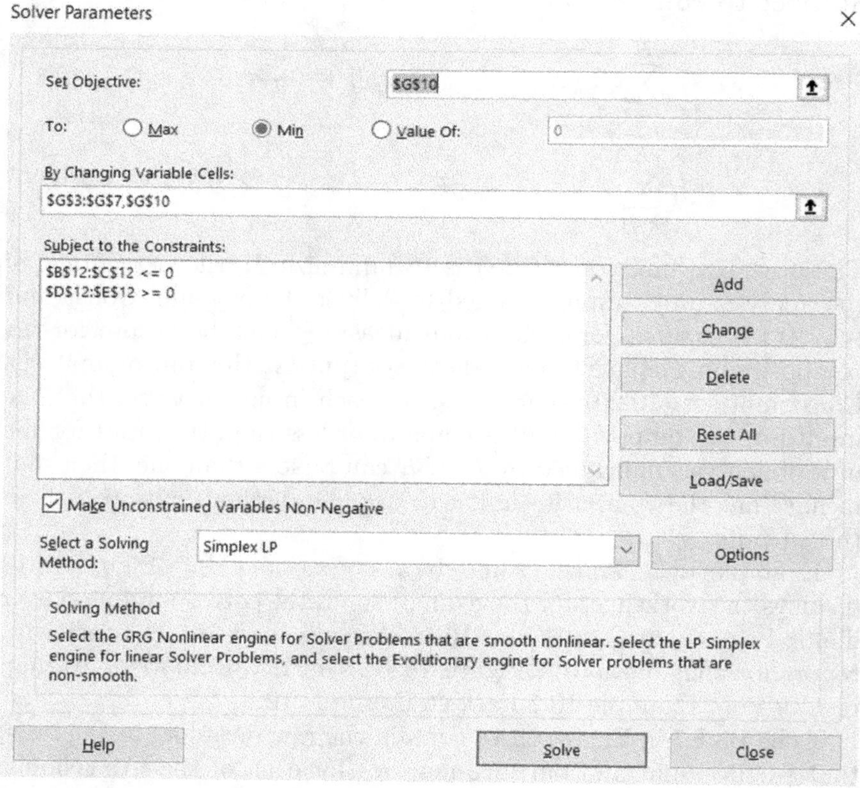

Solver Parameters   ✕

Set Objective: $G$10   ⬆

To: ○ Max   ● Min   ○ Value Of: 0

By Changing Variable Cells:

$G$3:$G$7,$G$10   ⬆

Subject to the Constraints:

$B$12:$C$12 <= 0   Add
$D$12:$E$12 >= 0
                                                                                                       Change

                                                                                                       Delete

                                                                                                       Reset All

                                                                                                       Load/Save

☑ Make Unconstrained Variables Non-Negative

Select a Solving   Simplex LP   Options
Method:

Solving Method

Select the GRG Nonlinear engine for Solver Problems that are smooth nonlinear. Select the LP Simplex
engine for linear Solver Problems, and select the Evolutionary engine for Solver problems that are
non-smooth.

        Help   Solve   Close

Figure 10.15.   Solver dialog box for Example 10.7.

6	Variable Cells						
7			Final	Reduced	Objective	Allowable	Allowable
8	Cell	Name	Value	Cost	Coefficient	Increase	Decrease
9	$G$3	School 1 Weights	0.0000000	0.1284414	0	1E+30	0.128441361
10	$G$4	School 2 Weights	0.0847605	0.0000000	0	0.136363636	0.086484811
11	$G$5	School 3 Weights	0.0000000	0.0554953	0	1E+30	0.055495339
12	$G$6	School 4 Weights	0.0000000	0.0302406	0	1E+30	0.030240624
13	$G$7	School 5 Weights	0.9036202	0.0000000	0	0.021693155	0.170454545
14	$G$10	E = Weights	0.8715586	0.0000000	⁻1	1E+30	1
15							
16	Constraints						
17			Final	Shadow	Constraint	Allowable	Allowable
18	Cell	Name	Value	Price	R.H. Side	Increase	Decrease
19	$B$12	Sums Faculty	0.0000000	-0.0349014	0	1.095238095	14.5952381
20	$C$12	Sums Budget	0.0000000	-0.0000114	0	13533.76623	1015.584416
21	$D$12	Sums Recruiters	0.0000000	0.0256341	0	1E+30	1.790391691
22	$E$12	Sums E =	4230.0672014	0.0000000	0	4230.067201	1E+30

Figure 10.16.   Sensitivity report for Example 10.7.

by constraining this cell to equal one, a user could shift the analysis from the assumption of *constant returns to scale* to the assumption of *variable returns to scale* (see Anderson *et al.*, 2019, Chapter 5). We assume constant returns to scale for consistency with Example 10.6. Cell G10 contains the efficiency decision variable, E.

Cells B12:C12 contain the sumproducts of the inputs and the weights in cells G3:G7 minus the efficiency in cell G10 times the input measure for the unit of analysis. (The worksheet in Figure 10.14 is configured for school 1 as the unit of analysis.) Cells B12:C12 are constrained to be equal to or less than zero in accordance with constraint set (10.20) as shown in the Solver dialog box in Figure 10.15.

Cells D12:E12 contain the sumproducts of the outputs and the weights in cells G3:G7 minus the output measure for the unit of analysis. These cells are constrained to equal or exceed zero in accordance with constraint set (10.19) as shown in the Solver dialog box.

When evaluating the optimal solution for school 1 in Figure 10.14, the first observation is that the value of the objective function, $E = 0.8716$, is identical to the value of the objective function for the primal formulation in Figure 10.11. Again, the implication is that school 1 is relatively inefficient.

The second observation concerning the solution in Figure 10.14 is that the weights for the schools are identical to the shadow (dual) prices for the primal problem that was observed in Figure 10.13.

Again, the implication here is that schools 2 and 5 are units that are not relatively inefficient, which shapes the relative inefficiency of school 1.

Finally, we turn to the sensitivity report for the dual formulation in Figure 10.16. Here, we observe that the shadow (dual) prices correspond precisely (except for sign differences for the input measures) to the weights on the input and output measures that we observed for the primal problem in Figure 10.11 (cells B9:E9).

Chapter 11

Sports

In this chapter, we consider several examples within the context of sports management. The first example pertains to league scheduling. In these problems, the goal is to determine when and where pairs of teams will play against one another during a season. Real-world applications of such problems abound in the literature and include the scheduling of a Chilean soccer league (Alarcôn *et al.*, 2017), a German basketball league (Westphal, 2014), and an Italian volleyball league (Cocchi, 2018). Although much smaller than these real-world applications, our example conveys some of the fundamental considerations that arise in such problems.

The second example focuses on the use of linear programming to determine when teams are mathematically eliminated from a division race. Traditionally, elimination has been based solely on the number of games a team trails behind the division leader and the number of games remaining to be played. However, when intra-division play is considered, it is possible that teams can be eliminated earlier. The particular model we use was proposed by Adler *et al.* (2002). Although our example focuses on elimination from a division race, the basic model can be extended to tackle problems such as elimination from playoff contention, division clinching, and playoff clinching.

The third example considers the ranking of n teams (or players) based on a tournament where each team plays every other team exactly once. The particular mathematical problem of interest is to find a permutation of the rows and columns of a matrix so as to maximize the sum of the elements above the main diagonal. In addition to ranking teams in a tournament, this problem has applications to electrical circuit theory (Lawler, 1964; Younger, 1963), triangularization of input–output matrices in economics (Korte & Oberhofer, 1971), and maximum likelihood paired comparison ranking (DeCani, 1969).

The fourth example also focuses on ranking. However, in this example, n teams or individuals are ranked based on their scores on a set of metric variables. We use a model proposed by Reinig and Horowitz (2018) that begins with the normalization of the variables. The normalized measures then serve as input to a linear programming model that seeks to minimize the maximum differences between weighted scores across all pairs of teams or individuals.

11.1   Example 11.1 — League Scheduling

Consider a small league with only four teams. The season consists of a pair of games played in each of six weeks (each team plays every week). Each team will play the other three teams twice during the six-week season, once *at home* and once *away*. The games are televised, and the league has estimated TV 'ratings' for each possible pairing of games each week at each location. The goal is to construct a season schedule that maximizes the overall TV ratings for the season. In addition to ensuring that each team plays exactly one game each week and plays every other team once at home and once away, the league would like to ensure that teams do not play each other over consecutive weeks. The integer linear programming formulation for this problem is as follows:

Parameters

C_{ijk} = the TV ratings if home team i plays visiting team j in week k, for all $1 \leq i \leq 4$, $1 \leq j \neq i \leq 4$, and $1 \leq k \leq 6$.

Decision variables

Y_{ijk} = 1 if home team i plays visiting team j in week k and 0 otherwise, for all $1 \leq i \leq 4$, $1 \leq j \neq i \leq 4$, and $1 \leq k \leq 6$. Note that there are 72 binary decision variables.

Objective function

$$\text{Maximize: } \sum_{i=1}^{4}\sum_{j\neq i}\sum_{k=1}^{6} C_{ijk}Y_{ijk}. \qquad (11.1)$$

Subject to

$$\sum_{j\neq i}(Y_{ijk} + Y_{jik}) = 1 \quad \text{for all } 1 \leq i \leq 4 \quad \text{and} \quad 1 \leq k \leq 6, \quad (11.2)$$

$$\sum_{k=1}^{6} Y_{ijk} = 1 \quad \text{for all } 1 \leq i \leq 4 \quad \text{and} \quad 1 \leq j \neq i \leq 4, \qquad (11.3)$$

$$Y_{ijk} + Y_{jik} + Y_{ij(k+1)} + Y_{ji(k+1)} \leq 1$$

$$\text{for all } 1 \leq i \leq 4, 1 \leq j > i \leq 4, \quad \text{and} \quad 1 \leq k \leq 5. \qquad (11.4)$$

The objective function in Equation (11.1) is to maximize the overall television ratings. Constraint set (11.2) ensures that each of the four teams will play exactly one game in each of the six weeks. Constraint set (11.3) guarantees that every home team i will play every other team (j) as the visiting team exactly once over the six-week season. Constraint set 11.4 requires that no pair of teams will play each other over consecutive weeks.

The TV ratings parameters for Example 1 are located in cells C2:H13 of Figure 11.1. The 72 decision variables are in cells C17:H28. Cell J4 contains the overall TV ratings objective function, which is the sumproduct of cells C2:H13 and C17:H28. Cell J4 is the 'Set Objective' cell in the Solver dialog box in Figure 11.2.

Cells C30:H33 of Figure 11.1 contain the sums of the decision variables in each column that correspond to games involving the team associated with the row label. For example, Cell C31 contains the formula $= C17 + C18 + \text{SUM}(C23:C26)$ because all of the cells in this formula involve team 2 playing in week 1. By constraining cells C30:H33 to equal one, constraint set 11.2 is ensured.

Cells J17:J28 contain the row sums of the matrix of decision variables in cells C17:H28. By constraining cells J17:J28 to equal one, we afford a guarantee that each team i will play every other team j once as a home team during the six-week season, as required by constraint set (11.3).

	Week 1	Week 2	Week 3	Week 4	Week 5	Week 6		
Home 1 Away 2	31.33	31.65	27.08	25.29	28.84	27.41		Total TV
Home 2 Away 1	34.16	27.99	29.19	28.05	34.13	27.56		Audience
Home 1 Away 3	33.79	31.10	33.39	27.58	25.21	34.18		Rating
Home 3 Away 1	25.64	33,77	31.48	30.56	34.23	32.90		386.94
Home 1 Away 4	30.01	32.37	32.49	27.28	27.16	27.59		
Home 4 Away 1	25.17	29.62	28.92	32.32	33.15	34.29		
Home 2 Away 3	30.88	25.34	33.86	31.97	29.55	28.72		
Home 3 Away 2	28.48	25.39	27.10	26.49	25.37	32.94		
Home 2 Away 4	25.49	30.61	26.67	32.49	30.16	29.43		
Home 4 Away 2	28.03	34.60	26.32	30.85	30.51	34.83		
Home 3 Away 4	25.46	29.65	28.79	26.69	25.66	30.58		
Home 4 Away 3	33.49	26.45	27.46	30.89	28.21	33.29		

							Once @	No Consecutive Weeks (ij)				
	Week 1	Week 2	Week 3	Week 4	Week 5	Week 6	Home	12	23	34	45	56
Home 1 Away 2	1	0	0	0	0	0	1					
Home 2 Away 1	0	0	0	0	1	0	1	1	0	0	1	1
Home 1 Away 3	0	1	0	0	0	0	1					
Home 3 Away 1	0	0	0	1	0	0	1	1	1	1	1	0
Home 1 Away 4	0	0	1	0	0	0	1					
Home 4 Away 1	0	0	0	0	0	1	1	0	1	1	0	1
Home 2 Away 3	0	0	1	0	0	0	1					
Home 3 Away 2	0	0	0	0	0	1	1	0	1	1	0	1
Home 2 Away 4	0	0	0	1	0	0	1					
Home 4 Away 2	0	1	0	0	0	0	1	1	1	1	1	0
Home 3 Away 4	0	0	0	0	1	0	1					
Home 4 Away 3	1	0	0	0	0	0	1	1	0	0	1	1
Team 1 Plays	1	1	1	1	1	1						
Team 2 Plays	1	1	1	1	1	1						
Team 3 Plays	1	1	1	1	1	1						
Team 4 Plays	1	1	1	1	1	1						

Figure 11.1.   Excel worksheet for Example 11.1.

The cells with values in the range spanning L17:P28 are the sums of variables for pairs of teams in successive weeks. For example, the formula in cell L18 is = SUM(C17:D18). The four cells in this sum are all possible home/away combinations for teams 1 and 2 in the first two weeks. In accordance with constraint set (11.4), we require this sum to be equal to or less than one. We could have teams 1 and 2 play each other in either week 1 or week 2, but we must forbid them from playing each other over consecutive weeks. So, by restricting L18 to be equal to or less than one, we are forbidding (i) team 1 from hosting team 2 in week 1 and team 2 from hosting team 1 in week 2, and also (ii) team 2 from hosting team 1 in week 1 and team 1 from hosting team 2 in week 2. The other cells in the L17:P28 accomplish the same for other pairs of teams and/or other pairs of consecutive weeks.

The optimal solution in Figure 11.1 provides a feasible schedule with overall TV ratings of 386.94. However, closer inspection reveals

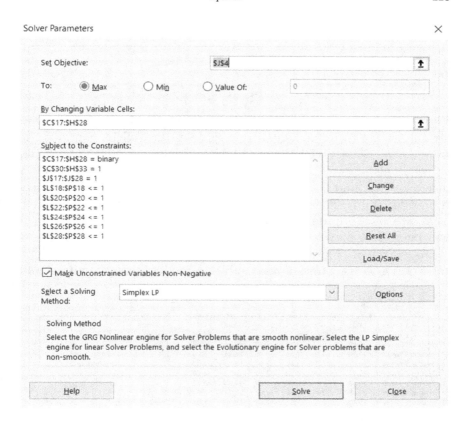

Figure 11.2. Solver dialog box for Example 11.1.

some aspects of the solution that might be undesirable. For example, we observe that C17, D19, and E21 are equal to one, which means that team 1 plays three consecutive home games (against teams 2, 3, and 4, respectively). Accordingly, team 1's last three games will be away. Team 2 also plays three consecutive home games: team 3 in week 3, team 4 in week 4, and team 1 in week 5 (see cells E23, F25, and G18). Team 3 plays three consecutive home games in weeks 4, 5, and 6 against teams 1, 4, and 2, respectively (see cells F20, G27, and H24).

It might be worthwhile to sacrifice some overall TV ratings to get a schedule that does not have bands of three consecutive home games. To evaluate this possibility, we incorporate the following constraint set to prevent each team from playing three consecutive home games:

$$\sum_{j \neq i}(Y_{ijk} + Y_{ij(k+1)} + Y_{ij(k+2)}) \leq 2$$

$$\text{for all } 1 \leq i \leq 4 \quad \text{and} \quad 1 \leq k \leq 4. \tag{11.5}$$

Figure 11.3 is basically a duplicate of Figure 11.1 but with constraint set (11.5) implemented in cells L9:O12. Cell L9 contains the formula = SUM(C17:E17) + SUM(C19:E19) + SUM(C21:E21), which is all possible home games that could be played by team 1 in weeks 1, 2, and 3. This sum must be equal to or less than 2. By constraining all cells in the L9:O12 range to be equal to or less than 2, as shown in the updated Solver dialog box in Figure 11.4, no team will be able to play more than two consecutive home games.

As shown in Figure 11.3, forbidding teams from playing three consecutive home games does result in a modest reduction in overall

	A	B	C	D	E	F	G	H	I	J	K	L	M	N	O	P
1			Week 1	Week 2	Week 3	Week 4	Week 5	Week 6		Total TV						
2	Home 1 Away 2		31.33	31.65	27.08	25.29	28.84	27.41		Audience						
3	Home 2 Away 1		34.16	27.99	29.19	28.05	34.13	27.56		Rating						
4	Home 1 Away 3		33.79	31.10	33.39	27.58	25.21	34.18		381.00						
5	Home 3 Away 1		25.64	33.77	31.48	30.56	34.23	32.90				No Stretch of 3				
6	Home 1 Away 4		30.01	32.37	32.49	27.28	27.16	27.59				Consecutive				
7	Home 4 Away 1		25.17	29.62	28.92	32.32	33.15	34.29				Home Games				
8	Home 2 Away 3		30.88	25.34	33.86	31.97	29.55	28.72				123	234	345	456	
9	Home 3 Away 2		28.48	25.39	27.10	26.49	25.37	32.94		Team 1		2	2	2	1	
10	Home 2 Away 4		25.49	30.61	26.67	32.49	30.16	29.43		Team 2		2	2	2	1	
11	Home 4 Away 2		28.03	34.60	26.32	30.85	30.51	34.83		Team 3		1	1	1	2	
12	Home 3 Away 4		25.46	29.65	28.79	26.69	25.66	30.58		Team 4		1	1	1	2	
13	Home 4 Away 3		33.49	26.45	27.46	30.89	28.21	33.29								
14																
15										Once @		No Consecutive Weeks (ij)				
16			Week 1	Week 2	Week 3	Week 4	Week 5	Week 6		Home		12	23	34	45	56
17	Home 1 Away 2		1	0	0	0	0	0		1						
18	Home 2 Away 1		0	0	0	0	1	0		1		1	0	0	1	1
19	Home 1 Away 3		0	0	0	1	0	0		1						
20	Home 3 Away 1		0	1	0	0	0	0		1		1	1	1	1	0
21	Home 1 Away 4		0	0	1	0	0	0		1						
22	Home 4 Away 1		0	0	0	0	0	1		1		0	1	1	0	1
23	Home 2 Away 3		0	0	1	0	0	0		1						
24	Home 3 Away 2		0	0	0	0	0	1		1		0	1	1	0	1
25	Home 2 Away 4		0	1	0	0	0	0		1						
26	Home 4 Away 2		0	0	0	1	0	0		1		1	1	1	1	0
27	Home 3 Away 4		0	0	0	0	1	0		1						
28	Home 4 Away 3		1	0	0	0	0	0		1		1	0	0	1	1
29																
30	Team 1 Plays		1	1	1	1	1	1								
31	Team 2 Plays		1	1	1	1	1	1								
32	Team 3 Plays		1	1	1	1	1	1								
33	Team 4 Plays		1	1	1	1	1	1								

Figure 11.3.　Updated worksheet for Example 11.1 with constraint set 11.5 added.

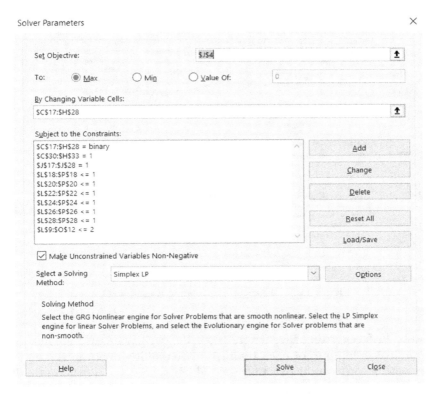

Figure 11.4.   Updated Solver dialog box after adding constraint set (11.5).

TV ratings from 386.94 to 381. Nevertheless, no team plays more than two consecutive home games and this desirable property might be deemed worth the reduction in ratings. For example, team 1's schedule is as follows: (week 1) home against team 2, (week 2) away against team 3, (week 3) home against team 4, (week 4) home against team 3, (week 5) away against team 2, and (week 6) away against team 4.

11.2   Example 11.2 — Who is Mathematically Eliminated?

Consider the five teams in the American League East in Major League Baseball and the hypothetical data reported in cells H1:L7 of

	A	B	C	D	E	F	G	H	I	J	K	L
1		Games Remaining Against Each Other									Games	Games
2		Baltimore	Boston	New York	Toronto	Tampa Bay			Won	Lost	Behind	Remaining
3	Baltimore	xxx						Baltimore	96	48	---	18
4	Boston	8	xxx					Boston	94	49	1.5	19
5	New York	4	4	xxx				New York	92	53	4.5	17
6	Toronto	3	0	3	xxx			Toronto	81	63	15	18
7	Tampa Bay	0	2	0	6	xxx		Tampa Bay	79	64	16.5	19
8												
9	Mininum # of Possible Wins for Division Champ					100						
10												
11												
12		Baltimore	Boston	New York	Toronto	Tampa Bay						
13	Baltimore	xxx	4	0	0	0		100				
14	Boston	4	xxx	0	0	2		100				
15	New York	4	4	xxx	0	0		100				
16	Toronto	3	0	3	xxx	6		93				
17	Tampa Bay	0	0	0	0	xxx		79				
18												
19		Baltimore	Boston	New York	Toronto	Tampa Bay						
20	Baltimore	xxx										
21	Boston	8	xxx									
22	New York	4	4	xxx								
23	Toronto	3	0	3	xxx							
24	Tampa Bay	0	2	0	6	xxx						

Figure 11.5. Excel worksheet for Example 11.2.

Figure 11.5 regarding the current record of each team and the number of games remaining for each team. We are coming down to the end of the season and would like to know which teams are mathematically eliminated from winning the division.

Based on the data in cells H1:L7 alone, no team is mathematically eliminated from winning the division. If the last-place team (Tampa Bay) won its remaining 19 games, they would finish with a record of 98 wins and 64 losses. Thus, 98 wins is the best Tampa Bay could achieve, but that is more than any other team in the division currently has, and thus it seems that Tampa Bay could still win the division if the other teams lost the rest of their games.

However, because some of the teams in the division have remaining games against each other, it is possible that the lower bound on the minimum number of wins for the division champion might exceed 98, and thus Tampa Bay would be mathematically eliminated now. Cells A1:F7 in Figure 11.5 provide the number of games remaining between each pair of teams of the division. Using this information, the linear programming model for determining the minimum number of possible wins for the division champions is as follows:

Parameters

$W_j =$ the current number of wins for team j,
$R_{jk} =$ the number of games remaining between teams j and k (for $1 \leq j \leq 4, 1 \leq k > j \leq 5$).

Decision variables

$M =$ the number of wins for the division champion,
$X_{jk} =$ the number of times team j beats team k in their remaining head-to-head games (for all $1 \leq j \leq 5$ and $1 \leq k \neq j \leq 5$).

Objective function

$$\text{Minimize: } M. \tag{11.6}$$

Subject to constraints

$$X_{jk} + X_{kj} = R_{jk} \qquad \text{(for } 1 \leq j \leq 4, 1 \leq k > j \leq 5\text{),} \tag{11.7}$$

$$M \geq W_j + \sum_{k \neq j} X_{jk} \qquad \text{(for all } 1 \leq j \leq 5\text{).} \tag{11.8}$$

The objective function in Equation (11.6) is to minimize the number of wins for the division champion. Constraint set (11.7) requires, for each pair of teams in the division, that the number of games won by each team in the pair sums to the total number of remaining games between the teams in the pair. Constraint set (11.8) requires, for each team, that the number of wins for the division champion equals or exceeds the current number of wins for the team plus the number of wins realized in the remaining games against other teams in the division.

The decision variable, M, is located in cell F9 of the worksheet displayed in Figure 11.5. This is the objective function and is the 'Set Objective' cell in the Solver dialog box shown in Figure 11.6. Cells B13:F17 contain the X_{jk} decision variables. For example, cell C13 is the number of times Baltimore beats Boston in their remaining eight games, whereas cell B14 is the number of times Boston

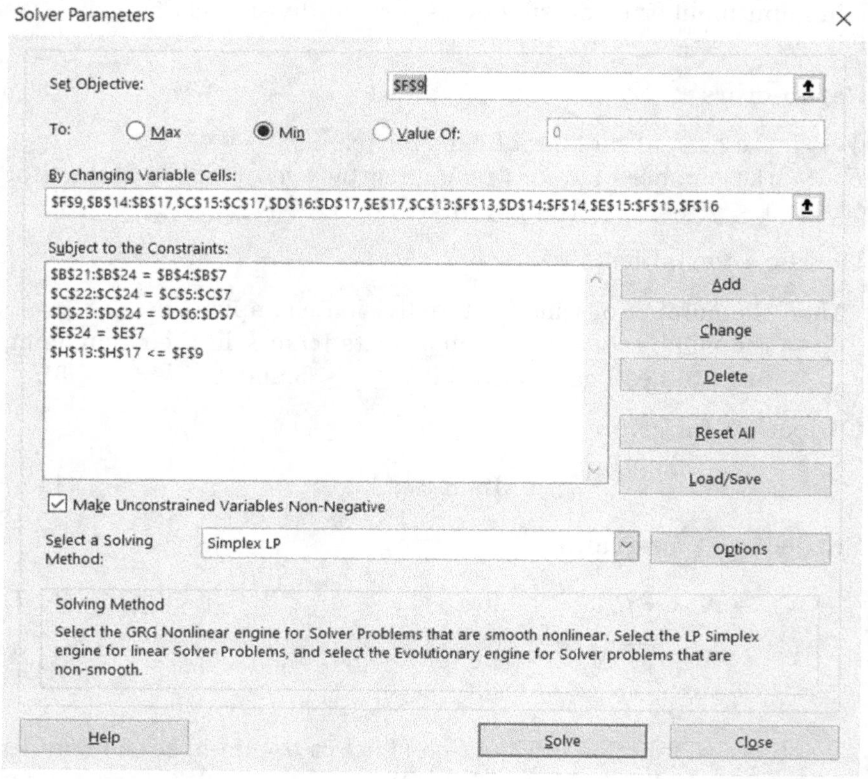

Figure 11.6.   Solver dialog box for Example 11.2.

beats Baltimore. Cell B21 contains the sum of cells B14 and C13, and cell B21 is constrained to equal the number of remaining games between Baltimore and Boston in cell B4. Similarly, constraint set (11.7) is established for all pairs of teams by constraining the lower triangular matrix in cells B20:F24 to equal the cells in the lower triangular matrix in cells B3:F7.

Cells H13:H17 are computed as the current number of wins for each team in cells I3:I7 plus the row sums of the matrix in cells B13:F17. Accordingly, cells H13:H17 are the right-hand sides of constraint set (11.8) and they are constrained to be equal to or less than the value of M in cell F9.

The optimal solution in Figure 11.5 reveals that, given the configuration of games remaining between teams in the division, the minimum number of wins possible for the division champion is 100.

In this particular solution, Baltimore, Boston, and New York all finish with 100 wins but there are alternative optimal solutions where only one or two teams finish with 100 wins.

As noted previously, the maximum number of possible wins for Tampa Bay is $79 + 19 = 98$, and, therefore, they are mathematically eliminated because they cannot reach 100 wins. Toronto is also mathematically eliminated because they currently have 81 wins but with only 18 games remaining their maximum number of wins is $81 + 18 = 99$.

11.3 Example 11.3 — Tournament Ranking

Suppose we have teams from six universities that played each other in a ping-pong tournament. Each team plays every other team once. At the end of the tournament, we want to establish a ranking of the teams. The data from the tournament are arranged in cells A1-J7 of Figure 11.7. Cells B2:B7 are the names of the universities and Cells A2:A7 are numerical identifiers for the universities, which will be used in the variable and parameter definitions for the model. A value of 1 in the body of the binary matrix in cells C2:H7 indicates that the team in the row defeated the team in the column, whereas a value of 0 indicates the row team lost to the column team. The row sums in cells J2:J7 are the total number of wins for each team in the tournament. To establish a ranking of the teams, the goal is to find

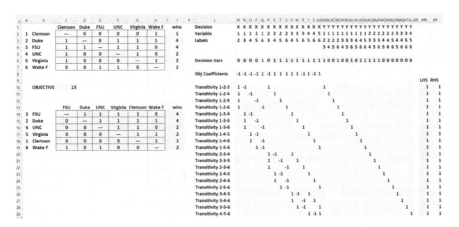

Figure 11.7.   Excel worksheet for Example 11.3.

a permutation of the rows and columns of the matrix such that the sum of the elements above the main diagonal of the reordered matrix is as large as possible. The integer linear programming formulation of this problem is as follows:

Parameters

$C_{ij} = 1$ if team i beats team j and 0 otherwise, for all $1 \leq i \leq 6$, and $1 \leq j \neq i \leq 6$.

Decision variables

$X_{ij} = 1$ if team i is ranked ahead of team j in the ordering, 0 otherwise, for all $1 \leq i \leq 5$, and $1 \leq j > i \leq 6$.

$Y_{ijk} =$ surrogate binary variables that are used to ensure transitivity in the ordering for all $1 \leq i \leq 4$, $1 \leq j > i \leq 5$, and $1 \leq k > j \leq 6$.

Objective function

$$\text{Maximize:} \quad \sum_{1 \leq i < j \leq 6} C_{ji} + \sum_{1 \leq i < j \leq 6} (C_{ij} - C_{ji}) X_{ij}. \qquad (11.9)$$

Subject to constraints

$$X_{ij} - X_{ik} + X_{jk} + Y_{ijk} = 1$$

$$\text{for all } 1 \leq i \leq 4, 1 \leq j > i \leq 5, \quad \text{and} \quad 1 \leq k > j \leq 6. \qquad (11.10)$$

The objective function in Equation (11.9) is the sum of elements above the main diagonal of the reordered matrix. The first summation term is a constant and can be dropped from the formulation and simply added to the objective function value at the end. However, the constant term is beneficial for understanding how the objective function works. If $X_{ij} = 0$, the j comes before i in the sequence and the C_{ji} collected in the constant term is gathered appropriately and $(C_{ij} - C_{ji})$ in the second term drops out because $X_{ij} = 0$. However, if $X_{ij} = 1$, then the C_{ji} contribution in the constant term is inappropriate and is removed in the second term because of the $-C_{ji}X_{ij}$ component. At the same time, $C_{ij}X_{ij}$ appropriately gathers this term because i precedes j in the sequence.

Constraint set (11.10) ensures transitivity in the ordering. For example, if team 1 precedes team 2 ($X_{12} = 1$) and team 2 precedes team 3 ($X_{23} = 1$), then team 1 must precede team 3 ($X_{13} = 1$). Constraint (11.10) would be $X_{12} - X_{13} + X_{23} + Y_{123} = 1$ or $1 - 1 + 1 + Y_{123} = 1$, so $Y_{123} = 0$ to make the constraint hold. Now, let us consider an obvious violation of transitivity: Team 1 precedes team 2 ($X_{12} = 1$), team 2 precedes team 3 ($X_{23} = 1$), and team 3 precedes team 1 ($X_{13} = 0$). Here, constraint (11.10) would be $X_{12} - X_{13} + X_{23} + Y_{123} = 1$ or $1 - 0 + 1 + Y_{123} = 1$. However, Y_{123} must be 0 or 1 and cannot be –1 to make the constraint hold, so such a violation of transitivity is prevented by the constraint.

The decision variables in Figure 11.7 are located in cells M6:AU6. The first 15 of the columns (M through AA) correspond to the X_{ij} variables and the remaining 20 (AB through AU) to the Y_{ijk} variables. The objective function coefficients ($C_{ij} - C_{ji}$) are located in cells M8:AA8 and the sumproduct of M6:AA6 is contained in cell C10, where it is added to the constant 7, which is the sum below the main diagonal in the matrix in cells C2:H7. Cell C10 is the objective function and the 'Set Objective' cell in the Solver dialog box in Figure 11.8.

Cells AW10:AW29 are equal to the sumproducts of the decision variables in M6:AU6 and the constraint coefficient in each row. Cells AW10:AW29 are constrained to equal the column of ones in cells AX10:AX29 to enforce the transitivity constraints in (11.10).

The optimal solution in Figure 11.7 returns 13 as the sum of the elements above the main diagonal in the optimally permuted matrix. The decision variable values in cells M6:AA6 can be used to establish the appropriate permutation of the teams. For example, because $X_{12} = X_{13} = X_{14} = X_{15} = 0$, we know that teams 2, 3, 4, and 5 all precede team 1 in the permutation. Team 1 does precede team 6 because $X_{16} = 1$, so we know that teams 1 and 6 are the last two teams in the permutation. Next, since $X_{23} = 0$ but $X_{24} = X_{25} = X_{26} = 1$, we can establish that team 3 is first in the permutation and team 2 is second. Continuing on, we find that the optimal ordering of teams is 3–2–4–5–1–6 (or FSU — Duke — UNC — Virginia — Clemson — Wake Forest). The permuted matrix is in cells C14:H19, and counting the number of 1s above the main diagonal yields 13.

It is not surprising that FSU and Duke are the top-ranked teams, as they both have four wins and no other team has more than two. But why is FSU first and Duke second? The reason is that FSU

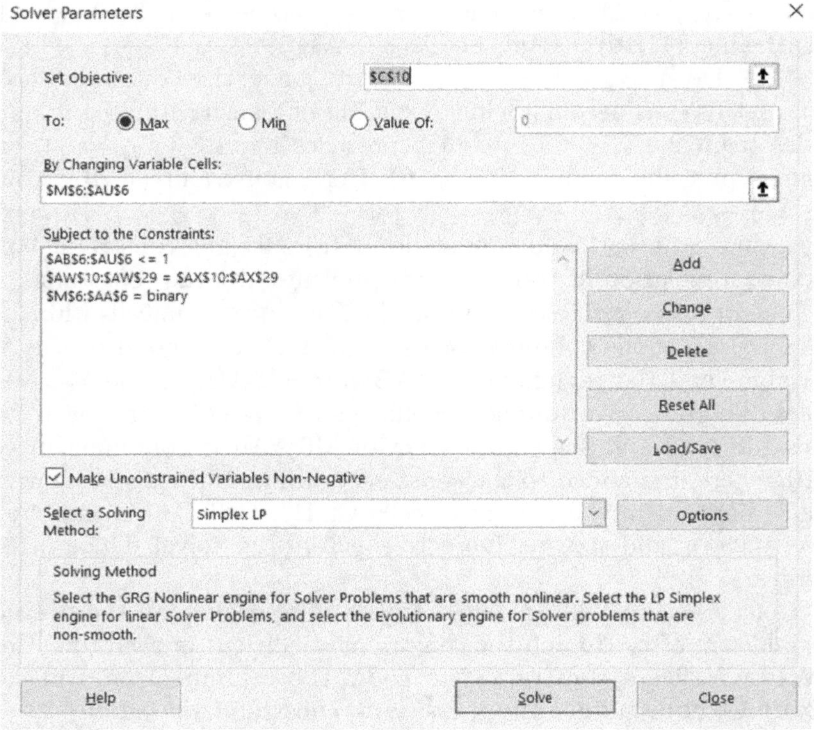

Figure 11.8.   Solver dialog box for Example 11.3.

won its ping-pong match against Duke. Moreover, if the order of
FSU and Duke is switched, then we lose the one above the main
diagonal in the FSU row, Duke column (nothing else above the main
diagonal would change), and the sum above the main diagonal drops
from 13 to 12. A similar logic is used to explain why UNC is ranked
ahead of Virginia even though they each have two wins. The fact that
Clemson defeated Wake Forest also explains why Clemson is ranked
ahead of Wake Forest even though Wake Forest has more wins.

11.4   Example 11.4 — Minimax Ranking Based on Metric Variables

Our final example focuses on ranking individuals based on a set
of metric variables. For this example, we obtained data pertaining

to 10 NFL quarterbacks with respect to four performance measures: (i) passing yards, (ii) completion percentage, (iii) touchdowns, and (iv) interceptions. The data were obtained from the website https://www.nfl.com/stats/player-stats/ on 12/4/2023. The quarterbacks were ranked in terms of passing yardage and we took the first 10 quarterbacks on the list. Cells A1:E12 in Figure 11.9 identify the quarterbacks and their performance measures. Interceptions are coded as negative values because, unlike the other three statistics, smaller values of this measure are preferable. This coding is in accordance with the ranking methodology we employ, which was developed by Reinig and Horowitz (2018).

Cells G3:G12 identify whether or not each quarterback is dominated by one or more quarterbacks. The principle of dominance, which is part of the Reinig and Horowitz (2018) methodology is defined as follows: Quarterback A dominates quarterback B if quarterback A's four performance measures equal or exceed the four performance measures for quarterback B and strictly exceed the measurement for quarterback B for at least one performance measure. For example, because Tagovailoa has more passing yards, a higher completion percentage, more touchdowns, and fewer interceptions (a larger negative interceptions value) than Allen, Tagovailoa dominates Allen. Prescott dominates Allen for the same reason.

The first step of the analysis is to normalize the raw measures by converting them to z-scores. The means and standard deviations

	Passing Yards	Completion Percentage	Touchdowns	(negative) Interceptions	Dominated?											
Quarterback																
C.J. Stroud	3540	63.4	20	-5	No, most yards				Maximum Difference in Weighted Scores =					1.0512		
Sam Howell	3466	65.8	18	-14	No, better % than Stroud											
Tua Tagovailoa	3457	70.1	24	-10	No, better % than Stroud & Howell			Sum of Weights				1				
Jared Goff	3288	67.7	20	-8	No, better % than Stroud & Howell, Fewer Ints than Tagovailoa											
Dak Prescott	3234	70.1	26	-6	No, most TDs							Dak Prescott	0.6429			
Josh Allen	3214	68.1	24	-13	Yes, by Tagovailoa & Prescott							Tua Tagovailoa	0.6429			
Brock Purdy	3185	70.2	23	-6	No, highest completion percentage							Brock Purdy	0.5591			
Patrick Mahomes	3127	67.8	22	-10	Yes, by Tagovailoa, Prescott, & Purdy							C.J. Stroud	0.2899			
Justin Herbert	3038	65.6	20	-6	Yes, by Prescott & Purdy							Jared Goff	0.1639			
Jalen Hurts	2995	66.5	19	-10	Yes, by Tagovailoa, Prescott, Purdy, & Mahomes (so the first three are redundant)							Patrick Mahomes	-0.3151			
												Sam Howell	-0.4059			
MEAN	3254.4	67.5	21.6	-8.8								Josh Allen	-0.4082			
STD DEV.	184.3	2.3	2.6	3.1								Justin Herbert	-0.4082			
												Jalen Hurts	-0.7613			

	Passing Yards	Completion Percentage	Touchdowns	(negative) Interceptions	Weighted Sum	Stroud	Howell	Tagovailoa	Goff	Prescott	Allen	Purdy	Mahomes	Herbert	Hurts
Quarterback															
C.J. Stroud	1.5496	-1.8348	-0.6176	1.2180	0.2899	xxxx	-0.6957	0.3531	-0.1260	0.3531	-0.6981	0.2692	-0.6049	-0.6981	-1.0512
Sam Howell	1.1481	-0.7686	-1.3896	-1.6668	-0.4059	0.6957	xxxx	1.0488	0.5698	1.0488	-0.0024	0.9649	0.0908	-0.0024	-0.3555
Tua Tagovailoa	1.0992	1.1417	0.9264	-0.3846	0.6429	-0.3531	-1.0488	xxxx	-0.4790	0.0000	-1.0512	-0.0839	-0.9580	-1.0512	-1.4043
Jared Goff	0.1823	0.0755	-0.6176	0.2564	0.1639	0.1260	-0.5698	0.4790	xxxx	0.4790	-0.5721	0.3952	-0.4790	-0.5721	-0.9252
Dak Prescott	-0.1107	1.1417	1.6985	0.8975	0.6429	-0.3531	-1.0488	0.0000	-0.4790	xxxx	-1.0512	-0.0839	-0.9580	-1.0512	-1.4043
Josh Allen	-0.2192	0.2552	0.9264	-1.3462	-0.4082	0.6981	0.0024	1.0512	0.5721	1.0512	xxxx	0.9673	0.0932	0.0000	-0.3531
Brock Purdy	-0.3765	1.1862	0.5404	0.8975	0.5591	-0.2692	-0.9649	0.0839	-0.3952	0.0839	-0.9673	xxxx	-0.8741	-0.9673	-1.3204
Patrick Mahomes	-0.6912	0.1199	0.1544	-0.3846	-0.3151	0.6049	-0.0908	0.9580	0.4790	0.9580	-0.0932	0.8741	xxxx	-0.0932	-0.4463
Justin Herbert	-1.1741	-0.8574	-0.6176	0.8975	-0.4082	0.6981	0.0024	1.0512	0.5721	1.0512	0.0000	0.9673	0.0932	xxxx	-0.3531
Jalen Hurts	-1.4074	-0.4576	-1.0036	-0.3846	-0.7613	1.0512	0.3555	1.4043	0.9252	1.4043	0.3551	1.3204	0.4463	0.3531	xxxx
Weights	0.3398	0.3365	0.0074	0.3163											
Z-score diffs	2.9570	3.0210	3.0881	2.8848											

Figure 11.9.   Excel worksheet for Example 11.4.

for each measure are computed in cells B14:E15. These values are then used to compute the z-scores in cells B19:E28. (We rounded the z-scores to four decimal places because using too much precision could sometimes cause problems with the Simplex LP engine of the Excel Solver.) Cells B32:E32 contain the maximum z-score minus the minimum z-score for each measure. These cells are not used in the analysis but indicate that the z-score range is rather consistent across the four measures.

Next, we turn to the linear programming model developed by Reinig and Horowitz (2018), which will try to find a set of weights for the (standardized) measures, so that when weighting the paired differences between teams for those standardized measures, we keep the largest of the weighted sums as small as possible. The formulation is as follows:

Decision variables

$M =$ the maximum difference between the pairwise weighted scores,
$X_j =$ the weight for measure j, $1 \leq j \leq 4$.

Parameters

$Z_{ij} =$ the z-score for quarterback i on measure j, for all $1 \leq i \leq 10$, and $1 \leq j \leq 4$.

Objective function

$$\text{Minimize: } M. \tag{11.11}$$

Subject to constraints

$$M \geq \sum_{j=1}^{4} X_j(Z_{ij} - Z_{kj}) \quad \text{for all } 1 \leq i \leq 9, \quad \text{and} \quad 1 \leq k > i \leq 10,$$
$$\tag{11.12}$$

$$M \geq \sum_{j=1}^{4} X_j(Z_{kj} - Z_{ij}) \quad \text{for all } 1 \leq i \leq 9, \quad \text{and} \quad 1 \leq k > i \leq 10,$$
$$\tag{11.13}$$

$$\sum_{j=1}^{4} X_j = 1. \tag{11.14}$$

The objective function in Equation (11.11) is to minimize the maximum difference between the pairwise weighted scores. Constraint sets (11.12) and (11.13) make sure that the maximum difference M equals or exceeds the pairwise weighted score differences (in both directions) for all pairs of quarterbacks. Constraint (11.14) requires the weights to sum to one.

Cell Q3 in Figure 11.9 contains the decision variable M, which is the objective function and the 'Set Objective' cell in the Solver dialog box in Figure 11.10. The remaining four decision variables (the weights on the measures) are located in cells B30:E30. The sum of these four cells is in cell O5 and this sum is constrained to equal one in accordance with constraint (11.14).

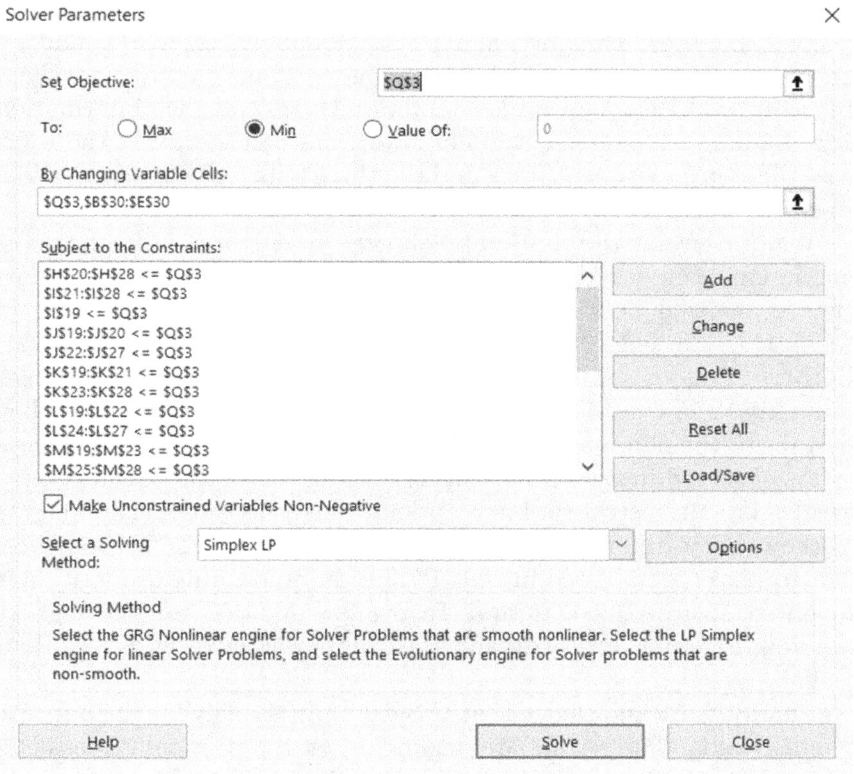

Figure 11.10.   Solver dialog box for Example 11.4.

Cells G19:G28 contain the sumproducts of the weights in cells B30:E30 and the z-scores for each of the 10 quarterbacks. Therefore, they are the weighted scores for the quarterbacks. Cells H19:Q28 contain the pairwise differences between the weighted scores for the quarterbacks. Specifically, the cell values contain the column quarterback's weighted score minus the row quarterback's weighted score.

Generally, the values in cells H19:Q28 are constrained to be equal to or less than the value of cell M in cell Q3 to accomplish constraint sets 11.12 and 11.13. However, in accordance with the approach described by Reinig and Horowitz (2018), we eliminate redundant dominant relationships from these constraint sets. Returning to the dominance information in cells G3:G12 in Figure 11.9, we see that Mahomes is dominated by Tagovailoa, Prescott, and Purdy. Later, we find that Hurts is dominated by Tagovailoa, Prescott, Purdy, and Mahomes, which is indicative of redundancy. In other words, if we know Tagovailoa, Prescott, and Purdy dominate Mahomes, and we know Mahomes dominates Hurts, then by transitivity Tagovailoa, Prescott, and Purdy must also dominate Hurts and such information is redundant. Therefore, we eliminate from consideration the pairwise difference between Hurts and the three quarterbacks Tagovailoa, Prescott, and Purdy.

We see from the optimal solution in Figure 11.9 that the maximum pairwise difference is 1.0512, and this pairwise difference is realized between several different pairs of quarterbacks in cells H19:Q28 (e.g., Stroud–Hurts, Tagovailoa–Allen, Tagovailoa–Herbert, Prescott–Allen, and Prescott–Herbert). The values larger than 1.0512 in cells J28, L28, and N28 can be ignored because they pertain to the eliminated pairwise differences based on redundancy.

As observed in cells B30:E30, all four measures have nonnegative weight, but the weight for interceptions is trivially small. The weights of the other three measures are very comparable. For convenience, quarterbacks are relisted in cells P7:P16 in descending order of their weighted scores. Prescott and Tagovailoa are tied for the highest ranking, followed by Purdy and then more distantly by Stroud and Goff.

Finally, we acknowledge that this example is solely for the purpose of illustration. The results are dependent on the particular measures selected for inclusion in the model, as well as the model itself.

Part II

Optimization Examples for Multivariate Statistical Methods Used in Predictive and Descriptive Analytics

Chapter 12

Regression

In this chapter, we focus on optimization issues that pertain to ordinary least squares (OLS) regression. In the first section, we focus on the nonlinear optimization problem underlying OLS regression and the derivation of the normal equation for estimating the regression coefficients. In the second section, using an example pertaining to the measurement of customer satisfaction with a restaurant experience (Brusco, 2019), we demonstrate how regression analysis can be accomplished in Excel using only a few basic commands. This example includes not only the estimation of the regression coefficients but also the decomposition of the sum of squares and the significance testing of the regression coefficients.

The third section of this chapter focuses on the important issues of model selection and the measurement of relative predictor importance. A VBA macro is used to implement all-possible-subsets regression by applying sweep operations to the correlation matrix. For each possible number of predictors, the macro obtains the subset that will minimize error sum of squares or, equivalently, maximize the proportion of explained variation (R^2). The selection of the subset size is based on Mallows's Cp index (Mallows, 1973). The macro also reports information pertaining to the relative importance of predictors, which can be used to rank the predictors in order of their contribution to the explanation of the dependent variable.

12.1   The Ordinary Least Squares Regression Model

The derivation of the normal equations for the OLS regression model uses the following definitions:

$n =$ the number of observations (e.g., survey respondents).

$p =$ the number of predictor variables.

$\mathbf{y} =$ an $n \times 1$ vector of dependent variable measurements.

$\mathbf{X} =$ an $n \times (p+1)$ matrix of variable measurements (including the intercept, which is just a column of ones). In actuality, there are p predictor variables but $p + 1$ parameters are estimated when counting the intercept.

$\boldsymbol{\beta} =$ a $(p + 1) \times 1$ vector of regression coefficients.

$\mathbf{X}\,\beta =$ an $n \times 1$ vector corresponding to the prediction of $\mathbf{y}$.

$\mathbf{e} =$ $\mathbf{e} = \mathbf{y} - \mathbf{X}\boldsymbol{\beta}$, is an $n \times 1$ vector of prediction errors.

$SSE =$ $SSE = \mathbf{e}'\mathbf{e}$ is a scalar representing the error (or residual) sum of squares.

$\bar{\mathbf{y}} =$ an $n \times 1$ vector of containing the mean of the dependent variable measurements.

$SST =$ $SST = (\mathbf{y} - \bar{\mathbf{y}})'(\mathbf{y} - \bar{\mathbf{y}})$ is a scalar representing the total sum of squares about the mean.

$SSR =$ $SST\text{–}SSE$ is a scalar representing the regression sum of squares.

$R^2 =$ $R^2 = \frac{SSR}{SST}$, the coefficient of determination, which is the proportion of variation in the dependent variable that is explained by the independent variables.

The optimization problem is to find $\boldsymbol{\beta}$ so as to minimize the SSE, which is equivalent to maximizing SSR and R^2.

$$\text{Minimize: } \mathbf{e}'\mathbf{e} = (\mathbf{Y} - \mathbf{X}\boldsymbol{\beta})^2 = \mathbf{Y}'\mathbf{Y} - 2\boldsymbol{\beta}'\mathbf{X}'\mathbf{Y} + \boldsymbol{\beta}'\mathbf{X}'\mathbf{X}\boldsymbol{\beta}. \quad (12.1)$$

Taking the derivative of Equation (12.1) with respect to β and setting it equal to zero yields

$$\partial L/\partial \beta = -2\mathbf{X}'\mathbf{Y} + 2\mathbf{X}'\mathbf{X}\boldsymbol{\beta} = 0. \quad (12.2)$$

Adding $2\,\mathbf{X}'\mathbf{Y}$ to both sides yields

$$2\,\mathbf{X}'\mathbf{X}\boldsymbol{\beta} = 2\,\mathbf{X}'\mathbf{Y}. \quad (12.3)$$

The 2's cancel and multiplying through by $(\mathbf{X'X})^{-1}$ yields

$$\boldsymbol{\beta} = (\mathbf{X'X})^{-1}\mathbf{X'Y}. \qquad (12.4)$$

Equation (12.4) is the normal equation for finding the slope coefficients for OLS regression. The coefficients in the beta vector $\boldsymbol{\beta} = [b_0, b_1, \ldots, b_p]$ can be used, in conjunction with predictor variable measurements, to establish a prediction, $\widehat{y}$, for the dependent variable using the following equation:

$$\widehat{y} = b_0 + b_1 x_1 + \cdots + b_p x_p. \qquad (12.5)$$

12.2 Example 12.1 — Illustration of Regression Mechanics

A marketing research firm is interested in determining the key drivers of overall customer satisfaction at a popular restaurant. Linear regression will be used to determine which drivers are most important. Suppose data were collected from $n = 60$ patrons of the restaurant. The metric-dependent variable, y, is overall satisfaction, which is measured on a scale of 0 to 100 with larger values indicating greater satisfaction. The firm is considering 10 candidate independent variables to predict overall satisfaction. Each candidate predictor variable is a Likert scale response to a statement ranging from $1 =$ strongly disagree to $7 =$ strongly agree. The 10 predictor statements are shown in cells A3:A12 of Figure 12.1. The dependent variable measurements are contained in cells B16:B75 and the independent variable measurements are contained in cells C16:M75.

The regression coefficients in cells I2:I12 are computed using Equation (12.4). The computation makes use of the three Excel matrix commands: (i) MMULT for multiplying two matrices together, (ii) TRANSPOSE for transposing a matrix, and (iii) MINVERSE for finding the inverse. The nested function in cell I2 to implement Equation (12.4) is as follows: = MMULT(MINVERSE(MMULT(TRANSPOSE(C16:M75),C16:M75)), MMULT(TRANSPOSE(C16:M75),B16:B75)).

Once the regression coefficients are obtained using the basic command in cell I2, it is desirable to obtain other information, such as the various sum-of-squares components and the significance of the regression coefficients. This is accomplished using the portion of the worksheet displayed in Figure 12.2. Cells O16:O75 in Figure 12.2 contain the squared deviation of each dependent variable measurement from its mean, and the inner product of these cells, *SST*, is located in Cell P2. Cells P16:P75 are computed as the matrix product (using MMULT) of cells C16:M75 and I2:I12, thus representing the predicted values for the dependent variable. The prediction errors in cells Q16:Q75 are the differences between the actual dependent variables in B16:B75 and the predicted values in P16:P75. The inner product of Q16:Q75 is *SSE* and is located in cell P3. The value of $SSR = SST - SSE$ is computed in cell P4 as P2–P3. R^2 is computed in cell P5 as P4/P2 (SSR/SST). The mean square error (MSE) is computed in cell P7 as $SSE/(n-p-1)$ (i.e., P3/49) because it will be used to compute standard errors (SE) for the coefficients.

Cells U2:AE12 in Figure 12.2 contain $(\mathbf{X'X})^{-1}$ using the MMULT, TRANSPOSE, and MINVERSE commands. The main diagonal of this matrix is necessary for the computation of the standard errors in cells J2:J12 in Figure 12.1. Specifically, the standard errors are computed as the square root of the products of the MSE in cell P7 and the main diagonal elements of the matrix in U2:AE12. The

	Beta	SE	t-stat	p-val
B0	26.845	11.663	2.30	.026
B1	2.746	1.531	1.79	.079
B2	2.811	1.672	1.68	.099
B3	2.231	1.667	1.34	.187
B4	0.004	2.055	0.00	.998
B5	3.182	1.677	1.90	.064
B6	-2.207	1.965	-1.12	.267
B7	0.764	1.664	0.46	.648
B8	-2.000	1.601	-1.25	.217
B9	-1.603	1.143	-1.40	.167
B10	3.891	1.214	3.21	.002

Y = overall satisfaction (0 to 100, highest = 100); X1 = the side dishes and salads are delicious; X2 = the entrées are delicious; X3 = the desserts are delicious; X4 = parking at the restaurant is ample; X5 = the ligthing in the restaurant is appropriate; X6 = the noise level in the restaurant is not distracting to me; X7 = the wait for beverage service is excessive; X8 = the wait for food service is excessive; X9 = the wait for the check is excessive; X10 = restaurant service personnel are courteous and friendly

Customer	Y = Overall Satisfaction	Intcpt	X1	X2	X3	X4	X5	X6	X7	X8	X9	X10
1	85	1	5	6	5	3	3	3	3	4	6	6
2	78	1	5	5	5	4	4	3	3	5	6	6
3	84	1	6	6	5	4	3	4	4	3	6	6
4	91	1	6	6	5	3	4	4	4	5	7	6
5	86	1	6	6	7	3	3	3	5	4	6	7
6	73	1	5	5	5	4	4	4	5	4	6	6
7	90	1	6	6	7	4	4	4	3	4	6	7
8	94	1	2	7	7	4	3	3	5	5	7	6
9	77	1	5	7	5	4	3	4	5	3	7	6
10	91	1	6	6	6	4	3	3	3	5	7	6
11	89	1	5	7	5	4	3	5	4	4	2	3
12	96	1	6	3	5	5	4	3	3	4	4	4
13	92	1	7	7	7	5	3	4	5	3	4	3
14	88	1	7	6	6	5	4	3	4	4	4	3
15	78	1	6	7	5	5	4	5	4	4	3	3

Figure 12.1.   A portion of the Excel worksheet for Example 12.1.

	O	P	Q	R	S	T	U	V	W	X	Y	Z	AA	AB	AC	AD	AE
1																	
2	SST =	15443.7					1.037	-0.064	0.019	-0.039	-0.028	-0.017	-0.026	-0.049	-0.008	-0.028	0.006
3	SSE =	6428.7					-0.064	0.018	-0.008	-0.003	0.000	0.000	0.003	0.005	-0.001	0.002	0.000
4	SSR =	9014.9					0.019	-0.008	0.021	-0.012	-0.002	0.005	-0.005	-0.002	0.001	-0.003	0.000
5	R^2 =	58.37%					-0.039	-0.003	-0.012	0.021	0.001	-0.001	0.003	0.004	-0.001	-0.001	-0.001
6							-0.028	0.000	-0.002	0.001	0.032	-0.010	-0.016	-0.004	0.002	0.003	0.000
7	MSE =	131.19883					-0.017	0.000	0.005	-0.001	-0.010	0.021	-0.010	0.004	-0.002	-0.002	-0.002
8							-0.026	0.003	-0.005	0.003	-0.016	-0.010	0.029	0.000	0.003	0.003	-0.004
9							-0.049	0.005	-0.002	0.004	-0.004	0.004	0.000	0.021	-0.015	-0.001	0.000
10							-0.008	-0.001	0.001	-0.001	0.002	-0.002	0.003	-0.015	0.020	-0.002	-0.001
11							-0.028	0.002	-0.003	-0.001	0.003	-0.002	0.003	-0.001	-0.002	0.010	-0.004
12							0.006	0.000	0.000	-0.001	0.000	-0.002	-0.004	0.000	-0.001	-0.004	0.011
13																	

	Mean Diff	Prediction	Error
14	Mean		
15	Diff	Prediction	Error
16	17.35	79.55	5.45
17	10.35	77.93	0.07
18	16.35	82.86	1.14
19	23.35	80.43	10.57
20	18.35	92.18	-6.18
21	5.35	79.25	-6.25
22	22.35	91.63	-1.63
23	26.35	76.51	17.49
24	9.35	82.09	-5.09
25	23.35	80.93	10.07
26	21.35	73.46	15.54
27	28.35	72.48	23.52
28	24.35	85.18	6.82
29	20.35	82.76	5.24
30	10.35	77.79	0.21

Figure 12.2.   A second portion of the Excel worksheet for Example 12.1.

t-statistics in cells K2:K12 are the coefficients in cells I2:I12 divided by their standard errors in cells J2:J12. The two-tailed p-values for the statistics in cells L2:L12 are computed using the TDIST function in Excel with $n - p - 1$ degrees of freedom.

When examining the results in Figure 12.2, we observed that the value of R^2 is 0.5837. Interestingly, only one of the 10 predictors (x_{10}) is statistically significant at the $\alpha = 0.05$ level. Moreover, the signs of some of the slope coefficients are discordant with theoretical expectations. For example, variable x_6 has a negative slope coefficient despite the fact that a positive coefficient should be expected (i.e., the more someone agrees with the statement that the noise level is not distracting, the higher satisfaction should be). The lack of predictor significance along with the atheoretical signs for the regression coefficients is a possible indication of *multicollinearity*, that is, high correlation among two or more predictor variables. The correlation matrix in Figure 12.3 provides further evidence of this problem. The correlations among the three food quality items (x_1, x_2, x_3) are all fairly high, as are the correlations among the three aesthetic items

	x1	x2	x3	x4	x5	x6	x7	x8	x9	x10
x1	1.000									
x2	0.681	1.000								
x3	0.672	0.777	1.000							
x4	-0.250	-0.169	-0.264	1.000						
x5	-0.309	-0.300	-0.306	0.754	1.000					
x6	-0.270	-0.163	-0.269	0.822	0.774	1.000				
x7	-0.481	-0.294	-0.393	0.095	0.032	0.028	1.000			
x8	-0.319	-0.199	-0.236	-0.029	-0.009	-0.087	0.791	1.000		
x9	-0.067	0.143	0.111	-0.202	-0.085	-0.187	0.285	0.358	1.000	
x10	-0.109	0.036	-0.009	0.274	0.342	0.337	0.203	0.227	0.382	1.000

Figure 12.3.　Correlation matrix for the predictors in Example 12.1.

(x_4, x_5, x_6) and two of the items that pertain to waiting (x_7, x_8). Thus, there are a few small groups of highly correlated variables and this is likely influencing the directional signs of some of the coefficients as well as the significance tests.

12.3　All-Possible-Subsets Regression

In this section, we focus on an all-possible-subsets regression analysis. As its name implies, all-possible-subsets regression focuses on the evaluation of all 2^p possible regression models (or $2^p - 1$ if the null model consisting of zero predictors is ignored). For example, for $p = 4$, the $2^4 = 16$ possible subsets are $\emptyset$, $\{1\}$, $\{2\}$, $\{3\}$, $\{4\}$, $\{1, 2\}$, $\{1, 3\}$, $\{1, 4\}$, $\{2, 3\}$, $\{2, 4\}$, $\{3, 4\}$, $\{1, 2, 3\}$, $\{1, 2, 4\}$, $\{1, 3, 4\}$, $\{2, 3, 4\}$, $\{1, 2, 3, 4\}$. The number of regression models increases rapidly as a function of p. For example, at $p = 20$, the number of possible subsets is just over one million. We use a VBA implementation (Brusco, 2019) of all-possible-subsets regression to complete our analyses, which is scalable for up to roughly 20 predictors. The use of the all-possible-subsets VBA macro will focus on two related but distinct objectives: (i) model selection and (ii) the measurement of predictor importance. Prior to our model selection and predictor importance demonstrations, however, we describe the principal engine of the VBA macro – the SWEEP algorithm.

12.3.1　*Example 12.2 — The SWEEP algorithm*

The primary engine for running APS regression is a SWEEP algorithm that operates on the $(p + 1) \times (p + 1)$ correlation matrix, **R**, associated with the p predictors and the dependent variable (by convention, the dependent variable corresponds to row $p + 1$ and column $p + 1$ of the matrix). SWEEP algorithms have been widely used for subset selection problems in regression (Brusco *et al.*, 2009; Furnival & Wilson, 1974; Garside, 1971), as well as other variable selection optimization problems in multivariate statistics (Brusco & Steinley, 2011; Duarte Silva, 2001). An excellent tutorial on the history and applications of the SWEEP algorithm is provided by Goodnight (1979).

The SWEEP algorithm transforms a given $(p + 1) \times (p + 1)$ matrix $\mathbf{R}$ into a new matrix $\mathbf{S}$ using the following process:

Step 1. Select one of the predictor variables $h(1 \leq h \leq p)$ for sweeping.
Step 2. Set the pivot element as $\lambda = 1/r_{hh}$.
Step 3. Set $s_{ij} = r_{ij} - \lambda r_{hj} r_{ih}$ for all $1 \leq i \neq h$, $j \neq h \leq p + 1$.
Step 4. Set $s_{hj} = \lambda r_{hj}$ for all $1 \leq j \neq h \leq p + 1$.
Step 5. Set $s_{ih} = -\lambda r_{ih}$ for all $1 \leq i \neq h \leq p + 1$.
Step 6. $s_{hh} = \lambda$.
Step 7. $\mathbf{R} = \mathbf{S}$.

A predictor variable, h, is selected for a SWEEP operation in Step 1. Step 2 defines the pivot element as the inverse of the main diagonal element of $\mathbf{R}$ corresponding to predictor h (i.e., $\lambda = 1/r_{hh}$). Step 3 computes all elements of $\mathbf{S}$ that do not correspond to either row h or column h. Step 4 computes the row h elements of $\mathbf{S}$, with the exception of the main diagonal element in that row (i.e., column h in row h). Step 5 computes the column h elements of $\mathbf{S}$, with the exception of the main diagonal element in that column (i.e., row h in column h). Step 6 sets the value for column h of row h in $\mathbf{S}$. Step 7 resets $\mathbf{R}$ to $\mathbf{S}$, which enables the exact same set of steps to be repeated for sweeping other variables in or out.

The SWEEP worksheet in the all-possible-subsets workbook (see Figure 12.4) provides a small numerical example to illustrate the algorithm. The problem consists of $n = 17$ respondents measured on $p = 3$ predictors (x_1, x_2, x_3) and a dependent variable (y). The raw data are provided in cells B2:E18 of the workbook. The means and standard deviations of the dependent variable and predictors are computed in cells B20:E21 to facilitate the computation of z-scores for each variable. The z-scores, which are obtained by differencing the raw variable measures from the mean and dividing by the standard deviation, are displayed in cells B24:E40. By defining the 17×4

	X1	X2	X3	Y
	2	3	8	16
	6	9	5	6
	8	3	1	15
	7	6	8	16
	4	2	5	15
	9	8	3	17
	1	5	6	1
	6	4	7	14
	8	7	4	17
	5	3	9	15
	9	2	5	25
	3	4	1	5
	1	1	6	5
	2	5	8	2
	7	8	3	4
	6	9	7	8
	4	6	5	8
MEAN	5.1765	4.8824	5.3529	11.1176
S.D.	2.6175	2.6095	2.3248	6.5069

ZX1	ZX2	ZX3	ZY
-1.2136	-1.4878	1.1386	.7503
.3146	1.5779	-.1518	-.7865
1.0787	-.7213	-1.8724	.5967
.6967	.4283	1.1386	.7503
-.4495	-1.1045	-.1518	.5967
1.4608	1.1947	-1.0121	.9040
-1.5956	.0451	.2783	-1.5549
.3146	-.3381	.7085	.4430
1.0787	.8115	-.5820	.9040
-.0674	-.7213	1.5687	.5967
1.4608	-1.1045	-.1518	2.1335
-.8315	-.3381	-1.8724	-.9402
-1.5956	-1.4878	.2783	-.9402
-1.2136	.0451	1.1386	-1.4012

SWEEP

CORRELATION MATRIX, R =

1.0000	.4078	-.3486	.6481
.4078	1.0000	-.1580	-.2902
-.3486	-.1580	1.0000	.0245
.6481	-.2902	.0245	1.0000

R2 = 1 - 1 = 0
Sweep IN X1 first

X1 has been swept in,
Model: Y on X1
R2 = 1 - .5800 = .4200
Now sweep IN X2

1.0000	.4078	-.3486	.6481
-.4078	.8337	-.0158	-.5545
.3486	-.0158	.8785	-.2504
-.6481	-.5545	.2504	.5800

X1 and X2 have now been swept in
Model: Y on X1 & X2
R2 = 1 - .2112 = .7888
Now sweep IN X3

1.1995	-.4892	-.3408	.9193
-.4892	1.1995	-.0190	-.6651
.3408	.0190	.8782	.2398
-.9193	.6651	.2398	.2112

X1, X2, and X3 have now been swept in
Model: Y on X1, X2, & X3
R2 = 1 - .1457 = .8543
Now sweep OUT X1

1.3317	-.4818	.3681	1.0124
-.4818	1.1999	.0216	-.6599
.3881	.0216	1.1387	.2731
-1.0124	.6599	-.2731	.1457

X1 swept OUT only X2 and X3 remain IN
Model: Y on X2, & X3
R2 = 1 - .9153 = .0847

.7509	-.3618	.2914	.7602
.3618	1.0256	.1620	-.2936
-.2914	.1620	1.0256	-.0219
.7602	.2936	.0219	.9153

SUMMARY OUTPUT (Y on X1)

Regression Statistics

Multiple R	.6481
R Square	.4200
Adjusted R Square	.3814
Standard Error	5.2754
Observations	17.0000

SUMMARY OUTPUT (Y on X1 & X2)

Regression Statistics

Multiple R	.8881
R Square	.7888
Adjusted R Square	.7586
Standard Error	3.2951
Observations	17

SUMMARY OUTPUT (Y on X1, X2, & X3)

Regression Statistics

Multiple R	0.9243
R Square	0.8543
Adjusted R Square	0.8207
Standard Error	2.8401
Observations	17

SUMMARY OUTPUT (Y on X2 & X3)

Regression Statistics

Multiple R	.2910
R Square	.0847
Adjusted R Square	-.0461
Standard Error	6.8599
Observations	17

Figure 12.4. Excel worksheet for Example 12.2.

matrix in these cells as $\mathbf{Z}$, the correlation matrix is computed as $\mathbf{R} = (1/17)\mathbf{Z}'\mathbf{Z}$. This is accomplished using the MMULT function in Excel, and the $(p+1) \times (p+1)$ correlation matrix, $\mathbf{R}$, is contained in cells H1:K4. The element in row $p+1$, column $p+1$ of the matrix (i.e., cell K4) is of particular interest because one minus this value is equal to R^2 for the regression model. Because none of the predictors have been swept at this point, cells H1:K4 corresponds to an absence of predictors in the model and, therefore, $R^2 = 1 - 1 = 0$.

The matrix in cells H6:K9 was obtained by setting $h = 1$ at Step 1 and applying the remaining steps of the SWEEP algorithm. Sweeping on the variable x_1 provides the R^2 value for the regression model for y using only x_1 as a predictor, which is equal to one minus the value in cell K9 (i.e., $1 - 0.5800 = 0.4200$). To verify this result, the data analysis toolpack in Excel was used to run regression analysis on the raw data and the results are displayed in cells M1:N8. The value of $R^2 = 0.4200$ in cell N5 confirms the results obtained by the SWEEP algorithm.

Next, the predictor x_2 was swept into the solution by setting $h = 2$ and applying the SWEEP algorithm to the matrix in cells H6:K9. The result is the matrix in cells H11:K14, which corresponds to the two-predictor model associated with x_1 and x_2. The value of R^2 for this model is found by taking one minus the value in cell K14 (i.e., $1 - 0.2112 = 0.7888$). The verification of this result was performed using the data analysis toolpack and the results are displayed in cells M10:N17. The value of $R^2 = 0.7888$ in cell N14 confirms the SWEEP algorithm results.

The predictor x_3 was swept into solution by setting $h = 3$ and applying the SWEEP algorithm to the matrix in cells H11:K14. The result is the matrix in cells H16:K19, which corresponds to the full regression model with all three predictors (x_1, x_2, x_3). The value of R^2 for the full regression model is obtained by taking one minus the value in cell K19 (i.e., $1 - 0.1457 = 0.8543$). The verification of this result was performed using the data analysis toolpack and the results are displayed in cells M19:N26. The value of $R^2 = 0.8543$ in cell N23 confirms the SWEEP algorithm results.

At this point, all three predictor variables have been swept into the model. Choosing one of these variables and applying the SWEEP algorithm again will sweep that variable out of the model. To illustrate, we selected variable x_1. The predictor $x1$ was swept out of the solution by setting $h = 1$ and applying the SWEEP algorithm to the matrix in cells H16:K19. The result is the matrix in cells H21:K24, which corresponds to the two-predictor model associated with x_2 and x_3. The value of R^2 for this model is found by taking one minus the value in cell K24 (i.e., $1 - 0.9153 = 0.0847$). The verification of this result was performed using the data analysis toolpack and the results are displayed in cells M28:N35. The value of $R^2 = 0.0847$ in cell N32 confirms the SWEEP algorithm results.

The numerical example in the SWEEP worksheet shows that the SWEEP algorithm is an efficient approach for obtaining the R^2 values associated with different regression models. Variables can be swept in (or out) of the model very rapidly. Such an approach is far less computationally intensive than running a separate regression analysis for each possible subset. The VBA macro in the APS workbook uses the SWEEP algorithm to conduct a systematic evaluation of all possible subsets in an efficient manner.

12.3.2 *Model selection*

Model selection (also known as subset selection or variable selection) in regression analysis is concerned with the choice of q

$(1 \leq q \leq p)$ predictors from the full set (P) of candidate predictors. Methods for model selection include stepwise regression (Efroymson, 1960), all-possible-subsets (APS) regression (Garside, 1971; LaMotte & Hocking, 1970; Smith, 1991), branch-and-bound programming (Brusco *et al.*, 2009; Furnival & Wilson, 1974), the garotte (Breiman, 1995), and the lasso (Tibshirani, 1996). Miller (2002) provides an extended treatment of many of these methods.

A desirable feature of all-possible-subsets regression is that it identifies the subset that maximizes R^2 for all subset sizes q on the interval $1 \leq q \leq p$. Branch-and-bound methods can afford this property and often accomplish it more efficiently because they can implicitly eliminate a large number of subsets. Stepwise methods commonly fail to identify the subset that maximizes R^2 for one or more values of q.

The playing field is level for subsets of the same size and, therefore, the subset yielding the largest value of R^2 can be identified as the 'best subset' for each $q(1 \leq q \leq p)$. However, because the addition of a variable to a regression model cannot possibly reduce R^2, we need some basis for comparing the best subsets across different q. One simple approach is to choose q based on inspection of the R^2 values. That is, a good value of q is one where the improvement in R^2 when moving from $q - 1$ to q is large but the improvement when moving from q to $q + 1$ is small. This type of ad hoc rule is often both impractical and suboptimal, but some formal indices are available. The particular measure used in the all-possible-subsets regression approach is Mallows's C_p (Mallows, 1973):

$$C_p(P_q) = [\text{SSE}(P_q)/\text{MSE}(P)] - (n - 2(q+1)), \quad \text{for } 1 \leq q \leq p,$$

$$(12.6)$$

where P_q is the best subset of q predictors, $\text{SSE}(P_q)$ is the error sum of squares for the regression using only the predictors in P_q, and $\text{MSE}(P)$ is the mean squared error associated with the full regression model using all candidate predictors. We adopt the convention used by Olejnik *et al.* (2000) that the subset yielding the minimum value of $C_p(P_q)$ is selected while recognizing that it is also desirable for $C_p(P_q)$ to be close to $q + 1$.

12.3.3 *Predictor importance*

Predictor importance is concerned with determining a ranking of the predictors with respect to their influence on the dependent variable. Numerous measures of predictor importance have been proposed. Bivariate correlations, partial correlations, and standardized regression coefficients are among the simplest measures, but more sophisticated indices are also available. Thorough evaluations of predictor importance measures have been provided by Nimon & Oswald (2013) and Grömping (2015). An index defined by the product of the bivariate correlation and the standardized regression coefficient, which was originally proposed by Hoffman (1960) and is often associated with the work of Pratt (1987), has been proposed and is favorably viewed by some researchers (Thomas *et al.*, 1998. 2018). The APS VBA macro allows for the use of a far richer measure, the general dominance index (Budescu, 1993; Lindeman *et al.*, 1980), which has been well received in the literature (Grömping, 2015; Nimon & Oswald, 2013).

12.3.4 *Example 12.3 — Restaurant data*

For our final example of this chapter, we apply the VBA macro for all-possible-subsets regression to the restaurant data in Example 12.1. We begin, however, by noting that Brusco (2019) ran stepwise regression on these data using SPSS. The stepwise approach resulted in a four-predictor subset (x_1, x_2, x_8, x_{10}) with $R^2 = .5176$. All four predictors were significant at the $\alpha = .05$ level and had the correct theoretical sign. However, as we shall see, this is not the four-predictor subset that maximizes R^2.

Figure 12.5 provides a screenshot of the Excel spreadsheet for APS regression. The number of candidate predictors is entered in cell E1 and the sample size is in cell E2. The sample measurements for the 10 predictors and the dependent variable must begin in cells B15 and V15, respectively. The button 'Run All-Possible-Subsets' runs the VBA macro for APS when clicked. The VBA macro reads n, p, the predictor variable data matrix ($\mathbf{X}$), and the dependent variable ($\mathbf{y}$) from the worksheet and constructs a $p + 1$ correlation matrix for the predictors and dependent variables. Sweep operations on the

	A	B	C	D	E	F	G	H	I	J	K
1	Number of Candidate Predictors				10						
2	Sample Size				60						

Run / All-Po[...]

Predictor importance row:

Predictor	sum =	.5837								
Importance	.1256	.1300	.1227	.0057	.0270	.0082	.0319	.0296	.0098	.0933

X1	X2	X3	X4	X5	X6	X7	X8	X9	X10	1111111112	Y
5	6	5	3	3	3	3	4	6	6		85
5	5	5	4	4	3	3	5	6	6		78
6	6	5	4	3	4	4	3	6	6		84
6	6	5	3	4	4	4	5	7	6		91
6	6	7	3	3	3	5	4	6	7		86
5	5	5	4	4	4	5	4	6	6		73
6	6	7	4	4	4	3	4	6	7		90
2	7	7	4	4	3	5	5	7	6		94
5	7	5	4	3	4	5	3	7	6		77
6	6	6	4	3	3	3	5	7	6		91
5	7	5	4	3	5	4	2	3			89
6	3	5	5	4	3	3	4	4	4		96
7	7	7	5	3	4	5	3	4	3		92
7	6	6	5	4	3	4	4	4	3		88
6	7	5	5	4	5	4	4	3	3		78
7	5	5	5	3	5	4	4	3	3		68

Cases: Case 1 – Case 16

All-possible-subsets results:

Subset Size	R^2	Mallows Cp	1	2	3	4	5	6	7	8	9	10
1	.3286	23.0268	0	1	0	0	0	0	0	0	0	0
2	.4345	12.5605	1	0	0	0	0	0	0	0	0	1
3	.4868	8.4124	1	0	1	0	0	0	0	0	0	1
4	.5223	6.2333	1	0	1	0	1	0	0	0	0	1
5	.5432	5.7720	1	0	1	0	1	0	0	1	0	1
6	.5582	6.0026	1	1	0	1	0	0	1	0	1	
7	.5685	6.7965	1	1	1	0	1	0	0	1	1	1
8	.5819	7.2177	1	1	1	0	1	1	0	1	1	1
9	.5837	9.0000	1	1	1	0	1	1	1	1	1	1
10	.5837	11.0000	1	1	1	1	1	1	1	1	1	1
11												
12												
13												
14												
15												
16												
17												
18												
19												
20												

Figure 12.5.   Excel worksheet for Example 12.3.

correlation matrix are used to compute the value of R^2 for all possible subsets.

Beginning in cells Y9 and Z9 of Figure 12.5, for each subset size, the R^2 and Mallows's C_p values are reported, respectively. Beginning in cell AA9, the selected variables are identified (1 if selected, 0 otherwise). For example, the best four-predictor subset consists of $\{x_1, x_3, x_5, x_{10}\}$. This four-predictor subset yields $R^2 = 0.5223$ and $Cp = 6.2333$. A plot of the number of predictors versus Mallows's C_p is shown in Figure 12.6. The general dominance measures of predictor importance begin in cell B9. The sum of these values is displayed in cell C8 and it is noted that this sum is equal to the R^2 value for the full (10-predictor) regression model (see cell Y18).

As noted, the SPSS stepwise regression solution consisted of four predictors; however, Mallows's C_p results from the all-possible-subsets regression analysis suggest that four is probably not the best number of predictors. The C_p index is minimized at $q = 5$ predictors. Moreover, the value of $C_p = 5.7720$ for $q = 5$ predictors is quite close to the number of parameters in the regression model $(5 + 1 = 6)$, which also lends support for $q = 5$ predictors. The all-possible-subsets results also reveal that, even if a four-predictor subset is desired, the stepwise regression subset $\{x_1, x_2, x_8, x_{10}\}$ is inferior to the APS subset $\{x_1, x_3, x_5, x_{10}\}$ with respect to explained variation. The stepwise regression subset yields $R^2 = 0.5176$, whereas the all-possible-subsets yields $R^2 = 0.5223$.

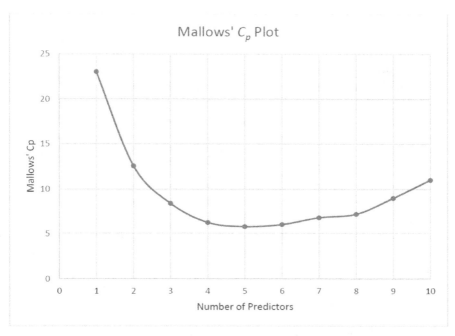

Figure 12.6. A plot of Mallows's Cp vs. number of predictors for Example 12.6.

To summarize, there are three salient findings from the model selection analysis. First, stepwise regression does not necessarily produce the maximum R^2 value for the number of predictors that it selects. Second, stepwise regression does not necessarily lead to the best choice for the number of predictors. Third, there is the potential for rather different subsets to yield very similar values of R^2. For example, the stepwise and all-possible-subsets results reveal that augmenting $\{x_1, x_{10}\}$ with $\{x_3, x_5\}$ results in an R^2 value that is only better by 0.0047 than the R^2 achieved when augmenting with $\{x_2, x_8\}$. Accordingly, model selection results alone are not sufficient to establish which variables are more important than others.

Turning to relative predictor importance, Brusco (2019) reported that, based on the absolute magnitude of the standardized regression coefficients from the full regression model, the rank order of the predictors from greatest to least with respect to importance is x_{10}, x_5, x_2, x_1, x_3, x_6, x_8, x_9, x_7, x_4. This is a somewhat perplexing ranking, as it would seem highly unlikely that the lighting in the restaurant (x_5) would be the second most important driver of satisfaction. The food quality items (x_2, x_1, x_3) are placed after lighting in the third through fifth positions. Another puzzling aspect of the predictor ranking is that the waiting time measures (x_8, x_9, x_7) occupy three of the last four positions in the ranking.

Contrastingly, based on the general dominance measures from the APS solution, the rank order of the predictors would be x_2, x_1, x_3, x_{10}, x_7, x_8, x_5, x_9, x_6, x_4. This ranking is far more logical. The food quality items occupy the first three positions in the ranking. The fourth position is occupied by the courteousness of the server measure (x_{10}). Three of the next four positions are occupied by the waiting time measures (x_7, x_8, x_9). Three of the last four positions are occupied by the aesthetic measures (x_5, x_6, x_4).

The key takeaways from the predictor importance analysis are twofold. First, it would be a mistake to rely solely on p-values or standardized regression coefficients from the full regression model to establish the relative importance of predictors. Second, all-possible-subsets regression provides a well-established measure for ranking predictors that, in this example, led to a far more interpretable ranking. The ranking obtained from the general dominance measures reflects the primacy of the food quality items (x_1, x_2, x_3) and the secondary importance of the service quality measures (courtesy measure x_{10} and promptness measures x_7, x_8, x_9). The aesthetic measures pertaining to parking (x_4), lighting (x_5), and noise level (x_6) are of lesser importance.

Chapter 13

Logistic Regression

Logistic regression is one of the most popular methodological tools in predictive analytics. Although logistic regression, broadly defined, encompasses a variety of models that allow for multiple (possibly ordinal) categories of a dependent variable, attention here is restricted to the case of a binary dependent variable. Thus, the term *logistic regression* when used in this chapter refers to what is sometimes called 'binary logistic regression'. Applications of the method abound in business-related contexts and include the prediction of bank failure (Zaghdoudi, 2013), loan default (Agbemava *et al.*, 2016), item purchase (Bejaei *et al.*, 2015), complaint behavior (Salamah & Ramayanti 2018), word-of-mouth communication regarding products (Alboqami *et al.*, 2015), and conversion of point differentials or point spreads into win–loss probabilities (Huggins *et al.*, 2020; Kvam & Sokol, 2004). Like ordinary least squares (OLS) regression in the case of a continuous dependent variable, logistic regression can be used for purely *predictive* purposes, for purely *explanatory* purposes to identify the key independent variables that drive the dependent variable, or for both predictive and explanatory purposes.

This chapter draws heavily from Brusco (2022c). It begins with a description of the logistic regression model and underlying optimization problem corresponding to maximum likelihood estimation. A small numerical example is included in this section. Next, we present a larger example, which serves as a resource for covering the important topics of model selection and relative predictor importance in subsequent sections.

13.1   Logistic Regression Model

13.1.1   *Mechanics*

Thorough treatments of logistic regression are provided by several authors including Hosmer *et al.* (2013) and Menard (2010). Here, the goal is to provide a brief description and a linkage to OLS regression. We define n as the number of observations in the training (or estimation) sample, v as the number of predictors, y_i as the binary dependent variable measurement for observation i, and x_{ij} as the independent variable measurement for observation i on predictor j. Moreover, it is assumed that predictor $j = 0$ is the intercept term and column 0 of the $n \times (v + 1)$ matrix, $\mathbf{X} = [x_{ij}]$, is just a column of 1's (i.e., $x_{i0} = 1$ for all $1 \leq i \leq n$). Denoting β_j as the regression coefficient for variable j (for all $0 \leq j \leq v$), a standard OLS regression model is

$$y_i = \sum_{j=0}^{v} \beta_j x_{ij}. \tag{13.1}$$

Two problems with the model are typically highlighted: (i) the potential for predicted values outside of the [0, 1] range for the dependent variable and (ii) violations of assumptions that facilitate testing of the regression coefficients, most notably the assumption of constant error variance. It is possible to use OLS regression for two-group discriminant analysis (see Ragsdale & Stam 1992); however, many researchers prefer logistic regression, which models the logarithm of the odds (i.e., logit) of the observations. Dropping the i subscripts momentarily to avoid notational clutter, an observation is modeled by replacing y on the left with the logarithm of its odds:

$$\log \frac{P(\mathbf{x})}{1 - P(\mathbf{x})} = \sum_{j=0}^{v} \beta_j x_j, \tag{13.2}$$

where $P(\mathbf{x})$ should be interpreted as the probability of $y = 1$ given the vector $\mathbf{x}$ of predictor variable measurements. Accordingly, the left side of (13.2) is the logarithm of the odds ratio (i.e., the probability of $y = 1$, $P(\mathbf{x})$, divided by the probability of $y = 0$, which is

$1 - P(\mathbf{x})$). Taking the exponent of both the left and right sides of (13.2) yields

$$\frac{P(\mathbf{x})}{1 - P(\mathbf{x})} = \exp\left(\sum_{j=0}^{v} \beta_j x_j\right) = \prod_{j=0}^{v} \exp(\beta_j x_j). \tag{13.3}$$

The far-right expression in (13.3) is included only to show the multiplicative (rather than additive) nature of the independent variables. In other words, for a one-unit increase in x_j, we would expect a change in the odds of the binary variable, assuming a value of one as being $\exp(\beta_j)$. Multiplying through by $1 - P(\mathbf{x})$ yields

$$P(\mathbf{x}) = \exp\left(\sum_{j=0}^{v} \beta_j x_j\right) - P(\mathbf{x}) \exp\left(\sum_{j=0}^{v} \beta_j x_j\right). \tag{13.4}$$

Collecting the $P(\mathbf{x})$ terms on the left,

$$P(\mathbf{x}) + P(\mathbf{x}) \exp\left(\sum_{j=0}^{v} \beta_j x_j\right) = \exp\left(\sum_{j=0}^{v} \beta_j x_j\right). \tag{13.5}$$

Factor out $P(\mathbf{x})$ and divide to obtain

$$P(\mathbf{x}) = \frac{\exp(\sum_{j=0}^{v} \beta_j x_j)}{1 + \exp(\sum_{j=0}^{v} \beta_j x_j)}. \tag{13.6}$$

Equation (13.6) is expressed in a slightly more compact way in the Excel spreadsheets:

$$P(\mathbf{x}) = \frac{\exp(\sum_{j=0}^{v} \beta_j x_j)}{1 + \exp(\sum_{j=0}^{v} \beta_j x_j)} \left(\frac{\exp -(\sum_{j=0}^{v} \beta_j x_j)}{\exp -(\sum_{j=0}^{v} \beta_j x_j)}\right)$$

$$= \frac{1}{1 + \exp -(\sum_{j=0}^{v} \beta_j x_j)}. \tag{13.7}$$

The goal in logistic regression is to estimate the β's so as to maximize the likelihood of the data across all observations. Therefore, because the likelihood is across all observations, it is necessary to bring back the i subscripts to write the likelihood function as follows:

$$L(\mathbf{X}|\boldsymbol{\beta}) = \prod_{i:y_i=1} P(\mathbf{x}_i) \prod_{i:y_i=0} (1 - P(\mathbf{x}_i)). \qquad (13.8)$$

It is much easier to work with the logarithm of the likelihood function, and the estimates that maximize one will also maximize the other:

$$LL(\mathbf{X}|\boldsymbol{\beta}) = \log(L(\mathbf{X}|\boldsymbol{\beta})) = \sum_{i:y_i=1} \log P(\mathbf{x}_i) + \sum_{i:y_i=0} \log(1 - P(\mathbf{x}_i)).$$

$$(13.9)$$

Estimating the parameters that maximize (13.9) requires partial differentiation. Unfortunately, unlike standard OLS regression, there is no closed-form solution. Fortunately, the GRG Nonlinear engine of the Excel Solver is typically effective for completing the estimation process, as is demonstrated in the Excel workbooks for the two examples presented later in this chapter. More specifically, the estimates of the coefficients obtained by the GRG Nonlinear engine are not overly sensitive to the initial starting values provided for those estimates.

Upon successful estimation, cases for the training sample, as well as a holdout (or validation) sample, are often made based on Equation (13.7) by assigning cases with $P(\mathbf{x}) > .5$ to the group associated with $y = 1$ and to the group associated with $y = 0$ otherwise. For predictive applications of logistic regression where the number of cases for the two groups is markedly imbalanced, and/or the costs of misclassification are asymmetric (i.e., it costs more to misclassify a group 1 case as group 2 than it does to misclassify a group 2 case as group 1, or vice versa), a probability other than 0.5 might be used for assigning cases (see, for example, Brownlee 2020).

Testing the significance of the estimated logistic regression coefficients, $\hat{\beta}_j$, requires estimates of their standard errors, $se(\hat{\beta}_j)$. The $se(\hat{\beta}_j)$ values are computed as the square roots of the main diagonal of the covariance matrix, $(\mathbf{X'VX})^{-1}$, where $\mathbf{V}$ is an $n \times n$ diagonal matrix with elements $P(\mathbf{x}_i)(1 - P\mathbf{x}_i)$ along the main diagonal. The Wald coefficients, $w_j = (\hat{\beta}_j/se(\hat{\beta}_j))^2$, and z-values, $z_j = (\hat{\beta}_j/se(\hat{\beta}_j))$, can then be computed and their significance tests made using the standard normal distribution.

13.1.2 *Example 13.1 — Insurance company data*

Our first logistic regression example (Example 13.1) uses a small dataset similar to one from Johnson and Wichern (2007, p. 578). Measurements are available for $n = 24$ respondents on one binary dependent variable and $p = 2$ predictor variables. The dependent variable is $y_i = 1$ if respondent i uses car insurance company X or $y_i = 0$ if the respondent does not use car insurance company X. The first predictor (x_1) is annual income in thousands of dollars and the second predictor (x_2) is age in years. The measurements for the dependent variable, intercept (x_0), and independent variables are contained in cells A7:D30 of Figure 13.1.

The logistic regression coefficients are located in cells C2:C4 of Figure 13.1. These coefficients, along with the data in cells A7:D30,

	A	B	C	D	E	F	G	H	I	J	K
1			β-hat	se(β-hat)	z	p-value	exp(β-hat)				
2		Intercept	-2.3431	3.4500	-0.6792	0.4970			11.9026	-0.0761	-0.1165
3		Income	-0.0339	0.0307	-1.1014	0.2707	0.9667		-0.0761	0.0009	0.0000
4		Age	0.1122	0.0512	2.1943	0.0282	1.1188		-0.1165	0.0000	0.0026
5											
6	Group (y)	x0	x1	x2	Prob	LL	Classify	V		-2LL =	24.5082
7	1	1	115.23	63	0.6956	-0.3630	1	0.2117		AIC =	30.5082
8	1	1	75.45	54	0.7619	-0.2719	1	0.1814		BIC =	34.0424
9	1	1	63.16	36	0.3915	-0.9377	0	0.2382			
10	1	1	92.22	66	0.8746	-0.1340	1	0.1097			
11	1	1	104.58	34	0.1123	-2.1870	0	0.0997			
12	1	1	78.14	68	0.9336	-0.0687	1	0.0620			
13	1	1	66.85	61	0.9038	-0.1012	1	0.0870			
14	1	1	72.44	45	0.5634	-0.5738	1	0.2460			
15	1	1	73.54	48	0.6352	-0.4539	1	0.2317			
16	1	1	66.60	57	0.8581	-0.1530	1	0.1218			
17	1	1	70.19	43	0.5266	-0.6412	1	0.2493			
18	1	1	83.22	55	0.7335	-0.3099	1	0.1955			
19	0	1	56.64	45	0.6878	-1.1642	1	0.2147			
20	0	1	94.51	37	0.1994	-0.2224	0	0.1596			
21	0	1	112.98	37	0.1176	-0.1251	0	0.1038			
22	0	1	88.67	62	0.8339	-1.7952	1	0.1385			
23	0	1	75.30	47	0.5945	-0.9027	1	0.2411			
24	0	1	119.55	34	0.0708	-0.0734	0	0.0658			
25	0	1	76.75	40	0.3889	-0.4924	0	0.2376			
26	0	1	101.47	32	0.1009	-0.1064	0	0.0907			
27	0	1	82.60	38	0.2943	-0.3485	0	0.2077			
28	0	1	85.35	36	0.2328	-0.2651	0	0.1786			
29	0	1	107.24	45	0.2843	-0.3345	0	0.2035			
30	0	1	86.89	35	0.2048	-0.2291	0	0.1628			

Figure 13.1.   Excel worksheet for Example 13.1.

are used to compute the probabilities of using insurance company X in cells E7:E30 using Equation (13.7). Respondents with a usage probability equaling or exceeding 0.5 are classified as users of car insurance company X in cells G7:G30, whereas those with a probability less than 0.5 are classified as non-users. In accordance with Equation (13.8), the log-likelihoods for each respondent are computed in cells F7:F30 and their sum is multiplied by -2 to establish the deviance measure in cell K6. The GRG engine of the Excel Solver is used to find the coefficients in cells C2:C4 that minimize the value in cell K6.

The main diagonal elements of the $\mathbf{V}$ matrix are computed in cells H7:H30. These measures are then used with the variable measurements in cells B7:D30 to compute $(\mathbf{X'VX})^{-1}$ in cells I2:K4 using the following formula: $=$ MINVERSE(MMULT(MMULT(TRANSPOSE(B7:D30), IF(H7:H30 $=$ TRANSPOSE(H7:H30), H7:H30, 0)), B7:D30)).

The standard errors of the coefficients are computed in cells D2:D4 using the matrix in I2:K4. The z values in cells E2:E4 are the ratios of the coefficients in cells C2:C4 to the standard errors in cells D2:D4. The p-values in cells F2:F4 are obtained using the NORMDIST function in Excel. Finally, cells G3:G4 are obtained from the exponentiation of the coefficients in C3:C4.

From cells F2:F4, we observe that only the predictor age is significant at the $\alpha = 0.05$ level. From cells G2:G4, we observe that for a one-unit increase in the income variable (effectively, a one-thousand-dollar increase because the variables are measured in thousands), we would expect a $1 - 0.9667 = 3.33\%$ *decrease* in the odds that the respondent will use company X. A stronger increase in the odds of using company X of 11.88% would be expected for a one-unit increase in the age variable.

From cells G7:G30, we observe that 19 of the 24 respondents (79.2%) are correctly classified. Cells G9 and G11 show that two users of insurance company X (respondents 3 and 5) were incorrectly classified as non-users, whereas cells G19, G22, and G23 show that three non-users of insurance company X (respondents 13, 16, and 17) were incorrectly classified as users of company X.

13.2 Example 13.2 — Restaurant Recommendation

13.2.1 *Data for the example*

The example depicts recommendation behavior in a services marketing context. The dependent variable assumes a value of $y_i = 1$ if customer i recommends the restaurant (i.e., engages in positive word-of-mouth behavior) and $y_i = 0$ otherwise. There are 10 predictor variables, each measured on a 7-point Likert scale (1 = strongly disagree to 7 = strongly agree). The predictor variable statements are displayed in cells A4:A13 of the MLE_Full worksheet in the Chapter 13 — LR_MSPI workbook. A screenshot of a portion of this worksheet is displayed in Figure 13.2. The training sample consists of $n = 1000$ observations for the dependent and predictor variables, which are contained in cells B17:M1016 of the worksheet. A correlation matrix for the predictors is provided in Figure 13.3.

In this type of application context, the goals of the logistic regression analysis might be both predictive and explanatory. Building a good predictive engine requires model selection. There are 500

Logistic Regression via Excel Solver to Maximize Log-Likelihood (Minimize Deviance)

	β-hat		
Y = 1 if recommended the resataurant, else 0			
Intercept Term	-5.1533	Deviance (-2LL)	1108.175
X1 = the side dishes and salads are delicious	.1705	Akaike's Information Criterion (AIC) =	1130.175
X2 = the entrées are delicious	.3899	Bayesian Information Criterion (BIC) =	1184.161
X3 = the desserts are delicious	-.1253		
X4 = parking at the restaurant is ample	.1561	% Correctly Classified (Training) =	78.0%
X5 = the lighting in the restaurant is appropriate	-.0490	% Correctly Classified (Holdout) =	75.2%
X6 = the noise level in the restaurant is not distracting to me	.1412		
X7 = the wait for beverage service is excessive	-.0026		
X8 = the wait for food service is excessive	-.4057		
X9 = the wait for the check is excessive	-.0787		
X10 = restaurant service personnel are courteous and friendly	.2444		

Customer	Y	X0	X1	X2	X3	X4	X5	X6	X7	X8	X9	X10	Prob Y = 1	LL	PRED	Correct?
1	0	1	7	7	7	6	6	6	5	4	6	5	.1818	-.2007	0	1
2	0	1	7	7	7	7	7	7	2	3	2	6	.4298	-.5617	0	1
3	0	1	7	7	7	5	4	7	2	3	2	7	.4493	-.5965	0	1
4	1	1	7	6	6	4	5	6	3	3	2	4	.1750	-1.7427	0	0
5	1	1	6	7	7	7	7	7	6	5	5	6	.1807	-1.7107	0	0
6	1	1	7	7	7	6	5	5	1	1	2	7	.6071	-.4990	1	1
7	1	1	4	5	4	7	7	7	3	3	3	2	.0947	-2.3566	0	0
8	0	1	7	7	7	6	7	7	1	1	1	7	.6678	-1.1020	1	0
9	1	1	7	7	7	7	7	7	1	1	1	7	.7015	-.3545	1	1
10	0	1	7	7	7	4	7	2	1	1	1	7	.4207	-.5460	0	1
11	0	1	7	6	7	7	7	7	2	3	3	4	.2244	-.2541	0	1
12	1	1	7	7	7	3	3	3	1	1	1	7	.4653	-.7650	0	0
13	1	1	7	7	6	7	6	4	4	4	3	5	.2200	-1.5143	0	0
14	0	1	7	7	7	6	6	6	3	2	2	4	.3505	-.4316	0	1
15	1	1	7	7	7	4	3	4	1	1	1	6	.4785	-.7371	0	0
16	1	1	7	7	7	5	6	6	1	1	1	7	.6106	-.4933	1	1
17	1	1	5	6	6	7	7	7	1	1	1	7	.5619	-.5765	1	1
18	1	1	2	2	2	7	7	3	5	4	5	5	.0195	-3.9367	0	0

Figure 13.2. Excel worksheet for Example 13.2.

	X1	X2	X3	X4	X5	X6	X7	X8	X9	X10
X1	1.000									
X2	0.846	1.000								
X3	0.850	0.841	1.000							
X4	0.157	0.124	0.148	1.000						
X5	0.151	0.089	0.121	0.636	1.000					
X6	0.116	0.079	0.100	0.617	0.612	1.000				
X7	-0.337	-0.318	-0.349	-0.103	-0.125	-0.144	1.000			
X8	-0.352	-0.316	-0.342	-0.136	-0.134	-0.148	0.825	1.000		
X9	-0.341	-0.315	-0.326	-0.140	-0.151	-0.175	0.836	0.821	1.000	
X10	0.507	0.522	0.500	0.114	0.112	0.101	-0.522	-0.513	-0.516	1.000

Figure 13.3. Correlation matrix for Example 13.2.

additional observations in cells B1017:M1516, which are used as a holdout sample to evaluate predictive performance.

13.2.2   *Maximum likelihood estimation (MLE)*

The maximum likelihood estimation of the coefficients for the full (i.e., using all 10 predictors) logistic regression model is accomplished in the MLE_Full worksheet of the Chapter 13 — LR_MSPI workbook shown in Figure 13.2. The coefficients are in cells H3:H13. These cells are used in conjunction with the predictor variable measurements to populate cells O17:O1516 with the $P(\mathbf{x})$ values associated with Equation (13.7). The log-likelihoods for each observation are computed in cells P17:P1516 using Equation (13.9), which requires the $P(\mathbf{x})$ values in O17:O1516 and the dependent variable measures in cells B17:B1516. The deviance of the model is computed in cell R3, using the training sample only, as –2*SUM(P17:P1016). Cell R3 is the objective function that the Excel Solver seeks to minimize by changing the values in H3:H13. The GRG Nonlinear engine of the Excel Solver is recommended. Initial values for H3:H13 can be established in various ways, although there is no guarantee of a globally optimal solution. To obtain the results shown in Figure 13.1, we initialized all of the values to zero and ran the algorithm. As a check, we obtained the same results using SPSS (IBM Corp., 2019).

The deviance measure in cell R3 is used to compute the AIC and BIC in cells R4 and R5, respectively. The predictions for the dependent variable are obtained in cells Q17:Q1516 by applying *if* statements to the $P(\mathbf{x})$ values in O17:O1516, and cells R17:R1516 assume values of one if the predictions match the observed y values in column B or zero if the predictions are wrong. Cells R7 and R8 contain the percentage of correct predictions for the training (78.0%) and holdout (75.2%) samples, respectively.

13.2.3 *Significance testing*

Although significance testing of the coefficients can be accomplished seamlessly using an Excel add-in, such as XLSTAT (Addinsoft, 2021), it is helpful to understand the mechanics of this process. The process is somewhat more involved than obtaining the coefficients themselves. For this reason, it is accomplished in a separate workbook, Chapter 13 — LR_SigTest. As shown in Figure 13.4, the ST_Full worksheet of this workbook is similar in appearance to the MLE_Full worksheet but has additional information in cells I3:M13 regarding the significance tests.

The values of the coefficients from cells H3:H13 of the MLE_Full worksheet should be copied and pasted to cells H3:H13 of the ST_Full worksheet. The $se(\hat{\beta}_j)$ values are computed in cells I3:I13. Their computation requires the diagonal elements of $\mathbf{V}$ in cells Q17:Q1016, which are used to obtain the covariance matrix, $(\mathbf{X'VX})^{-1}$, in cells O3:Y13. The matrix computation is accomplished using the MINVERSE, MMULT, and TRANSPOSE functions, as well as an *if* statement that produces the diagonal matrix. To ensure that the

Logistic Regression Significance Testing for the Full Model

Y = 1 if recommended the restaurant, else 0

	p-hat	se(β-hat)	Wald	z	p-value	exp(β-hat)
Intercept Term	-5.1533	.7700	44.7868	-6.6923	.0000	
X1 = the side dishes and salads are delicious	.1705	.1251	1.8582	1.3632	.1728	1.1859
X2 = the entrées are delicious	.3899	.1240	9.8861	3.1442	.0017	1.4768
X3 = the desserts are delicious	-.1253	.1234	1.0316	-1.0157	.3098	.8822
X4 = parking at the restaurant is ample	.1561	.0802	3.7950	1.9481	.0514	1.1690
X5 = the lighting in the restaurant is appropriate	-.0490	.0794	.3808	-.6171	.5372	.9522
X6 = the noise level in the restaurant is not distracting to me	.1412	.0778	3.2907	1.8140	.0697	1.1516
X7 = the wait for beverage service is excessive	-.0026	.1212	.0005	-.0216	.9828	.9974
X8 = the wait for food service is excessive	-.4057	.1207	11.3044	-3.3622	.0008	.6665
X9 = the wait for the check is excessive	-.0787	.1201	.4300	-.6557	.5129	.9243
X10 = restaurant service personnel are courteous and friendly	.2444	.0779	9.8466	3.1379	.0017	1.2768

Covariance Matrix Estimation for Computing se(β-hat)

.5930	-.0080	-.0095	-.0110	-.0079	-.0099	-.0112	-.0104	-.0076	-.0133	-.0288
-.0080	.0157	-.0065	-.0069	.0000	-.0010	.0004	-.0007	.0004	.0005	-.0004
-.0095	-.0065	.0154	-.0083	.0001	.0007	-.0002	-.0007	-.0001	.0000	-.0015
-.0110	-.0069	-.0083	.0152	-.0004	.0003	.0000	.0014	-.0001	-.0004	-.0003
-.0079	.0000	.0001	-.0004	.0064	-.0025	-.0023	-.0005	.0003	-.0001	.0001
-.0099	-.0010	.0007	.0003	-.0025	.0063	-.0022	.0000	.0003	.0001	.0000
-.0112	.0004	-.0002	.0000	-.0023	-.0022	.0061	.0004	-.0005	.0004	.0000
-.0104	-.0007	-.0007	.0014	-.0005	.0000	.0004	.0147	-.0063	-.0064	.0011
-.0076	.0004	-.0001	-.0001	.0003	.0003	-.0005	-.0063	.0146	-.0055	.0003
-.0133	.0005	.0000	-.0004	-.0001	.0001	.0004	-.0064	-.0055	.0144	.0009
-.0288	-.0004	-.0015	-.0003	.0001	.0000	.0000	.0011	.0003	.0009	.0081

Deviance (-2LL)	1108.175
Akaike's Information Criterion (AIC) =	1130.175
Bayesian Information Criterion (BIC) =	1184.161
% Correctly Classified (Training) =	78.0%
% Correctly Classified (Holdout) =	75.2%

Customer	Y	X0	X1	X2	X3	X4	X5	X6	X7	X8	X9	X10	Y = 1	LL	V	PRED	Correct?
1	0	1	7	7	7	6	6	5	4	6	5		.1818	-.2007	.1488	0	1
2	0	1	7	7	7	7	7	2	3	2	6		.4298	-.5617	.2451	0	1
3	0	1	7	7	5	4	7	2	3	2	7		.4493	-.5965	.2474	0	1
4	1	1	7	6	6	4	5	6	3	3	2	4	.1750	-1.7427	.1444	0	0
5	1	1	6	7	7	7	7	6	5	5	6		.1807	-1.7107	.1481	0	0
6	1	1	7	7	7	6	5	5	1	1	2	7	.6071	-.4990	.2385	1	1
7	1	1	4	5	4	7	7	3	3	3	2		.0947	-2.3566	.0858	0	0
8	0	1	7	7	7	6	7	7	1	1	1	7	.6678	-1.1020	.2218	1	0
9	1	1	7	7	7	7	7	1	1	1	1	7	.7015	-.3545	.2094	1	1
10	0	1	7	7	7	4	7	2	1	1	1	7	.4207	-.5460	.2437	0	1
11	1	1	7	6	7	7	7	2	3	3	4		.2244	-.2541	.1741	0	1
12	1	1	7	7	7	3	3	3	1	1	1	7	.4653	-.7650	.2488	0	0

Figure 13.4. Excel significance testing worksheet for Example 13.2.

V matrix is diagonal, the elements in Q17:Q1016 are perturbed via a small random error to guarantee that they are unique.

Once the $se(\hat{\beta}_j)$ values are obtained, the w_j and z_j values can be computed in cell ranges J3:J13 and K3:K13, respectively. The p-values for the coefficients are computed using the NORMDIST function in cells L3:L13. Only three predictors (x_2, x_8, and x_{10}) are significant at $\alpha = .05$ (x_4 and x_6 are borderline). Finally, the exponents of each regression coefficient are provided in cells M3:M13 for easy interpretation.

13.2.4 *Substantive interpretation*

An examination of the solution reveals some interesting findings. For example, for a one-unit increase in the response to the statement 'the side dishes and salads are delicious', we would expect an 18.6% increase in the odds that the individual will engage in positive word of mouth by recommending the restaurant. A stronger increase in the odds of 47.7% would be expected for a one-unit increase in the response to the statement 'the entrées are delicious'. Contrastingly, in light of its negative coefficient, we would expect a decrease of 11.8% for a one-unit increase in the response to the statement 'the desserts are delicious'. This is somewhat surprising. We would expect that greater agreement with the statements regarding side dishes and salads, entrées, and desserts (x_1, x_2, and x_3) would have a positive effect on recommendation behavior.

Most likely, the antithetical sign for desserts is attributable to the problem of multicollinearity. Specifically, Figure 13.3 reveals that the pairwise correlations between x_1, x_2, and x_3 all exceed 0.8. An antithetical sign is also observed for x_5 — 'the lighting in the restaurant is appropriate'. In light of these results, the interpretation and testing of regression coefficients are likely impaired by the multicollinearity that exists. Moreover, although the predictive results for the training and holdout samples are reasonably good, there is also the potential for overfitting in light of the relatively weak coefficients (and lack of statistical significance) for many of the variables. These issues will be explored in greater detail in the model selection section.

13.3 Model Selection

As in the case of OLS Regression, there are different possible approaches to model selection in logistic regression. One approach would be to use all-possible-subsets logistic regression in a manner similar to how we applied all-possible-subsets OLS regression in Chapter 12. This is apt to be computationally demanding in Excel and therefore is not considered here. However, there is no reason that we cannot apply all-possible-subsets OLS regression as a heuristic approach for selecting a subset of variables, and that approach will be evaluated. A second approach that we consider is the lasso.

13.3.1 *The all-possible-subsets method*

The first approach is to use all-possible-subsets regression (Miller, 2002). With this approach, all 2^v regression models are estimated and the model obtaining the best fit is stored for each subset size. For OLS regression, the measure of fit is typically the residual sum of squares, whereas for logistic regression, the *deviance* ($-2LL$) is the common measure. For both models, the measure of fit generally improves (decreases) as the subset size increases. Therefore, it is necessary to employ criteria that can facilitate the selection of the appropriate subset size. As in the case for OLS regression, two of the most popular criteria for logistic regression are Akaike's information criterion (AIC $= -2LL + 2(v + 1)$), (Akaike, 1973, 1974) and the Bayesian information criterion (BIC $= -2LL + (v + 1)\log(n)$), Schwartz, 1978). The best subset size and corresponding model would be selected based on the minimum value for the AIC or BIC criterion.

Unlike the lasso method, the all-possible-subsets approach can be used to establish the 'best subset' for each number of predictors on the interval $1 \leq u \leq v$. Accordingly, it explicitly includes or excludes variables for each subset size. We used the APS worksheet of the Chapter 13 — LR_MSPI workbook (see Figure 13.5) to obtain the results for an all-possible-subsets OLS regression for the 10 predictors. The dependent variable measures for the OLS regression are constructed by 'nudging' the binary outcomes, as described by Schield (2017). Specifically, $y_i = 1$ is replaced by $y_i = 1 - \varepsilon$ and

		A	B	C	D	E	F	G	H	I	J	K	V	Y	Z	AA	AB	AC	AD	AE	AF	AG	AH	AI	AJ

Number of Candidate Predictors: 10
Sample Size: 1000 — Run All-Possible-Subsets

Predictor	sum = .1932										Subset Size	R^2	Mallows C_p	1	2	3	4	5	6	7	8	9	10
Importance	.0225	.0300	.0169	.0084	.0031	.0089	.0194	.0328	.0212	.0300	1	.1157	87.9749	0	0	0	0	0	0	0	1	0	0
											2	.1689	24.7399	0	1	0	0	0	0	0	1	0	0
											3	.1807	12.3298	0	1	0	0	0	1	0	1	0	0
	X1	X2	X3	X4	X5	X6	X7	X8	X9	X10	4	.1892	3.8906	0	1	0	0	0	1	0	1	0	1
											5	.1913	3.3550	0	1	0	1	0	1	0	1	0	1
											6	.1918	4.7159	0	1	1	1	0	1	0	1	0	1

(Y column header: 1111111112 Y)

Case	X1	X2	X3	X4	X5	X6	X7	X8	X9	X10	Y	Subset Size	R^2	C_p	1	2	3	4	5	6	7	8	9	10
Case 1	7	7	7	6	6	6	5	4	6	5	-6.907	7	.1927	5.5888	1	1	1	1	0	1	0	1	0	1
Case 2	7	7	7	7	7	7	2	3	2	6	-6.907	8	.1931	7.1634	1	1	1	1	1	1	0	1	0	1
Case 3	7	7	7	5	4	7	2	3	2	7	-6.907	9	.1932	9.0031	1	1	1	1	1	1	0	1	1	1
Case 4	7	6	6	4	5	6	3	3	2	4	6.907	10	.1932	11.0000	1	1	1	1	1	1	1	1	1	1
Case 5	6	7	7	7	7	7	6	5	5	6	6.907	11												
Case 6	7	7	7	6	5	5	1	1	2	7	6.907	12												
Case 7	4	5	4	7	7	7	3	3	3	2	6.907	13												
Case 8	7	7	7	6	7	1	1	1	1	7	-6.907	14												
Case 9	7	7	7	7	7	7	1	1	1	7	6.907	15												
Case 10	7	7	7	4	7	2	1	1	1	7	-6.907	16												
Case 11	7	6	7	7	7	7	2	3	3	4	-6.907	17												
Case 12	7	7	7	3	3	3	1	1	1	7	6.907	18												
Case 13	7	7	6	7	6	4	4	4	3	5	6.907	19												

Figure 13.5.   All-possible-subsets OLS regression worksheet for Example 13.2.

$y_i = 0$ is replaced by $y_i = \varepsilon$ and the dependent variable measures are $\log \frac{y_i}{(1-y_i)}$. We used $\varepsilon = 0.001$ for the analyses. All-possible-subsets are evaluated by clicking on the button 'Run All-Possible-Subsets', which runs a VBA macro that performs sweep operations (Goodnight, 1979) on the correlation matrix for the training sample to complete the analysis (see, for example, Brusco, 2019). The macro reads the number of predictors (cell E1), the number of training sample observations (cell E2), and the training sample measures (B15:K1014 and V15:V1014). Upon evaluation of the all-possible-subsets results, the macro writes the best R^2 value for each subset size to cells Y9:Y18. The predictors associated with the best subset for each subset size are marked by a '1' in the cell range AA9:AJ18. The selection of the best subset size can be facilitated by the Mallows's C_p values (Mallows, 1973) in cells Z9:Z18. Mallows's C_p is similar to the AIC; in fact, the selection of subset size based on the minimum value of Mallows's C_p is equivalent to selection based on the minimum AIC.

The five-predictor subset $\{x_2, x_4, x_6, x_8, x_{10}\}$ yields the minimum value (3.35) of Mallows's C_p in Figure 13.5; however, the four-predictor subset $\{x_2, x_6, x_8, x_{10}\}$ is very close (3.89). Maximum likelihood estimation for the five-predictor and four-predictor subsets was completed in the worksheets BestSub5 and BestSub4 of the

Chapter 13 — LR_MSPI workbook, respectively. The worksheets ST_BestSub5 and ST_BestSub4 of the Chapter 13 — LR_SigTest workbook provide the results of the significance tests, and screenshots of the five- and four-predictor subset results are provided in Figures 13.6 and 13.7, respectively.

Logistic Regression Significance Testing for the 5-Predictor Model

Y = 1 if recommended the restaurant, else 0	β-hat	se(β-hat)	Wald	z	p-value	exp(β-hat)	Covariance Matrix					
Intercept Term	-5.3287	.7239	54.1924	-7.3615	.0000		.5240	-.0236	-.0135	-.0131	-.0232	-.0271
X2 = the entrées are delicious	0.4261	.0754	31.9446	5.6520	.0000	1.531	-.0236	.0057	-.0001	.0001	.0001	-.0020
X4 = parking at the restaurant is ample	0.1376	.0735	3.5067	1.8731	.0610	1.148	-.0135	-.0001	.0054	-.0031	.0000	.0001
X6 = the noise level in the restaurant is not distracting to me	0.1287	.0721	3.1892	1.7858	.0741	1.137	-.0131	.0001	-.0031	.0052	.0002	-.0001
X8 = the wait for food service is excessive	-0.4624	.0769	36.1530	-6.0127	.0000	0.630	-.0232	.0001	.0000	.0002	.0059	.0019
X10 = restaurant service personnel are courteous and friendly	0.2566	.0761	11.3635	3.3710	.0007	1.293	-.0271	-.0020	.0001	-.0001	.0019	.0058

Deviance (-2LL)	1111.0646
Akaike's Information Criterion (AIC) =	1123.0646
Bayesian Information Criterion (BIC) =	1152.5111
% Correctly Classified (Training Sample) =	78.2%
% Correctly Classified (Holdout Sample) =	75.4%

Customer	Y	X0	X2	X4	X6	X8	X10	Prob Y = 1	LL	V	PRED	Correct?
1	0	1	7	6	6	4	5	.2117	-.2379	.1669	0	1
2	0	1	7	7	7	3	6	.4184	-.5421	.2433	0	1
3	0	1	7	5	7	3	7	.4139	-.5343	.2426	0	1
4	1	1	7	6	6	3	4	.1406	-1.9616	.1209	0	0
5	1	1	7	7	7	5	6	.2220	-1.5049	.1727	0	0
6	1	1	7	6	5	1	7	.6123	-.4905	.2374	1	1
7	1	1	5	7	7	3	2	.0991	-2.3120	.0892	0	0
8	0	1	7	6	7	1	7	.6714	-1.1129	.2206	1	0
9	1	1	7	7	7	1	7	.7010	-.3552	.2096	1	1
10	0	1	7	4	2	1	7	.4491	-.5962	.2474	0	1
11	0	1	6	7	7	3	4	.2195	-.2479	.1713	0	1
12	1	1	7	3	3	1	7	.4469	-.8054	.2472	0	0
13	1	1	7	7	4	4	5	.1924	-1.6479	.1554	0	0
14	0	1	7	6	6	2	4	.3438	-.4213	.2256	0	1

Figure 13.6. Worksheet for five-predictor subsets for Example 13.2.

Logistic Regression Significance Testing for the 4-Predictor Model

Y = 1 if recommended the restaurant, else 0	β-hat	se(β-hat)	Wald	z	p-value	exp(β-hat)	Covariance Matrix				
Intercept Term	-5.0074	.6986	51.3693	-7.1672	.0000		.4881	-.0238	-.0208	-.0233	-.0269
X2 = the entrées are delicious	0.4313	.0753	32.8521	5.7317	.0000	1.539	-.0238	.0057	.0001	.0000	-.0020
X6 = the noise level in the restaurant is not distracting to me	0.2095	.0581	12.9871	3.6038	.0003	1.233	-.0208	.0001	.0034	.0002	-.0001
X8 = the wait for food service is excessive	-0.4634	.0769	36.3363	-6.0280	.0000	0.629	-.0233	.0000	.0002	.0059	.0019
X10 = restaurant service personnel are courteous and friendly	0.2549	.0762	11.2065	3.3476	.0008	1.290	-.0269	-.0020	-.0001	.0019	.0058

Deviance (-2LL)	1114.615
Akaike's Information Criterion (AIC) =	1124.615
Bayesian Information Criterion (BIC) =	1149.154
% Correctly Classified (Training Sample) =	79.1%
% Correctly Classified (Holdout Sample) =	76.8%

Customer	Y	X0	X2	X6	X8	X10	Prob Y = 1	LL	V	PRED	Correct?
1	0	1	7	6	4	5	.2125	-.2389	.1673	0	1
2	0	1	7	7	3	6	.4056	-.5202	.2411	0	1
3	0	1	7	7	3	7	.4682	-.6316	.2490	0	1
4	1	1	6	6	3	4	.1776	-1.7284	.1460	0	0
5	1	1	7	7	5	6	.2127	-1.5481	.1674	0	0
6	1	1	7	5	1	7	.5940	-.5208	.2412	1	1
7	1	1	5	7	3	2	.0941	-2.3634	.0852	0	0
8	0	1	7	7	1	7	.6899	-1.1708	.2139	1	0
9	1	1	7	7	1	7	.6899	-.3712	.2139	1	1
10	0	1	7	2	1	7	.4384	-.5769	.2462	0	1
11	0	1	6	7	3	4	.2103	-.2360	.1661	0	1
12	1	1	7	3	1	7	.4904	-.7125	.2499	0	0
13	1	1	7	4	4	5	.1507	-1.8923	.1280	0	0
14	0	1	7	6	2	4	.3457	-.4241	.2262	0	1

Figure 13.7. Worksheet for four-predictor subset for Example 13.2.

The five-predictor model has a slightly lower AIC than the four-predictor model; however, the four-predictor model has a slightly lower BIC. The five-predictor model provides only slightly better prediction of the training and holdout samples than the full logistic regression model. The predictive performance of the four-predictor model is appreciably better. Another advantage of the four-predictor model is that all predictors are significant ($\alpha = 0.05$) in the four-predictor model, but x_4 and x_6 are not significant in the five-predictor model.

13.3.2   *The lasso method*

A second approach is to constrain or penalize the coefficients of the model in some fashion. This is the principle used in the ridge regression (Hoerl and Kennard 1970) and l_1-regularized (lasso: Tibshirani 1996) methods. Here, attention is restricted to the latter method because of its immense popularity. Using the constrained approach, the goal is to maximize Equation (13.9) or, equivalently, minimize the deviance measure, subject to the restriction that the sum of the absolute values of the model coefficients does not exceed some threshold, λ. More formally, this constraint is specified as follows:

$$\sum_{j=1}^{v} |\hat{\beta}_j| \leq \lambda. \tag{13.10}$$

We used the Lasso worksheet displayed in Figure 13.8 to investigate the effect on prediction for the holdout sample at different values of λ. The Lasso worksheet of the Chapter 13 — LR_MSPI workbook is very similar to the MLE_Full worksheet. The primary differences are that (i) the significance testing of the coefficients is omitted in the Lasso worksheet because the testing process used in MLE_Full is no longer viable and (ii) cell P1 is included to contain the user-specified tuning parameter, λ. In addition, the sum of the absolute values of the predictor-variable coefficients is computed and stored in cell P4. A constraint is added in the Excel Solver to ensure that cell P4 is equal to or less than cell P1.

Relative to the MLE_Full worksheet, the results obtained by the GRG Nonlinear engine of the Excel Solver for the Lasso worksheet are more sensitive to the initial starting values of the coefficients. One approach to this problem is, for each value of λ evaluated, to

	L1-regularized Logistic Regression via Excel Solver								Tuning parameter, λ =					0.8
1	L1-regularized Logistic Regression via Excel Solver								Tuning parameter, λ =					0.8
2	Y = 1 if recommended the resataurant, else 0							p-hat	exp(p-hat)					
3	Intercept Term							-2.8962		Deviance (-2LL)					1142.835
4	X1 = the side dishes and salads are delicious							.0426	1.0436	Absolute sum of coefficients =					.800
5	X2 = the entrées are delicious							.2304	1.2591						
6	X3 = the desserts are delicious							.0000	1.0000						
7	X4 = parking at the restaurant is ample							.0310	1.0315						
8	X5 = the lighting in the restaurant is appropriate							.0002	1.0002	% Correctly Classified (Training) =					80.6%
9	X6 = the noise level in the restaurant is not distracting to me							.0339	1.0345	% Correctly Classified (Holdout) =					77.4%
10	X7 = the wait for beverage service is excessive							-.0025	.9975						
11	X8 = the wait for food service is excessive							-.2792	.7564						
12	X9 = the wait for the check is excessive							-.0133	.9868						
13	X10 = restaurant service personnel are courteous and friendly							.1669	1.1816						

Customer	Y	X0	X1	X2	X3	X4	X5	X6	X7	X8	X9	X10	Prob Y = 1	LL
1	0	1	7	7	7	6	6	6	5	4	6	5	.2751	-.3217
2	0	1	7	7	7	7	7	7	2	3	2	6	.4020	-.5142
3	0	1	7	7	7	5	4	7	2	3	2	7	.4274	-.5575
4	1	1	7	6	6	4	5	6	3	3	2	4	.2514	-1.3806
5	1	1	6	7	7	7	7	7	6	5	5	6	.2596	-1.3485
6	1	1	7	7	7	6	5	5	1	1	2	7	.5576	-.5841
7	1	1	4	5	4	7	7	7	3	3	3	2	.1585	-1.8417
8	0	1	7	7	7	6	7	7	1	1	1	7	.5776	-.8618
9	1	1	7	7	7	7	7	7	1	1	1	7	.5852	-.5359
10	0	1	7	7	7	4	7	2	1	1	1	7	.5203	-.7346
11	0	1	7	6	7	7	7	7	2	3	3	4	.2740	-.3202

Figure 13.8.   Lasso worksheet for Example 13.2.

run the GRG Nonlinear engine of the Solver multiple times using different (e.g., random) initial values for the coefficients. A second approach, which we used here, is to set the initial values for the logistic regression coefficients equal to those obtained for the unconstrained solution using the MLE_Full worksheet.

In this example, the sum of the absolute values of the predictor variable coefficients for the maximum likelihood logistic regression solution in Figure 13.2 is 1.764. Therefore, we began with an initial value of $\lambda = 1.7$ in Step 2 and systematically decreased the value by 0.1 in Step 3. The most interesting range of solutions was over the interval $0.8 \leq \lambda \leq 1.3$. Values of $\lambda > 1.3$ led to results comparable to those for the unconstrained model, whereas values of $\lambda < 0.8$ led to inferior results. Figure 13.9 displays the results for $\lambda = 1.3, \lambda = 1.1, \lambda = 0.9$, and $\lambda = 0.8$. Although global optimality is not guaranteed for any of these solutions, the monotonic improvement in prediction accuracy for both the training and holdout samples over this range is useful for elucidating the potential benefits of the lasso method to students.

The results in Figure 13.9 pertain to a single run of the lasso method for a given training and holdout sample. It is important to explain to students that a better approach to the selection of the appropriate value of λ would be accomplished using some type of sampling approach, such as bootstrapping or k-fold cross-validation.

		Best Subsets		l_1-regularized logistic regression (lasso)			
	Full model	Bestsub-5	Bestsub-4	$\lambda = 1.3$	$\lambda = 1.1$	$\lambda = 0.9$	$\lambda = 0.8$
b0	-5.1533	-5.3287	-5.0074	-4.8078	-4.0224	-3.2807	-2.8962
b1	.1705			.0921	.0682	.0514	.0426
b2	.3899	.4261	.4314	.3214	.2871	.2488	.2304
b3	-.1253			.0000	.0000	.0001	.0000
b4	.1561	.1376		.1137	.0814	.0482	.0310
b5	-.0490			.0000	.0000	.0000	.0002
b6	.1412	.1287	.2095	.1067	.0778	.0493	.0339
b7	-.0026			-.0026	-.0026	-.0025	-.0025
b8	-.4057	-.4624	-.4634	-.3730	-.3274	-.2986	-.2792
b9	-.0787			-.0654	-.0546	-.0224	-.0133
b10	.2444	.2566	.2549	.2251	.2010	.1787	.1669
-2LL	1108.175	1111.065	1114.615	1110.861	1118.108	1132.551	1142.835
AIC	1130.175	1123.065	1124.615	n/a	n/a	n/a	n/a
BIC	1184.161	1152.511	1149.154	n/a	n/a	n/a	n/a
Training (% correct)	78.0%	78.2%	79.1%	78.4%	79.5%	80.1%	80.6%
Holdout (% correct)	75.2%	75.4%	76.8%	75.2%	76.4%	76.6%	77.4%

Figure 13.9.   Summary of model selection results for Example 13.2.

This would not be easy to implement in Excel because it is computationally demanding. For each value of λ evaluated, it would be necessary to use the GRG Nonlinear engine to estimate the lasso model for each sample and obtain an average measure of predictive performance for that value of λ.

In addition, it should be noted that the lasso does not directly include or exclude variables, but merely restricts the sum of absolute values of their coefficients. From the solution in Figure 13.7 for $\lambda = 0.8$, it does appear that the coefficients for x_3 and x_5 have been driven to near zero, such that they might be safely excluded from the model. Other variables have very small coefficients yet must be retained in the model.

13.3.3   Summary of results

From Figure 13.9, it is clear that the lasso yielded the best prediction of the holdout sample, correctly classifying 77.4% of the cases.

The four-predictor best-subsets model was second best, correctly classifying 76.8% of the cases. It is quite likely that extensions of the lasso, such as the adaptive Lasso (Zou, 2006), might perform even better; however, our goal here is primarily to illustrate the optimization process in Excel.

13.4 Relative Predictor Importance

As we observed in Chapter 12, establishing the relative importance of predictors can be accomplished using all-possible-subsets regression. In particular, the measure of general dominance, which is based on R^2 shares, has proved useful for OLS regression (Budescu 1993, Lindeman *et al.* 1980). The R^2 share for a predictor corresponds to its average contribution to explained variation across all possible orderings for sequential inclusion of the predictors in the model (see Grömping 2015, p. 143 for details). An especially desirable property of this measure is that the sum of the R^2 shares across all predictors is equal to R^2 for the full regression model using all predictors. Extensions of this approach to logistic regression have been developed by Azen and Traxel (2009).

Given that the all-possible-subsets method can be used for both model selection and relative predictor importance, it provides a useful framework for both OLS and logistic regression. However, as noted, the computational demand for logistic regression is appreciably greater than it is for OLS regression and, accordingly, the feasibility of an Excel spreadsheet implementation of an all-possible-subsets analysis is less likely to be viable for logistic regression. However, we can use the results in Figure 13.5 for the all-possible-subsets OLS regression as a surrogate for the all-possible-subsets logistic regression.

Running the all-possible-subsets algorithm in the APS worksheet also generates the measures of relative predictor importance. The R^2-shares for each variable are shown in cells B9:K9 of Figure 13.5. The R^2-shares sum to the R^2 value of .1932 for the full regression model, as shown in cell B8. The three most important explanatory variables based on the rank order of R^2 shares (from least to greatest) are x_8 ('the wait for food service is excessive'), x_2 ('the entrées are delicious'), and x_{10} ('restaurant service personnel are courteous and friendly'). This is not surprising given their importance in all of the predictive models displayed in Figure 13.9.

The last three explanatory variables in the rank order of R^2 shares are x_4 ('parking at the restaurant is ample'), x_6 ('the noise level in the restaurant is not distracting to me'), and x_5 ('the lighting in the restaurant is appropriate'). It is not surprising that these three aesthetic items would have less explanatory value for recommendation behavior than the food quality items (x_1, x_2, x_3) and the service waiting time and quality items (x_7, x_8, x_9, x_{10}).

Once again, there is a critical practical observation that should not be overlooked. A variable's exclusion from a predictive model does not necessarily mean that it is less important than some variable that is included in the model. For example, although it is less important as an explanatory variable, the variable x_6 (and also x_4) was often a more important 'predictor' of recommendation behavior than x_1, x_3, x_7, and x_9 (notice that none of these four variables was selected for the four- or five-predictor subset). The reason is attributable to multicollinearity. For example, although x_1 and x_3 are good explanatory variables, because of their high correlation to x_2, they add little predictive benefit when x_2 is included in the model. The same holds true for x_7 and x_9 when x_8 is already included in the predictive model. However, x_6 and/or x_4 are clearly useful for augmenting the predictive model.

Chapter 14

Linear Discriminant Analysis

This chapter focuses on Fisher's (1936, 1938) method for linear discriminant analysis (LDA). Although other sophisticated classification methods have been proposed over the past several decades, LDA remains an important tool for both two-group and multigroup (i.e., three or more groups) classification problems.

We begin by describing the underlying optimization problem and mechanics associated with LDA. Although some presentations begin with a description of the special case of the two-group problem and then generalize to more than two groups, we have elected to work with the general formulation. Our presentation closely follows the structure used by Welling (2006).

Following the section on mechanics, we present a two-group example of LDA. In particular, we revisit the insurance brand example that we used to illustrate logistic regression in Chapter 13. In this section, we also address the simplified computations that are available for the two-group case. We finish the chapter with a multigroup example pertaining to the classic Fisher's iris dataset (Anderson, 1935; Fisher, 1936, 1938).

14.1 Linear Discriminant Analysis Mechanics

Our presentation of the LDA model and method uses the following notation:

$\mathbf{X} =$ an $n \times p$ matrix of measurements for n observations on p metric variables. The $1 \times p$ row vectors of $\mathbf{X}$ correspond to the

$p \times 1$ vectors denoted by $\mathbf{x}_i$ (for $1 \leq i \leq n$) in the remaining notation;

$g =$ the number of categories or groups;

$n_k =$ the number of observations in group k;

$\mathbf{y} =$ an $n \times 1$ vector of group measurements $(1, 2, \ldots, g)$ for n observations;

$\mathbf{a} =$ a $p \times 1$ 'projection' vector of discriminant coefficients;

$\boldsymbol{\mu}_k =$ a $p \times 1$ vector of variable means for group k (for $k = 1$ to g);

$\boldsymbol{\mu} =$ a $p \times 1$ vector of (grand) variable means across all groups;

$\mathbf{T} =$ the total sum-of-squares and cross-product matrix (or total scatter matrix); $\mathbf{T} = \sum_{i=1}^{n}(\mathbf{x}_i - \boldsymbol{\mu})(\mathbf{x}_i - \boldsymbol{\mu})^T$;

$\mathbf{H}_k =$ the within-group sum-of-squares and cross-product matrix (or scatter matrix) for group k; $\mathbf{H}_k = \sum_{i:y_i=k}(\mathbf{x}_i - \boldsymbol{\mu}_k)(\mathbf{x}_i - \boldsymbol{\mu}_k)^T$;

$\mathbf{W} =$ the within-group scatter matrix; $\mathbf{W} = \sum_{k=1}^{g}\mathbf{H}_k$;

$\mathbf{B} =$ the between-group scatter matrix; $\mathbf{B} = \sum_{k=1}^{g}n_k(\boldsymbol{\mu}_k - \boldsymbol{\mu})(\boldsymbol{\mu}_k - \boldsymbol{\mu})^T$.

The sum-of-squares decomposition among the $p \times p$ scatter matrices is given by the relationship $\mathbf{T} = \mathbf{B} + \mathbf{W}$. Fisher's LDA seeks a projection vector $\mathbf{a}$ that maximizes the ratio of between-category scatter to within-category scatter. That is, find $\mathbf{a}$ so as to

$$\text{maximize: } F(\mathbf{a}) = \frac{\mathbf{a}^T \mathbf{B} \mathbf{a}}{\mathbf{a}^T \mathbf{W} \mathbf{a}}. \qquad (14.1)$$

As noted by Welling (2004), a critical aspect of Equation (14.1) is that it is invariant with respect to $\mathbf{a}$ and, accordingly, LDA can be reframed as the following constrained optimization problem:

$$\text{minimize: } F_2(\mathbf{a}) = -\frac{1}{2}\mathbf{a}^T \mathbf{B} \mathbf{a}, \qquad (14.2)$$

$$\mathbf{a}^T \mathbf{W} \mathbf{a} = 1. \qquad (14.3)$$

The Lagrangian for the problem posed by Equations (14.2) and (14.3) is

$$L = -\frac{1}{2}\mathbf{a}^T \mathbf{B} \mathbf{a} + \frac{1}{2}\lambda(\mathbf{a}^T \mathbf{W} \mathbf{a} - 1). \qquad (14.4)$$

The first-order Karush–Kuhn–Tucker conditions yield

$$\mathbf{B}\mathbf{a} = \lambda \mathbf{W}\mathbf{a}, \quad \text{or,} \qquad (14.5)$$

$$\mathbf{W}^{-1}\mathbf{B}\mathbf{a} = \lambda \mathbf{a}, \qquad (14.6)$$

which is a generalized eigenproblem.

We consider two ways of tackling the generalized eigenproblem in Equation (14.6). One approach is to apply the power method directly to the asymmetric $\mathbf{W}^{-1}\mathbf{B}$ matrix, using Wielandt deflation to adjust the eigenvectors. The second approach is to capitalize on the fact that $\mathbf{B}$ is a positive definite matrix and make use of the fact that $\mathbf{B} = \mathbf{B}^{1/2}\mathbf{B}^{1/2}$. This process begins by adding a small positive constant to the main diagonal of $\mathbf{B}$, which is generally not full rank, and then obtaining its eigendecomposition:

$$\mathbf{B} = \mathbf{U}\mathbf{\Lambda}\mathbf{U}', \tag{14.7}$$

where $\mathbf{U}$ is a $p \times p$ matrix of eigenvectors for $\mathbf{B}$, and $\mathbf{\Lambda}$ is a diagonal matrix containing the eigenvalues of $\mathbf{B}$. This enables the computation of $\mathbf{B}^{1/2}$ as follows:

$$\mathbf{B}^{1/2} = \mathbf{U}\mathbf{\Lambda}^{1/2}\mathbf{U}'. \tag{14.8}$$

Setting $\mathbf{v} = \mathbf{B}\mathbf{a}$ allows Equation (14.6) to be rewritten as

$$\mathbf{B}^{1/2}\mathbf{W}^{-1}\mathbf{B}^{1/2}\mathbf{v} = \lambda\mathbf{v}, \tag{14.9}$$

Because $\mathbf{B}^{1/2}\mathbf{W}^{-1}\mathbf{B}^{1/2}$ is a positive definite symmetric matrix, it is now a standard eigenvalue problem for which we find the eigenvector $\mathbf{v}_1$ corresponding to the largest eigenvalue λ_1. We can then map $\mathbf{v}_1$ back to $\mathbf{a}_1$ via the following relationship:

$$\mathbf{a}_1 = \mathbf{B}^{-1/2}\mathbf{v}_1. \tag{14.10}$$

A maximum of $q = \min\{g - 1, p\}$ projections can be extracted from the leading eigenvalues and eigenvectors of (14.9).

14.2   Example 14.1 — Two-Group Insurance Data

Our first LDA example (Example 14.1) uses the same data that we used for Example 13.1, which was motivated by the lawnmower example from Johnson and Wichern (2007, p. 578). Measurements are available for $n = 24$ respondents on one binary dependent variable and $p = 2$ predictor variables. This classification problem has only $g = 2$ groups, where the dependent variable is $y_i = 1$ if respondent i uses car insurance brand X or $y_i = 0$ if the respondent does not use car insurance brand X. The first predictor (x_1) is annual income in thousands of dollars and the second predictor (x_2) is age in years. The measurements for the dependent variable and two independent variables are contained in cells B3:D26 of the 'Insurance 1' worksheet of the Chapter 14 — LDA workbook in Figure 14.1.

The first step is to obtain the grand means and group means for the predictor variables in cells C28:D28 and C30:D31, respectively. The differences between the grand means and the group means are located in cells C33:D34. The (grand-)mean-differenced variable measurements are in cells F3:G26. The total scatter matrix, $\mathbf{T}$, is computed in cells L5:M6 as the product of the transpose of F3:G26 and itself. The between-groups scatter matrix, $\mathbf{B}$, is computed in cells L2:M3 using the differences between the grand and group means in cells C33:D34 and the number of respondents in each group. The within-groups scatter matrix, $\mathbf{W}$, is computed in cells L8:M9 via $\mathbf{T} - \mathbf{B}$, that is, subtracting the elements in L2:L3 from their corresponding values in cells L5:M6. The pooled covariance matrix, $\mathbf{S}$, is obtained in cells P8:Q9 by dividing the elements in $\mathbf{W}$ (cells L8:M9) by $n - g$.

The $\mathbf{W}^{-1}\mathbf{B}$ matrix is located in cells L11:M12. Because $g = 2$, only one projection can be extracted for the lawnmower data.

Case #	y	X1	X2	Mean Differences X1	Mean Differences X2	Disc Scores	Classify
1	1	115.23	63.0	29.8	16.4	0.7123	1
2	1	75.45	54.0	-9.9	7.4	0.9128	1
3	1	63.16	36.0	-22.2	-10.6	-0.3833	0
4	1	92.22	66.0	6.8	19.4	1.5606	1
5	1	104.58	34.0	19.2	-12.6	-1.6068	0
6	1	78.14	68.0	-7.3	21.4	2.0943	1
7	1	66.85	61.0	-18.5	14.4	1.7545	1
8	1	72.44	45.0	-13.0	-1.6	0.1857	1
9	1	73.54	48.0	-11.9	1.4	0.4256	1
10	1	66.60	57.0	-18.8	10.4	1.4038	1
11	1	70.19	43.0	-15.2	-3.6	0.0640	1
12	1	83.22	55.0	-2.2	8.4	0.8060	1
13	0	56.64	45.0	-28.8	-1.6	0.5843	1
14	0	94.51	37.0	9.1	-9.6	-1.0851	0
15	0	112.98	37.0	27.6	-9.6	-1.5511	0
16	0	88.67	62.0	3.3	15.4	1.2932	1
17	0	75.30	47.0	-10.1	0.4	0.2920	1
18	0	119.55	34.0	34.2	-12.6	-1.9845	0
19	0	76.75	40.0	-8.6	-6.6	-0.3693	0
20	0	101.47	32.0	16.1	-14.6	-1.7068	0
21	0	82.60	38.0	-2.8	-8.6	-0.6953	0
22	0	85.35	36.0	0.0	-10.6	-0.9432	0
23	0	107.24	45.0	21.8	-1.6	-0.6923	0
24	0	86.89	35.0	1.5	-11.6	-1.0713	0
Grand Mean		85.4	46.6				
Group 1 Mean		80.1	52.5		2.6631		
Group 2 Mean		90.7	40.7		1.3415		
Grand - Group 1		5.3	-5.9				
Grand - Group 2		-5.3	5.9				

Matrices (right side of worksheet):

$$\mathbf{B} = \begin{bmatrix} 664.970 & -747.453 \\ -747.453 & 840.167 \end{bmatrix}$$

$$\mathbf{T} = \begin{bmatrix} 6962.829 & -930.793 \\ -930.793 & 2995.833 \end{bmatrix}$$

$$\mathbf{W} = \begin{bmatrix} 6297.659 & -183.340 \\ -183.340 & 2155.667 \end{bmatrix} \qquad \mathbf{S} = \begin{bmatrix} 288.257 & -8.334 \\ -8.334 & 97.985 \end{bmatrix}$$

$$\mathbf{W}^{-1}\mathbf{B} = \begin{bmatrix} 0.0957 & -0.1076 \\ -0.3386 & 0.3806 \end{bmatrix}$$

	Approx. u		Ru	Normal Ru	Eigenvalue
	0.0957	-0.1076	-0.2721	-0.1296	-0.2721	0.4763
	-0.3386	0.3806	0.9623	0.4584	0.9623	

0.4763

NORMALIZATION
u'Su
116.2846   -0.0252
            0.0892

Figure 14.1.   Excel worksheet for Example 14.1.

Therefore, standard eigendecomposition was used to extract the first eigenvalue and eigenvector using the Rayleigh coefficient and the power method. This approach will be described more fully in Chapter 15. A brief explanation is provided here. We begin with an initial approximation for **a** in cells O16:O17. The product of $\mathbf{R} = \mathbf{W}^{-1}\mathbf{B}$ and **a** is obtained in cells P16:P17. The square root of the sum of the squared values in cells P16:P17 is computed in cell P19 and this is used to normalize the vector in P16:P17 to unit length in cells Q16:Q17. The eigenvalue is estimated in cell S16 as the ratio of $\mathbf{aRa}/(\mathbf{a}'\mathbf{a})$. Estimation is accomplished by successively copying the values in cells Q16:Q17 and pasting them (using 'paste special' — values only in Excel) over O16:O17 until convergence. For this 2×2 matrix and starting with values of 1 in cells O16:O17, convergence is extremely rapid.

We follow standard practice and normalize the eigenvectors based on the relationship $\mathbf{a}'\mathbf{S}\mathbf{a} = 1$. In cell M22, we compute $\mathbf{a}'\mathbf{S}\mathbf{a}$ and then divide the eigenvector estimate in cells O16:O17 by the square root of this value to obtain the normalized coefficients in O22:O23. These coefficients are then multiplied by the group means in cells C30:D31 to get the discriminant scores at the group means, which are displayed in cells H30:H31. The assignment of individual cases to groups is accomplished in cells H3:I26 by computing the discriminant score for the case and assigning it to the group associated with its nearest group mean discriminant score. Specifically, discriminant scores for the cases are computed in cells H3:H26 by multiplying the predictor variables in C3:D26 by the coefficients in O22:O23 and subtracting the midpoint between the two group discriminant scores. With this approach, cases are classified in cells I3:I26 simply based on the sign of the discriminant score (positive = group 1 and negative = group 0).

The results in cells I3:I26 of Figure 14.1 reveal that 19 of the 24 (79.2%) respondents are classified correctly. Two users of insurance brand X (respondents 3 and 5) are misclassified as non-users and three of the non-users (cases 13, 16, and 17) are misclassified as users. Figure 14.2 provides a visual representation of the projection. The row axis is income (in thousands of $) and the column axis is age. The diamonds indicate the coordinates of the users of brand X and the squares indicate the coordinates of the non-users. The two diamonds below the line and three squares above the line are the misclassified cases.

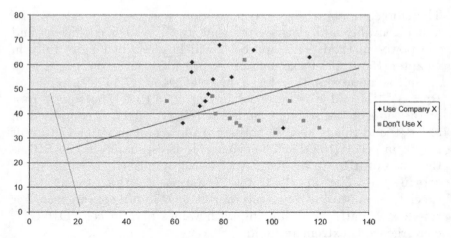

Figure 14.2. Discriminant function for Example 14.2.

At this juncture, we note that the establishment of the LDA classifier in cells O22:O23 can be obtained more efficiently for the two-group case. In particular, for $g = 2$, it is more common to specify **B** using sum-of-squares differences between the means of the two groups rather than differences between group means and grand means. That is, $\mathbf{B} = (\boldsymbol{\mu}_1 - \boldsymbol{\mu}_2)(\boldsymbol{\mu}_1 - \boldsymbol{\mu}_2)'$. The solution of the eigenproblem is no longer necessary and the discriminant function coefficients can be obtained via $\mathbf{a} = \mathbf{W}^{-1}(\boldsymbol{\mu}_1 - \boldsymbol{\mu}_2)$ or, more commonly, $\mathbf{a} = \mathbf{S}^{-1}(\boldsymbol{\mu}_1 - \boldsymbol{\mu}_2)$.

The worksheet 'Insurance2' displayed in Figure 14.3 shows how the same results can be obtained without computing the lead eigenvalue and eigenvector of $\mathbf{W}^{-1}\mathbf{B}$. The mean vectors for the users and non-users of brand X are computed in cells A17:B17 and D17:E17, respectively. The mean difference vector computed from A17–D17 and B17–E17 is located in cells A25:A26.

The mean-differenced data for users of brand X are located in cells G4:H15 and the mean-differenced data for non-users are in cells J4:K15. From the mean-corrected data, sample covariance matrices for the user and non-user groups are obtained in cells G18:H19 and J18:K19, respectively. The pooled covariance matrix is computed in cells J22:K23 and the inverse of this matrix is obtained using the MINVERSE function in cells J25:K26. The unstandardized discriminant function is computed

	A	B	C	D	E	F	G	H	I	J	K
1	Discriminant Analysis - Two Group Example						Mean differenced data				
2	Use Company X			Don't Use X			Use Company X			Don't Use X	
3	x1	x2		x1	x2		x1	x2		x1	x2
4	115.2	63.0		56.6	45.0		35.0950	10.5000		-34.0225	4.3333
5	75.5	54.0		94.5	37.0		-4.6850	1.5000		3.8475	-3.6667
6	63.2	36.0		113.0	37.0		-16.9750	-16.5000		22.3175	-3.6667
7	92.2	66.0		88.7	62.0		12.0850	13.5000		-1.9925	21.3333
8	104.6	34.0		75.3	47.0		24.4450	-18.5000		-15.3625	6.3333
9	78.1	68.0		119.6	34.0		-1.9950	15.5000		28.8875	-6.6667
10	66.9	61.0		76.8	40.0		-13.2850	8.5000		-13.9125	-0.6667
11	72.4	45.0		101.5	32.0		-7.6950	-7.5000		10.8075	-8.6667
12	73.5	48.0		82.6	38.0		-6.5950	-4.5000		-8.0625	-2.6667
13	66.6	57.0		85.4	36.0		-13.5350	4.5000		-5.3125	-4.6667
14	70.2	43.0		107.2	45.0		-9.9450	-9.5000		16.5775	4.3333
15	83.2	55.0		86.9	35.0		3.0850	2.5000		-3.7725	-5.6667
16											
17	80.135	52.5		90.6625	40.66667						
18	↑			↑			260.0144	30.6636		312.5000	-47.3309
19	x_1	These are vectors of variable means		x_2			30.6636364	128.636364	S_2	-47.330909	67.333333
20							S_1				
21											
22	Group Mean						S = pooled covariance matrix			286.257	-8.334
23	Difference									-8.334	97.985
24	Vector										
25	-10.5275		Scores at Centroids				S^{-1} = inverse of pooled matrix			.003502	.000298
26	11.8333		Group 1	Group 2						.000298	.010231
27			2.663078101	1.3415116							
28	-.025230		Midpoint	2.0022949			Discriminant function			-.033343	
29	.089236		0.660783232	-0.660783			$a = S^{-1}(x_1-x_2)$			.117931	
30	↑										
31	Normalized Discriminant Function						Normalization index a'Sa = 1			1.74653792	

Figure 14.3. Alternative LDA worksheet for the two-group example.

in cells J28:J29 by multiplying the inverse of the pooled covariance matrix in cells J25:K26 by the mean difference vector in cells A25:A26.

In cell J31, we compute **a'Sa** and then divide the unstandardized coefficients in cells J28:J29 by the square root of this value to obtain the standardized coefficients in A28:A29. Using these standardized coefficients, the discriminant scores at the group means for users and non-users are computed in cells C27 and D27, respectively. The midpoint of the discriminant scores is computed in cell D28. Discriminant scores for each respondent are computed by multiplying the standardized coefficients by the variable measurements for the respondent and subtracting the midpoint. Respondents with positive scores are classified as users and respondents with negative scores are classified as non-users. The discriminant scores are computed in cells C42:C65 (not shown in Figure 14.3), but they match the scores that are displayed in Figure 14.1.

14.3   Example 14.2 — Iris Data

Our second example is a classic multigroup classification problem. The data were originally collected by Anderson (1935) and subsequently analyzed by Fisher (1936, 1938) in his pioneering development of LDA. Specifically, the data were obtained from 50 samples of each of $g = 3$ species of iris flowers (Setosa, Versicolor, and Virginica), resulting in a total of $n = 150$ plants. Each plant was measured on $p = 4$ variables: (a) sepal length; (b) sepal width; (c) petal length; and (d) petal width. These measurements are located in cells B12:E161 of Figure 14.4.

14.3.1   *Analysis using the* $\mathbf{B} = \mathbf{B}^{1/2}\mathbf{B}^{1/2}$ *relationship*

As in Example 14.1, our first step was to compute the grand means (cells B163:E163), group means (cells B165:E167), and differences between grand means and group means (cells B169:E171). Next, we

		Sepal Length	Sepal Width	Petal Length	Petal Width										

Fisher's Linear Discriminant Analysis
Applied to the 3-Group Iris Data
 - Group 1 - Setosa (cases 1-50)
 - Group 2 - Versicolor (cases 51-100)
 - Group 3 - Virginica (cases 101-150)

Mean Centered Data used to compute the total sum of squares and cross-products matrix, T.

Computation of Relevant Input Matrices T and B computed directly. W = T - B.
$S_{pool} = (1/147)(W)$

Case #	X1	X2	X3	X4		X1	X2	X3	X4						
1	5.1	3.5	1.4	0.2		-0.7433	0.4427	-2.3580	-0.9993	B =	63.2121	-19.9527	165.2484	71.2793
2	4.9	3	1.4	0.2		-0.9433	-0.0573	-2.3580	-0.9993		-19.9527	11.3449	-57.2396	-22.9327
3	4.7	3.2	1.3	0.2		-1.1433	0.1427	-2.4580	-0.9993		165.2484	-57.2396	437.1028	186.7740
4	4.6	3.1	1.5	0.2		-1.2433	0.0427	-2.2580	-0.9993		71.2793	-22.9327	186.7740	80.4133
5	5	3.6	1.4	0.2		-0.8433	0.5427	-2.3580	-0.9993					
6	5.4	3.9	1.7	0.4		-0.4433	0.8427	-2.0580	-0.7993	T =	102.1683	-6.3227	189.8730	76.9243
7	4.6	3.4	1.4	0.3		-1.2433	0.3427	-2.3580	-0.8993		-6.3227	28.3069	-49.1188	-18.1243
8	5	3.4	1.5	0.2		-0.8433	0.3427	-2.2580	-0.9993		189.8730	-49.1188	464.3254	193.0458
9	4.4	2.9	1.4	0.2		-1.4433	-0.1573	-2.3580	-0.9993		76.9243	-18.1243	193.0458	86.5699
10	4.9	3.1	1.5	0.1		-0.9433	0.0427	-2.2580	-1.0993					
11	5.4	3.7	1.5	0.2		-0.4433	0.6427	-2.2580	-0.9993	W =	38.9562	13.6300	24.6246	5.6450
12	4.8	3.4	1.6	0.2		-1.0433	0.3427	-2.1580	-0.9993		13.6300	16.9620	8.1208	4.8084
13	4.8	3	1.4	0.1		-1.0433	-0.0573	-2.3580	-1.0993		24.6246	8.1208	27.2226	6.2718
14	4.3	3	1.1	0.1		-1.5433	-0.0573	-2.6580	-1.0993		5.6450	4.8084	6.2718	6.1566
15	5.8	4	1.2	0.2		-0.0433	0.9427	-2.5580	-0.9993					
16	5.7	4.4	1.5	0.4		0.1433	1.3427	-2.2580	-0.7993	W⁻¹ =	.0738	-.0366	-.0612	.0232
17	5.4	3.9	1.3	0.4		-0.4433	0.8427	-2.4580	-0.7993		-.0366	.0968	.0182	-.0606
18	5.1	3.5	1.4	0.3		-0.7433	0.4427	-2.3580	-0.8993		-.0612	.0182	.1006	-.0606
19	5.7	3.8	1.7	0.3		-0.1433	0.7427	-2.0580	-0.8993		.0232	-.0606	-.0606	.2502
20	5.1	3.8	1.5	0.3		-0.7433	0.7427	-2.2580	-0.8993					
21	5.4	3.4	1.7	0.2		-0.4433	0.3427	-2.0580	-0.9993	W⁻¹B =	-3.0584	1.0814	-8.1119	-3.4586
22	5.1	3.7	1.5	0.4		-0.7433	0.6427	-2.2580	-0.7993		-5.5616	2.1782	-14.9646	-6.3077
23	4.6	3.6	1	0.2		-1.2433	0.5427	-2.7580	-0.9993		8.0074	-2.9427	21.5116	9.1421
24	5.1	3.3	1.7	0.5		-0.7433	0.2427	-2.0580	-0.6993		10.4971	-3.4199	27.5485	11.8459
25	4.8	3.4	1.9	0.2		-1.0433	0.3427	-1.8580	-0.9993					
26	5	3	1.6	0.2		-0.8433	-0.0573	-2.1580	-0.9993	S_pool =	.2650	.0927	.1675	.0384
27	5	3.4	1.6	0.4		-0.8433	0.3427	-2.1580	-0.7993		.0927	.1154	.0552	.0327
28	5.2	3.5	1.5	0.2		-0.6433	0.4427	-2.2580	-0.9993		.1675	.0552	.1852	.0427
29	5.2	3.4	1.4	0.2		-0.6433	0.3427	-2.3580	-0.9993		.0384	.0327	.0427	.0419

Figure 14.4.   Excel worksheet for Example 14.2 — part 1.

obtained the grand-mean-corrected measurements of the four variables in cells G12:J161.

The total scatter matrix, $\mathbf{T}$, is computed in cells M17:P20 as the product of the transpose of G12:G161 and itself. The between-groups scatter matrix, $\mathbf{B}$, is computed in cells M12:P15 using the differences between the grand and group means in cells B169:E171 and the number of plants in each group. The within-groups scatter matrix, $\mathbf{W}$, is computed in cells M22:P25 via $\mathbf{T} - \mathbf{B}$, that is, subtracting the elements in M12:P15 from their corresponding values in cells M17:P20. The matrix $\mathbf{W}^{-1}$ is located in cells M27:M30 and was obtained from $\mathbf{W}$ using the MINVERSE command. The $\mathbf{W}^{-1}\mathbf{B}$ matrix is displayed in cells M32:P35 and we see that it is asymmetric. The pooled covariance matrix, $\mathbf{S}$, is obtained in cells M37:P40 by dividing the elements in $\mathbf{W}$ (cells L8:M9) by $n - g$.

In Figure 14.5, we implement the process described in Equations (14.7)–(14.9). The matrix in cells R17:U20 is obtained by adding the small positive constants from the matrix in cells R12:U15 to the main diagonal of $\mathbf{B}$ in cells M12:P15. The first eigenvalue and eigenvector of this slightly augmented version of $\mathbf{B}$ are

Eigendecomposion of $\mathbf{W}^{-1}\mathbf{B}$ Using the Decomposition of $\mathbf{B}$ into $\mathbf{B}^{1/2}\mathbf{B}^{1/2}$

Thus, the first step is a spectral decomposition of $\mathbf{B}$. Because $\mathbf{B}$ is NOT full rank (it is singular), we add error perturbation to main diagonal of .00001. The results matched those obtained by applying the power method with Wielandt deflation directly to $\mathbf{W}^{-1}\mathbf{B}$.

E =					Approx. u	Ru	Normal Ru	Eigenvalues		U =	.3267	.3312	.8852	.0000
.00001											.1118	.8885	-.2912	.3366
	.00001										.8628	-.1336	-.2685	.4069
		.00001									.3692	.2882	-.2441	-.8492
			.00001											
63.21214	-19.95267	165.24840	71.27933		0.3267	191.7781	0.3267	587.00026	$\Lambda^{1/2} =$		24.2281	.0000	.0000	.0000
-19.95267	11.34494	-57.23960	-22.93267		-0.1118	-65.6413	-0.1118				.0000	2.2523	.0000	.0000
165.24840	-57.23960	437.10281	186.77400		0.8628	506.4843	0.8628				.0000	.0000	.0032	.0000
71.27933	-22.93267	186.77400	80.41334		0.3692	216.6918	0.3692				.0000	.0000	.0000	.0032
						587.0003			$\mathbf{B}^{1/2} =$		2.8357	-.2231	6.7294	3.1363
											-.2231	2.0816	-2.6043	-.4241
.55657	1.49292	-.22443	.48423		0.3312	1.6803	0.3312	5.07296			6.7294	-2.6043	18.0784	7.6295
1.49292	4.00461	-.60200	1.29890		0.8885	4.5072	0.8885				3.1363	-.4241	7.6295	3.4911
-.22443	-.60200	.09050	-.19526		-0.1336	-0.6776	-0.1336							
.48423	1.29890	-.19526	.42131		0.2882	1.4619	0.2882		$\mathbf{B}^{1/2}\mathbf{W}^{-1}\mathbf{B}^{1/2} =$		3.2126	-1.3745	8.7624	3.6554
											-1.3745	.8941	-4.0567	-1.5898
						5.0730					8.7624	-4.0567	24.2092	9.9962
											3.6554	-1.5898	9.9962	4.1614
.00001	.00000	.00000	.00000		0.8852	0.0000	0.8852	.00001						
.00000	.00000	.00000	.00000		-0.2912	0.0000	-0.2912		$\mathbf{B}^{1/2} =$		247.8333	-81.3805	-75.1616	-68.2758
.00000	.00000	.00000	.00000		-0.2685	0.0000	-0.2685				-81.3805	62.9938	67.9813	-67.8021
.00000	.00000	.00000	.00001		-0.2441	0.0000	-0.2441				-75.1616	67.9813	75.1988	-88.5561
											-68.2758	-67.8021	-88.5561	246.9148
						0.0000								
									$\mathbf{B} - \mathbf{B}^{1/2}\mathbf{B}^{1/2} =$		63.2121	-19.9527	165.2484	71.2793
.00000	.00000	.00000	.00000		0.0000	0.0000	0.0000	.00001			-19.9527	11.3449	-57.2396	-22.9327
.00000	.00000	.00000	.00000		0.3366	0.0000	0.3366				165.2484	-57.2396	437.1028	186.7740
.00000	.00000	.00000	.00000		0.4069	0.0000	0.4069				71.2793	-22.9327	186.7740	80.4133
.00000	.00000	.00000	.00001		-0.8492	0.0000	-0.8492		The above matrix computation is unnecessary. It is just a check.					

Figure 14.5. Excel worksheet for Example 14.2 — part 2.

extracted in cells W17:AA22 using the power method as described in Example 14.1. Matrix $\mathbf{B}$ is then deflated by removing the variation explained by the first eigenvalue/eigenvector, and the updated matrix is displayed in cells R24:U27. The power method is then applied to this matrix to extract the second eigenvalue/eigenvector. This process is then repeated two more times so that we obtain $p = 4$ eigenvalues and eigenvectors.

The four eigenvectors are copied and pasted (values only) into cells AD12:AG15 to create the matrix $\mathbf{U}$. The square roots of the eigenvalues are placed in cells AD17, AE18, AF19, and AG20 to create the diagonal matrix $\mathbf{\Lambda}^{1/2}$. Next, we use Equation (14.8) to create $\mathbf{B}^{1/2} = \mathbf{U}\mathbf{\Lambda}^{1/2}\mathbf{U}'$ in cells AD22:AG25 via the MMULT and TRANSPOSE functions. We are then able to compute the matrix $\mathbf{B}^{1/2}\mathbf{W}^{-1}\mathbf{B}^{1/2}$ in cells AD27:AG30 using Equation (14.9) via the MMULT function. We also use the MINVERSE function to obtain $\mathbf{B}^{-1/2}$ in cells AD32:AG35 from $\mathbf{B}^{1/2}$ in cells AD22:AD25.

Figure 14.6 presents the final stages of the process, which begin with the extraction of the $g - 1 = 2$ eigenvalues and eigenvectors from the positive definite symmetric matrix $\mathbf{B}^{1/2}\mathbf{W}^{-1}\mathbf{B}^{1/2}$ that has been copied into cells AJ17:AM20. (Because $g = 3$, a maximum of two discriminant functions can be extracted.) The first

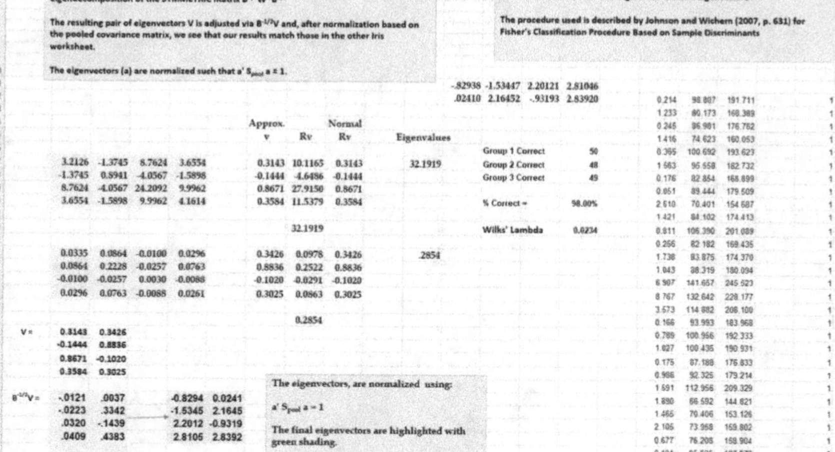

Figure 14.6.   Excel worksheet for Example 14.2 — part 3.

eigenvalue and eigenvector are obtained in cells AO17:AS22 using the power method and the matrix is subsequently deflated in cells AJ24:AM27. The power method is then reapplied to extract the second eigenvalue/eigenvector.

The eigenvectors in cells AQ17:AQ20 and AQ24:AQ27 were copied into cells AJ30:AK33. This matrix was then multiplied by $\mathbf{B}^{-1/2}$ in cells AD32:AG35 to obtain the eigenvector matrix $\mathbf{A} = [\mathbf{a}_1\ \mathbf{a}_2]$ in cells AJ35:AK38, which are in the original space in accordance with Equation (14.10). Cells AJ40 and AK40 are obtained by pre- and post-multiplying the pooled covariance matrix $\mathbf{S}$ (cells M37:P40) by the vectors in cells AJ35:AJ38 and AK35:AK38, respectively. The square roots of the values in the AJ40 and AK40 are then used to create the normalized eigenvectors in cells AM35:AN38 in accordance with $\mathbf{a}'\mathbf{Sa}$. The transpose of the normalized eigenvectors is copied into cells AT11:AW12, where it will be used to establish the classifiers for the cases.

We used the classification rule for Fisher's LDA described by Johnson and Wichern (2007, p. 631) to assign the plants to groups. Specifically, by defining γ as the number of discriminant functions extracted, which in this example is $\min\{p, g - 1\} = 2$, case i is assigned to group k if

$$\sum_{l=1}^{\gamma}[\mathbf{a}_l(\mathbf{x}_i - \boldsymbol{\mu}_k)]^2 \leq \sum_{l=1}^{\gamma}[\mathbf{a}_l(\mathbf{x}_i - \boldsymbol{\mu}_h)]^2 \quad \text{for } 1 \leq h \neq k \leq g. \quad (14.11)$$

For this example, the value of $\sum_{l=1}^{\gamma}[\mathbf{a}_l(\mathbf{x}_i - \boldsymbol{\mu}_k)]^2$ is computed for each plant $1 \leq i \leq 150$ and each group $1 \leq k \leq 3$ in cells AY12:BA161 of Figure 14.6. Cells BC12:BC161 use IF statements to assign the plant to the group for which the distance measure is minimized.

Cells AW16:AW18 count the number of correctly classified plants (out of 50 possible) for each group. All 50 Setosa plants were correctly classified. Forty-eight of the 50 Versicolor plants were correctly classified and the two that were misclassified were incorrectly assigned to the Virginica group. Forty-nine of the 50 Virginica plants were correctly classified and the one that was misclassified was incorrectly assigned to the Versicolor group. Cell AW20 sums the number of correct classifications in AW16:AW18 and divides that number by $n = 150$, revealing that 98% of the plants were correctly classified.

For good measure, we also report the value of Wilks's lambda in cell AW22, which is computed as follows:

$$\text{Wilks}'\lambda = \frac{|\mathbf{W}|}{|\mathbf{T}|}. \qquad (14.12)$$

This ratio of the determinant of $\mathbf{W}$ to the determinant of $\mathbf{T}$ is computed using the MDETERM function in Excel. The range of possible values for Wilks's lambda is 0 to 1, with values near zero suggesting greater classificatory efficacy. The obtained value of 0.0234 is very good.

14.3.2   *Analysis using* $\mathbf{W}^{-1}\mathbf{B}$ *directly with Wielandt deflation*

In this section, we repeat the analyses in Section 14.3.1, but, instead of the approach that uses the decomposition of $\mathbf{B}$, we apply the power method with Wielandt deflation directly to the asymmetric $\mathbf{W}^{-1}\mathbf{B}$ matrix. The worksheet for this approach differs only in the method and, therefore, we show only the portion of the worksheet that implements the Wielandt deflation in Figure 14.7.

The input for this approach is the asymmetric $\mathbf{W}^{-1}\mathbf{B}$ matrix in cells R16:U19 in Figure 14.7. We then apply the power method to this matrix just as we did to the positive definite symmetric matrix $\mathbf{B}^{1/2}\mathbf{W}^{-1}\mathbf{B}^{1/2}$ in Figure (14.6) and find that the lead eigenvalue is the same but the eigenvector is different. However, if the lead eigenvector for $\mathbf{B}^{1/2}\mathbf{W}^{-1}\mathbf{B}^{1/2}$ is multiplied by $\mathbf{B}^{-1/2}$ and scaled to unit length, then it is identical to the lead eigenvector we find in cells Y16:Y19 of Figure 14.7.

Removing the variation explained by the lead eigenvalue and eigenvector for $\mathbf{W}^{-1}\mathbf{B}$ results in the matrix found cells R24:R27. When we apply the power method to this matrix, we find that the second eigenvalue is also the same as the second eigenvalue for $\mathbf{B}^{1/2}\mathbf{W}^{-1}\mathbf{B}^{1/2}$ in Figure 14.6 but the eigenvector is different. Moreover, multiplying the second eigenvector for $\mathbf{B}^{1/2}\mathbf{W}^{-1}\mathbf{B}^{1/2}$ by $\mathbf{B}^{-1/2}$ and scaling to unit length does *not* yield the second eigenvector displayed for the $\mathbf{W}^{-1}\mathbf{B}$ matrix in cells Y24:Y27. Wielandt deflation is necessary to correct the shifting of the second eigenvector of the $\mathbf{W}^{-1}\mathbf{B}$ matrix, which stems from the application of the power rule to an asymmetric matrix.

	R	S	T	U	V	W	X	Y	Z	AA	AB	AC	AD	AE
11											-.8294	-1.5345	2.2012	2.8105
12											-.0241	-2.1645	.9319	-2.8392
13	Compute Lead Eigenvalue/Eigenvector				Approx.			Normal						
14						u	Ru	Ru		Eigenvalues				
15													Group 1 Correct	50
16	-3.0584	1.0814	-8.1119	-3.4586		-.2087	-6.7198	-.2087		32.1919			Group 2 Correct	48
17	-5.5616	2.1782	-14.9646	-6.3077		-.3862	-12.4326	-.3862					Group 3 Correct	49
18	8.0774	-2.9427	21.5116	9.1421		.5540	17.8347	.5540						
19	10.4971	-3.4199	27.5485	11.8459		.7074	22.7710	.7074					% Correct =	98.00%
20														
21							32.1919						Wilks' Lambda	0.0234
22	Compute Second Eigenvalue/Eigenvector*													
23														
24	0.0000	0.0000	0.0000	0.0000		.0000	.0000	.0000		.2854				
25	0.0968	0.1775	0.0437	0.0913		.5711	.1630	.5711						
26	-0.0396	-0.0727	-0.0179	-0.0374		-.2338	-.0667	-.2338						
27	0.1334	0.2446	0.0602	0.1258		.7869	.2246	.7869						
28														
29							.2854							
30														
31	.45313	-.16092	1.20717	.51470										
32	-0.209	-0.386	0.554	0.7074										
33	1.000				.0000	.4551	-.2072	-.0065						
34					.5711	-.1609	-18.6052	-.5866						
35	Using Wielandt deflation here to				-.2338	1.2072	8.0104	.2526						
36	compute the second eigenvalue				.7869	.5147	-24.4043	-.7695						
37	and eigenvector.													
38							31.7165	1.0000						
39														
40														
41										The eigenvectors are normalized using:				
42	-.2087	-.0065		-0.8294	-0.0241		-.8294	-.0241						
43	-.3862	-.5866		-1.5345	-2.1645		-1.5345	-2.1645		a' S_pool a = 1				
44	.5540	.2526		2.2012	0.9319		2.2012	.9319						
45	.7074	-.7695		2.8105	-2.8392		2.8105	-2.8392		The final eigenvectors are highlighted with green shading.				
46														
47	.0633	.0734												

Figure 14.7.   Excel worksheet for Example 14.2 using Wielandt deflation.

The first step of the deflation process is to define η' as the $1 \times p$ vector corresponding to the first row of $\mathbf{W}^{-1}\mathbf{B}$, λ_1 and λ_2 as the largest (lead) and second-largest eigenvalues, respectively, $\mathbf{a}_1$ and $\mathbf{a}_2$ as the eigenvectors corresponding to the two largest eigenvalues, and $\mathbf{a}_1(1)$ as the first element of the lead eigenvector $\mathbf{a}_1$. We begin by computing the following:

$$\omega = \frac{1}{\lambda_1 \mathbf{a}_1(1)}\eta. \tag{14.13}$$

The ω' vector is computed in cells R31:U31 of Figure 14.7 and ω is subsequently stored in cells W33:W36. The next step is to adjust the second eigenvector, which was copied into cells V33:V36 for convenience. The update is accomplished as follows:

$$\delta = (\lambda_2 - \lambda_1)\mathbf{a}_2 + \lambda_1 \mathbf{a}_1(1)\sum_{j=1}^{p}\mathbf{a}_2(j)\omega(j), \tag{14.14}$$

where δ will replace $\mathbf{a}_2$ as the second eigenvector.

The computation of δ is implemented in cells X33:X36 using the following formula in cell X33 and copying through X36:

= (AA\$24–AA\$16)*V33 + AA\$16*Y16*SUMPRODUCT(V\$33: V\$36,W\$33:W\$36).

The vector δ is normalized to unit length in cells Y33:Y36. The first eigenvector is pasted into cells R42:R45 and the adjusted second eigenvector $a_2 = \delta$ is pasted into cells S42:S45. Cells R47:S47 contain $a_1 Sa_1$ and $a_2 Sa_2$, respectively. The square roots of the values in these cells are used to obtain the normalized eigenvectors in cells U42:U45. Except for the arbitrary reversal of the signs on the second eigenvector, these eigenvectors match those found in Figure 14.6. The subsequent classification results are identical.

Chapter 15

Factor Analysis

Factor analysis encompasses a broad range of methods that seek to reduce the dimensionality of data. That is, rather than dealing with a large number of variables, the goal is to identify a small number of factors that explain most of the variation in the dataset. One important basis for categorizing factor analysis is whether it is confirmatory or exploratory.

Exploratory factor analysis methods do not typically prespecify the number of factors or which variables are associated with each factor. Principal component analysis (PCA) is the foundation for exploratory factor analysis and is covered in Section 15.1. The underlying optimization problem is to identify a small number of components (factors) that explain the most variation in the data. Several important aspects of the process are described, including (i) the mechanics of extracting eigenvalues and eigenvectors from the correlation matrix, (ii) choosing the number of components, and (iii) rotating the component loadings to improve interpretability.

Confirmatory factor analysis prespecifies the number of factors and the variables that can load on each factor. The underlying optimization is based on maximum likelihood. Confirmatory factor analysis examples are covered in Section 15.2.

15.1   Principal Component Analysis

15.1.1   *Overview of Example 15.1*

Principal component analysis (PCA) is a well-known multivariate statistical technique that is used for data reduction; an excellent treatment of the topic is provided by Jolliffe (2002). Given a data matrix, $\mathbf{X} = [x_{ij}]$, consisting of measurements for n observations on p variables, the goal of PCA is to select a few linear combinations (i.e., components) of the variables that explain most of the variation in the full data matrix. PCA is closely related to several other multivariate techniques, such as singular value decomposition (Eckart & Young, 1936), exploratory factor analysis (Spearman, 1904), correspondence analysis (Greenacre, 1984), and biplots (Gower & Hand, 1996).

PCA Example 15.1 draws heavily from Brusco (2018). The Excel workbook for PCA consists of three worksheets used for Example 15.1. The first worksheet, which is described in Section 15.1.2, is used to obtain the correlation matrix from the raw data. The second worksheet, explained in Section 15.1.3, is used to extract the eigenvectors from the correlation matrix. This is accomplished using the power method for finding the dominant eigenvector and corresponding eigenvalue of a matrix. Approaches for choosing the number of eigenvectors to retain are also described in this section. The selected eigenvectors and their corresponding eigenvalues are used to compute the principal component loadings (correlations between each variable and each component), and principal component scores for the respondents in the sample are also computed. The third worksheet, described in Section 15.1.4, is provided for the rotation of the component loadings to improve their interpretability.

15.1.2   *Obtaining the correlation matrix from the raw data*

The first worksheet 'HSM_Data' of the Excel workbook Chapter 15 — PCA is shown in Figure 15.1. The worksheet contains the raw data in the form of an $n \times p$ data matrix, $\mathbf{X} = [x_{ij}]$. In the example, the data are 7-point Likert scale measurements for $n = 30$ respondents for each of $p = 6$ variables (or scale item statements) pertaining to hedonic shopping motivations. More specifically, the measurements

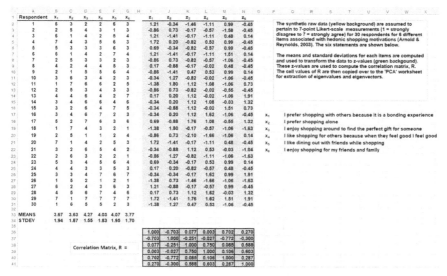

Figure 15.1. Excel worksheet for Example 15.1 — obtaining correlation matrix.

correspond to each respondent's level of agreement with each of the six statements in Figure 15.1, where the measurements range from $1 =$ strongly disagree to $7 =$ strongly agree. The raw data occupy cells B2:G31 and are shown in Figure 15.1. Although the data are synthetic, they are based on actual constructs and questionnaire items associated with an extensive study of hedonic shopping motivations conducted by Arnold and Reynolds (2003). The means $(\bar{x}_j)$ and standard deviations (s_j) of each of the variables $(1 \leq j \leq p)$ are computed (they are displayed in cells B33:G34 in the worksheet). The means and standard deviations are used to transform the raw data to an $n \times p$ matrix of z-scores, $\mathbf{Z} = [z_{ij}]$, which is accomplished as follows:

$$z_{ij} = (x_{ij} - \bar{x}_j)/s_j, \quad \forall \; 1 \leq i \leq n \quad \text{and} \quad 1 \leq j \leq p. \tag{15.1}$$

The z-scores, which occupy cells I2:N31 in Figure 15.1, are then used to compute the correlation matrix, $\mathbf{R}$, as follows:

$$\mathbf{R} = (1/(n-1))\mathbf{Z}^{\mathrm{T}}\mathbf{Z}. \tag{15.2}$$

The correlation matrix (contained in cells I36:N41 in Figure 15.1) is computed with the aid of the MMULT function based on the z-score

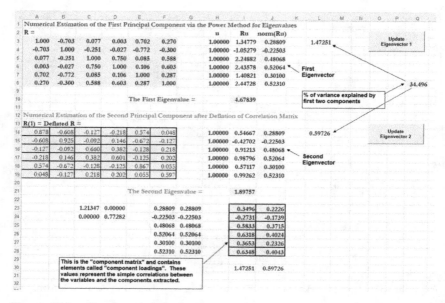

Figure 15.2.   Excel worksheet for Example 15.1 — extract components part 1.

columns. The correlation matrix could have been obtained directly using the Data Analysis Toolpack capabilities of Excel. However, the use of the z-scores and MMULT function to compute $\mathbf{R}$ is concordant with the goals of providing a thorough and detailed presentation of all of the computation aspects of PCA. The correlation matrix $\mathbf{R}$ is copied (cell values only, not the formulas) to the top left corner of the second worksheet 'extract'.

15.1.3   *Extraction of the principal components*

In the 'extract' worksheet displayed in Figure 15.2, the correlation matrix $\mathbf{R}$ is in cells A3:F8. The $p \times p$ correlation matrix is a real symmetric positive semidefinite matrix and, therefore, has real non-negative eigenvalues that sum to the trace of $\mathbf{R}$. Because the main diagonal elements of the correlation matrix are all one, the trace of $\mathbf{R}$ is equal to the number of variables, p. Moreover, based on the principles of eigendecomposition, also known as spectral decomposition, the correlation matrix can be written as a function of its eigenvalues $(\lambda_1, \ldots, \lambda_p)$ and eigenvectors $(\mathbf{u}_1, \ldots, \mathbf{u}_p)$ as follows:

$$\mathbf{R} = \lambda_1 \mathbf{u}_1 \mathbf{u}_1^T + \lambda_2 \mathbf{u}_2 \mathbf{u}_2^T + \cdots + \lambda_p \mathbf{u}_p \mathbf{u}_p^T = \mathbf{U}\boldsymbol{\Lambda}\mathbf{U}^T, \qquad (15.3)$$

where Λ is a $p \times p$ diagonal matrix containing the eigenvalues of $\mathbf{R}$ and $\mathbf{U} = [\mathbf{u}_1, \ldots, \mathbf{u}_p]$ is the $p \times p$ matrix of corresponding eigenvectors. Without loss of generality, we will assume that the eigenvalues are sequenced in non-increasing order of magnitude (i.e., $\lambda_1 \geq \lambda_2, \ldots, \geq \lambda_{p-1} \geq \lambda_p$). The eigenvalues are measures of the explained variation in the correlation matrix and, accordingly, the eigenvalue–eigenvector pairs associated with the largest eigenvalues are those that make the greatest contribution to the decomposition of $\mathbf{R}$.

Rather than a complete decomposition of $\mathbf{R}$ as in Equation (15.3), in PCA we would like a low-rank ($q << p$) approximation that explains the greatest amount of variation in $\mathbf{R}$. The underlying optimization problem associated with the first principal component, $\mathbf{u}$, is as follows:

$$\text{Maximize: } \mathbf{u}^{\mathrm{T}}\mathbf{R}\mathbf{u}. \tag{15.4}$$

$$\text{Subject to: } \mathbf{u}^{\mathrm{T}}\mathbf{u} = 1. \tag{15.5}$$

The Lagrangian function associated with the optimization problem is as follows:

$$\text{Maximize: } L = \mathbf{u}^{\mathrm{T}}\mathbf{R}\mathbf{u} - \lambda(\mathbf{u}^{\mathrm{T}}\mathbf{u} - 1). \tag{15.6}$$

The first order condition is as follows:

$$\partial L/\partial \mathbf{u} = 2\mathbf{R}\mathbf{u} - 2\lambda\mathbf{u} = \mathbf{0} \quad \text{or} \quad \mathbf{R}\mathbf{u} = \lambda\mathbf{u}, \tag{15.7}$$

which is an eigenstructure where $\mathbf{0}$ is a $p \times 1$ vector of zeros, the Lagrange multiplier λ is the eigenvalue, and $\mathbf{u}$ is its corresponding eigenvector.

Although we know that there are typically p eigenvalues that solve the determinant polynomial corresponding to Equation (15.7), we also know from the eigendecomposition in Equation (15.3) that it is the largest eigenvalue and its corresponding eigenvector will explain the most variation in $\mathbf{R}$. Therefore, we want to find the largest (or *dominant*) eigenvalue/ eigenvector pair associated with Equation (15.7), that is, $\lambda = \lambda_1$ and $\mathbf{u} = \mathbf{u}_1$. Once this pair is identified, the correlation matrix can be deflated by removing the contribution from λ_1 and $\mathbf{u}_1$ using Equation (15.3). We use the notation $\mathbf{R}(q)$ to denote the deflated correlation matrix associated with the elimination of variation stemming from the first q principal components.

After the first eigenvalue/eigenvector pair is extracted, the correlation matrix is deflated via the following:

$$\mathbf{R}(1) = \mathbf{R} - \lambda \mathbf{u}_1 \mathbf{u}_1^T. \tag{15.8}$$

The second eigenvalue/eigenvector pair $(\lambda_2, \mathbf{u}_2)$ associated with $\mathbf{R}$ is the dominant eigenvalue associated with $\mathbf{R}(1)$ and would then be extracted in a similar fashion. Thus, the process of extracting principal components from $\mathbf{R}$ is sequential (Lattin *et al.*, 2003, p. 98). After the extraction of $(\lambda_2, \mathbf{u}_2)$, deflation occurs by setting

$$\mathbf{R}(2) = \mathbf{R}(1) - \lambda \mathbf{u}_2 \mathbf{u}_2^T, \tag{15.9}$$

and the third eigenvalue/eigenvector pair $(\lambda_3, \mathbf{u}_3)$ associated with $\mathbf{R}$ would be extracted as the dominant eigenvalue associated with $\mathbf{R}(2)$. We use this process in our Excel spreadsheet to sequentially extract all of the eigenvalue/eigenvector pairs for $\mathbf{R}$. Each stage of the sequential extraction process is accomplished using successive approximation via the *power method* (Lay *et al.*, 2015).

The power method is one of the most conceptually straightforward approaches for finding the dominant eigenvector and corresponding eigenvalue of a correlation matrix (as well as other symmetric matrices). Using the power method, the eigenvector at iteration $k + 1$ ($\mathbf{u}^{k+1}$) is estimated from the multiplication of the eigenvector at iteration $k(\mathbf{u}^k)$ by $\mathbf{R}$. That is, $\mathbf{u}^{k+1} = \mathbf{R}\mathbf{u}^k$. To illustrate why this process will converge to the dominant eigenvector, let us assume that the initial estimate $(\mathbf{u}^0)$ is expressed as a linear combination (with coefficients α_j, for $1 \leq j \leq p$) function of the p eigenvectors of $\mathbf{R}$ as follows: $\mathbf{u}^0 = \alpha_1 \mathbf{u}_1 + \alpha_2 \mathbf{u}_2 + \cdots + \alpha_p \mathbf{u}_p$. For the first iteration, we would have $\mathbf{u}^1 = \mathbf{R}\mathbf{u}^0 = \alpha_1 \mathbf{R}\mathbf{u}_1 + \alpha_2 \mathbf{R}\mathbf{u}_2 + \cdots + \alpha_p \mathbf{R}\mathbf{u}_p$ and, because of the eigenstructure relationship $\mathbf{R}\mathbf{u} = \lambda\mathbf{u}$, we can rewrite this as $\mathbf{u}^1 = \mathbf{R}\mathbf{u}^0 = \alpha_1 \lambda_1 \mathbf{u}_1 + \alpha_2 \lambda_2 \mathbf{u}_2 + \cdots + \alpha_p \lambda_p \mathbf{u}_p$. Recalling that we have defined λ_1 as the largest eigenvalue, it is helpful to rewrite this as $\mathbf{u}^1 = \mathbf{R}\mathbf{u}^0 = \lambda_1 [\alpha_1 \mathbf{u}_1 + \alpha_2 (\lambda_2/\lambda_1)\mathbf{u}_2 + \cdots + \alpha_p (\lambda_p/\lambda_1)\mathbf{u}_p]$. After $m + 1$ iterations of the power method, we have $\mathbf{u}^{m+1} = \mathbf{R}^m \mathbf{u}^0 = (\lambda_1)^m [\alpha_1 \mathbf{u}_1 + \alpha_2 (\lambda_2/\lambda_1)^m \mathbf{u}_2 + \cdots + \alpha_p (\lambda_p/\lambda_1)^m \mathbf{u}_p.]$. In the limit, as $m \to \infty$, the terms $(\lambda_2/\lambda_1)^m, \ldots, (\lambda_p/\lambda_1)^m$ go to zero as long as λ_1 is strictly larger than all of the other eigenvalues. Assuming $\alpha_1 \neq 0$, the result is convergence to an eigenvector that is a multiple of the dominant eigenvector: That is, $\mathbf{u}^{m+1} = \mathbf{R}^m \mathbf{u}^0 = \alpha_1 (\lambda_1)^m \mathbf{u}_1$.

Faster convergence is generally achievable if the updated eigenvector is normalized to unit length using the following equation:

$$\mathbf{u}^{k+1} = \frac{\mathbf{Ru}^k}{\sqrt{\left(\mathbf{Ru}^k\right)^T \left(\mathbf{Ru}^k\right)}}. \tag{15.10}$$

Thus, assuming λ_1 is strictly greater than all other eigenvalues, the power method estimates the eigenvector via an iterative process of multiplying by $\mathbf{R}$ and normalizing the result. Upon convergence of the estimation process to the eigenvector $\mathbf{u}$, the corresponding eigenvalue for an eigenvector can be obtained via the Rayleigh quotient as follows:

$$\lambda = \frac{\mathbf{u}^T \mathbf{Ru}}{\mathbf{u}^T \mathbf{u}}. \tag{15.11}$$

The Rayleigh quotient makes use of the facts that $\mathbf{u}$ is an eigenvector of $\mathbf{R}$ and $\mathbf{Ru} = \lambda \mathbf{u}$. Thus, the numerator of Equation (15.11) could be rewritten as $\lambda \mathbf{u}^T \mathbf{u}$, which implies $\lambda = \lambda$. If the eigenvector is unit length, then the denominator is one and the eigenvalue is equal to the numerator, which is the objective function in Equation (15.4) and indicates that $\lambda = \mathbf{u}^T \mathbf{Ru}$ is the variance extracted.

For successive extraction of eigenvalues from correlation matrices, the convergence of the power method to the dominant eigenvectors is generally rapid; however, convergence can take more iterations as the ratio of the second largest eigenvalue to the largest eigenvalue approaches one. For example, when computing the dominant eigenvalue for $\mathbf{R}$, the ratio of λ_2/λ_1 may have an effect on the rapidity of convergence. When computing the dominant eigenvalue for $\mathbf{R}(1)$, it is the ratio of λ_3/λ_2 that affects convergence.

An initial estimate of the first eigenvector ($\mathbf{u}$) is placed in cells H3:H8. We usually just start with all ones in these cells. Cells I3:I8 contain the vector formed by the $\mathbf{Ru}$ product. The value in cell I10 is the constant $\mathbf{u}^T \mathbf{Ru}$. Again, if $\mathbf{u}$ is normalized, then this quantity is equal to the eigenvalue estimate. Cells J3:J8 contain $\mathbf{Ru}$ after it has been normalized to unit length. This is the updated eigenvector. Clicking on the button 'Update Eigenvector 1' simply reads the cell values in J3:J8 and re-pastes them in cells H3:H8 to obtain the updated eigenvector for the next iteration. As we continue to tap this button, we can see the eigenvalue get larger. After 15–20 taps, the

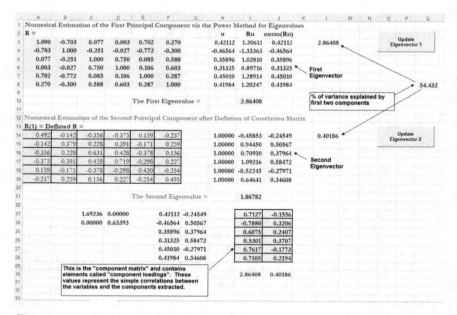

Figure 15.3. Excel worksheet for Example 15.1 — extract components part 2.

changes tend to be small. We can continue tapping until the values in H3:H8 and J3:J8 have stabilized (in other words, the values in these two cell ranges do not change). The results after convergence are shown in Figure 15.3. The first eigenvalue is $\lambda = 2.86408$ and its corresponding eigenvector is shown in cells J3:J8.

Now that the first eigenvalue and eigenvector have been extracted, the correlation matrix is deflated using Equation (15.8). The deflated correlation matrix is shown in cells A14:F19. Once again, the initial estimates for the eigenvector are all ones in cells H14:H19. Tapping the 'Update Eigenvector 2' button will lead to convergence of the estimation of the second eigenvalue and eigenvector in a manner similar to the first. The result is shown in Figure 15.4. The second eigenvalue is $\lambda = 1.93877$ and its corresponding eigenvector is shown in cells J14:J19.

Continuing in this manner, the remaining four eigenvalue/ eigenvector pairs are extracted in rows 34 to 76 of the worksheet. Cells A80:F85 show the fully deflated correlation matrix after extraction of all six components. The six eigenvalues are copied into cells U2:U7 so as to facilitate the production of a scree plot, which is

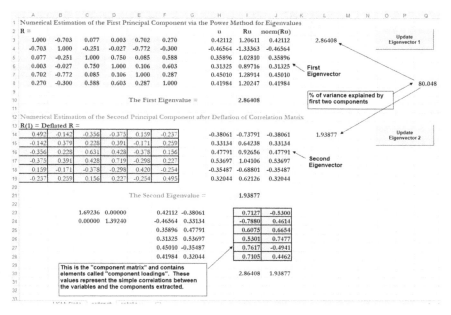

Figure 15.4.   Excel worksheet for Example 15.1 — extract components part 3.

shown in Figure 15.5. The sharp elbow in the plot suggests that two components should be extracted because of the huge drop in the size of the third eigenvalue in comparison to the second. Two components would also be chosen based on the popular default rule (e.g., in SPSS) for correlation matrices of selecting all components with eigenvalues greater than one. The first two components explain just over 80% of the variation in the dataset.

The 2×2 diagonal matrix (**D**) in cells C23:D24 contains the square roots of the eigenvalues on the main diagonal. The two columns of the 6×2 matrix ($\mathbf{U}_2$) in cells F23:G28 are the two eigenvectors. The matrix product $\mathbf{G} = \mathbf{U}_2\mathbf{D}$ yields the 6×2 matrix of *component loadings* in cells I23:J28. These (unrotated) loadings, which are interpreted as correlations between the six items and the two components, are visually displayed in the correlation circle plot in Figure 15.6. Five of the six variables (all but x_2) have fairly high positive loadings on component 1. Similarly, four of the six variables (all but x_1 and x_5) have fairly high positive loadings on component 2. Moreover, the component loadings all fall between 0.4 and 0.8 in absolute value.

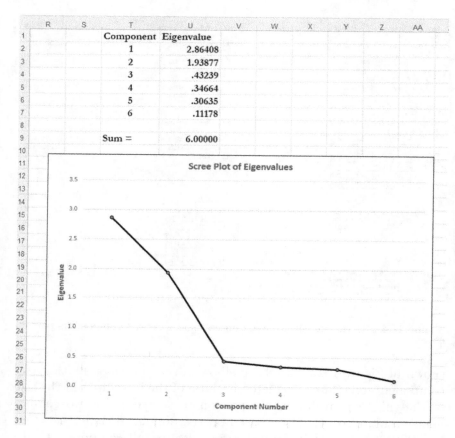

	Component	Eigenvalue
1		
2	1	2.86408
3	2	1.93877
4	3	.43239
5	4	.34664
6	5	.30635
7	6	.11178
9	Sum =	6.00000

Figure 15.5.   Excel worksheet for Example 15.1 — scree plot.

Ideally, we would like to have a *simple structure* whereby the values in each column are close (in absolute value) to either zero or one. This makes it easy to ascertain which variables correspond most heavily to each component. Therefore, in Section 15.1.4, we will copy these component loadings into the next worksheet ('rotate') and rotate them to see if we can improve interpretability.

15.1.4   *Rotation of the component loadings*

In the 'rotate' worksheet displayed in Figure 15.7, the matrix of unrotated component loadings (**G**) has been pasted into cells E6:F11. We seek to find a rotation of these component loadings to

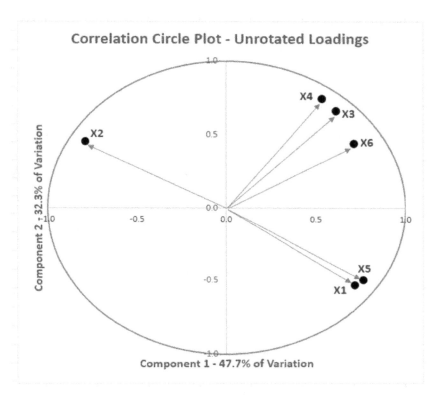

Figure 15.6. Correlation circle plot of unrotated loadings for Example 15.1.

a new set of coordinates, $\mathbf{H} = \mathbf{GW}$, that is more interpretable. The counterclockwise angle of rotation (θ, in degrees) is placed in cell A5. The 2×2 rotation matrix, $\mathbf{W}$, is in cells A7:B8. To understand the rationale for the rotation matrix, $\mathbf{W}$, it is helpful to think of a coordinate pair (g_1, g_2) for the unrotated loadings in terms of polar coordinates (d, φ), where $d = \sqrt{g_1^2 + g_2^2}$ and $\varphi = \arccos(d/g_1)$ if $g_2 \geq 0$ and $\varphi = -\arccos(d/g_1)$ if $g_2 < 0$. The unrotated component loading coordinates can then be expressed in terms of the polar coordinates as

$$g_1 = d\cos\varphi, \tag{15.12}$$

$$g_2 = d\sin\varphi. \tag{15.13}$$

Likewise, recognizing that the counterclockwise angle of rotation will be θ, the coordinate pair (h_1, h_2) for the rotated loadings in terms of polar coordinates will be

$$h_1 = d\cos(\varphi - \theta), \qquad (15.14)$$

$$h_2 = d\sin(\varphi - \theta). \qquad (15.15)$$

Based on the trigonometric rules for compound angles, (15.14) and (15.15) can be rewritten as

$$h_1 = d\cos\varphi\cos\theta + d\sin\varphi\sin\theta, \qquad (15.16)$$

$$h_2 = d\sin\varphi\cos\theta - d\cos\varphi\sin\theta. \qquad (15.17)$$

Equations (15.12) and (15.13) can be used to simplify Equations (15.16) and (15.17) as follows:

$$h_1 = g_1\cos\theta + g_2\sin\theta, \qquad (15.18)$$

$$h_2 = -g_1\sin\theta + g_2\cos\theta. \qquad (15.19)$$

More generally, in matrix notation,

$$[h_1 \; h_2] = [g_1 \; g_2] \begin{bmatrix} \cos\theta & -\sin\theta \\ \sin\theta & \cos\theta \end{bmatrix}. \qquad (15.20)$$

The 2×2 matrix on the right side of Equation (15.20) is the rotation matrix in cells A7:B8. The matrix of rotated component loadings, $\mathbf{H} = \mathbf{GW}$, is computed using this matrix and the values are displayed in cells H6:I11. Because the angle of rotation in cell A5 is set to zero in Figure 15.7, $\mathbf{H} = \mathbf{G}$. Plots of the unrotated and rotated loadings are displayed at the bottom of Figure 15.7.

The user can manually change the angle of rotation in cell A5 and the rotated loadings in $\mathbf{H}$ will be updated automatically. Alternatively, an optimal selection of the angle of rotation can be made based on the well-known *varimax* criterion developed by Kaiser (1958):

$$V = \frac{1}{p}\sum_{l=1}^{q}\left[\sum_{j=1}^{p}h_{jl}^4 - \frac{1}{p}\left(\sum_{j=1}^{p}h_{jl}^2\right)^2\right], \qquad (15.21)$$

where q is the number of selected components (in this example, $q = 2$). Kaiser (1958) developed the *raw varimax criterion* in

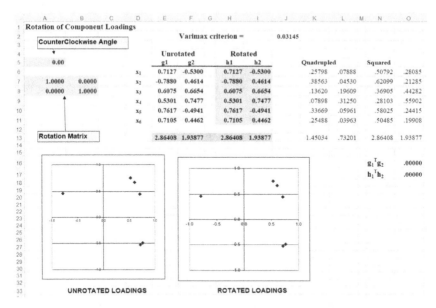

Figure 15.7.   Excel worksheet for rotation of component loadings — part 1.

Equation (15.21) with the goal of producing an orthogonal rotation of the component axes so as to induce a *simple structure* in the rotated loadings. For this criterion, simplicity is operationalized *as the variance of the squared loadings.* Maximizing this quantity has a propensity to drive the loadings toward zero or one in absolute value, which makes it easier to ascertain the variables that correlate strongly with each component.

Kaiser also proposed a normalized varimax criterion that is particularly relevant in the context of factor analysis. The normalized version adjusts the formula in Equation (15.21) to account for the *communality* of each variable (i.e., the variance in the variable that is accounted for by the common factor). Because PCA makes no distinction between common and specific factors, communality and the normalized varimax criterion are not relevant in the context of PCA. Although other options for rotation of the loadings are available, varimax is one of the most common and easiest to implement. The values in cells K6:O13 are scratch computations used to facilitate the computation of the value of the varimax criterion, which is in cell J2.

Effectively, we have another nonlinear optimization problem associated with the rotation of the loadings, which we will accomplish

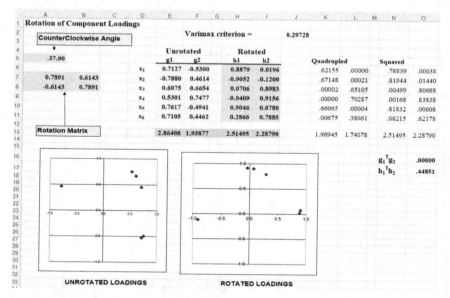

Figure 15.8.   Excel worksheet for rotation of component loadings — after rotation.

using the GRG Nonlinear engine of the Solver. The Excel Solver (using $\theta = 0°$ as the initial value in cell A5) was used to find the angle of rotation that maximizes the value of the varimax criterion in cell J2. The results are displayed in Figure 15.8. The optimal counterclockwise angle of rotation is $\theta = -37.9°$ (i.e., a clockwise rotation of $\theta = 37.9°$). The rotated loadings have changed dramatically, and the correlation circle plot of these loadings in Figure 15.9 now shows that the variables tend to be associated with one of the two components but not both. Cells O16:O17 show that unrotated loading vectors are orthogonal ($\mathbf{g}_1^T \mathbf{g}_2 = 0$) but the rotated loadings are not ($\mathbf{h}_1^T \mathbf{h}_2 \neq 0$). Moreover, it is noteworthy that, although the total variation explained by the two components is unchanged, the relative contribution of the explained variation for the two components is more equally distributed. That is, for the unrotated loadings, it is noted that $2.86048 + 1.93877 = 4.80285$, and for the rotated loadings, $2.51495 + 2.28790 = 4.80285$.

When examining the rotated loadings, it is evident that the variables x_1, x_2, and x_5 have large absolute values on component 1 and small absolute values on component 2. The reverse is true for the

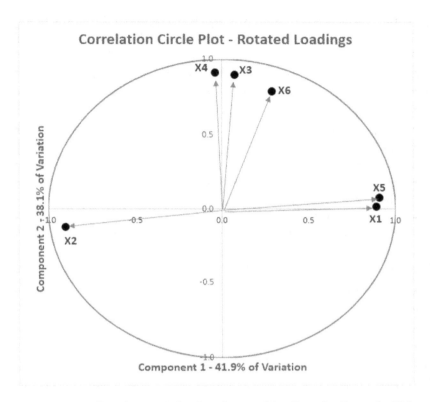

Figure 15.9.   Correlation circle plot of rotated loadings for Example 15.1.

other three variables x_3, x_4, and x_6. Accordingly, the interpretation of the rotated components is much cleaner. As observed from Figure 15.1, Likert scale items x_1, x_2, and x_5 all pertain to shopping *with* other people, whereas items x_3, x_4, and x_6 are related to shopping *for* other people. In the language used by Arnold and Reynolds (2003), component 1 would be identified as a *social shopping* construct and component 2 as a *role shopping* construct.

15.2   Confirmatory Factor Analysis

15.2.1   *Overview of the MLE model*

Miles (2005) discussed the value of using Excel spreadsheets to better understand the mechanics of confirmatory factor analysis. The focus was limited to a small example in the context of a unidimensional (i.e., one factor) model. In this section, we build on Miles's (2005) work by considering the unidimensional model and both uncorrelated and correlated multifactor models. To verify our results, we use the `lavaan` R package. Our description of confirmatory factor analysis uses the following notation:

n = the number of observations,
p = the number of variables,
q = the number of factors,
$\mathbf{S}$ = the $p \times p$ sample covariance (or, possibly, correlation matrix),
$\mathbf{\Sigma}$ = the $p \times p$ implied or estimated covariance matrix,
$\mathbf{\Lambda}$ = the $p \times q$ matrix of factor loadings,
$\mathbf{\Phi}$ = the $q \times q$ matrix of correlations between factors,
$\mathbf{\Theta}$ = the $p \times p$ matrix of error terms,
df = degrees of freedom for model estimation.

The model of interest is

$$\mathbf{\Sigma} = \mathbf{\Lambda}\mathbf{\Phi}\mathbf{\Lambda}' + \mathbf{\Theta}. \tag{15.22}$$

The discrepancy function to be minimized in accordance with maximum likelihood estimation of the $\mathbf{\Lambda}$, $\mathbf{\Phi}$, and $\mathbf{\Theta}$ parameters is as follows:

$$F_{\text{MLE}} = \ln |\mathbf{\Sigma}| + tr\left(\mathbf{S}\mathbf{\Sigma}^{-1}\right) - \ln |\mathbf{S}| - p. \tag{15.23}$$

The total number of parameters estimated in the implied covariance matrix is $p(p+1)/2$ and constraints are commonly applied to $\mathbf{\Lambda}$ and $\mathbf{\Phi}$ to ensure that a sufficient number of degrees of freedom remain for the model to be identified.

The measures of fit considered in our analyses are as follows:

(i) Chi-square:

$$\chi^2 = n \times F_{\text{MLE}}. \tag{15.24}$$

(ii) Root mean square error of approximation (RMSEA):

$$\text{RMSEA} = \sqrt{\max\left\{0, \frac{\left(\frac{\chi^2}{df} - 1\right)}{n}\right\}}. \tag{15.25}$$

(iii) Comparative fit index (CFI):

$$\text{CFI} = 1 - \frac{\max\left\{0, \chi^2 - df\right\}}{\max\left\{0, \chi^2_{\text{null}} - df_{\text{null}}\right\}}, \qquad (15.26)$$

where χ^2_{null} and df_{null} are the chi-square and degrees of freedom for the null model, respectively.

(iv) Composite reliability for each factor k (CR_k):

$$\text{CR}_k = \frac{\left(\sum_{j=1}^{p} \lambda_{jk}\right)^2}{\left(\left(\sum_{j=1}^{p} \lambda_{jk}\right)^2 + \sum_{j \in V_k} \theta_{jj}\right)} \qquad \text{for } 1 \le k \le q, \quad (15.27)$$

where V_k is the set of variables with unconstrained loadings on factor k.

(v) Average variance extracted for each factor k (AVE_k):

$$\text{AVE}_k = \frac{\left(\sum_{j \in V_k} \lambda_{jk}^2\right)}{\left(\left(\sum_{j \in V_k} \lambda_{jk}^2\right) + \sum_{j \in V_k} \theta_{jj}\right)} \qquad \text{for } 1 \le k \le q. \quad (15.28)$$

15.2.2 *Example 15.2 — Customer satisfaction data*

To illustrate the implementation of confirmatory factor analysis in Excel, we return to the customer satisfaction data used for logistic regression in Section 13.2. The raw data consist of $n = 1000$ measurements on $p = 9$ Likert scale variables. The variables and the correlation matrix are shown in Equation 15.10. We hypothesize that the underlying structure of these data consists of $q = 3$ factors with three variables each: Factor 1 — food quality (variables x_1, x_2, and x_3); Factor 2 — aesthetics (variables x_4, x_5, and x_6); and Factor 3 — service efficiency (variables x_7, x_8, and x_9).

15.2.2.1 *Null model*

In Figure 15.11, we display the results for the null model because it is necessary for the computation of our performance measures. The sample correlation matrix, $\mathbf{S}$, from Figure 15.10 is pasted into cells A2:I10 of Figure 15.11. The transpose ($\mathbf{\Lambda'}$) of the factor loadings

Confirmatory Factor Analysis

X1 = the side dishes and salads are delicious
X2 = the entrées are delicious
X3 = the desserts are delicious
X4 = parking at the restaurant is ample
X5 = the lighting in the restaurant is appropriate
X6 = the noise level in the restaurant is not distracting to me
X7 = the wait for beverage service is excessive
X8 = the wait for food service is excessive
X9 = the wait for the check is excessive

	X1	X2	X3	X4	X5	X6	X7	X8	X9
X1	1.000								
X2	0.846	1.000							
X3	0.850	0.841	1.000						
X4	0.157	0.124	0.148	1.000					
X5	0.151	0.089	0.121	0.636	1.000				
X6	0.116	0.079	0.100	0.617	0.612	1.000			
X7	-0.337	-0.318	-0.349	-0.103	-0.125	-0.144	1.000		
X8	-0.352	-0.316	-0.342	-0.136	-0.134	-0.148	0.825	1.000	
X9	-0.341	-0.315	-0.326	-0.140	-0.151	-0.175	0.836	0.821	1.000

Customer	X1	X2	X3	X4	X5	X6	X7	X8	X9
1	7	7	7	6	6	6	5	4	6
2	7	7	7	7	7	7	2	3	2
3	7	7	7	5	4	7	2	3	2
4	7	6	6	4	5	6	3	3	2
5	6	7	7	7	7	7	6	5	5
6	7	7	7	6	5	5	1	1	2
7	4	5	4	7	7	7	3	3	3
8	7	7	7	6	7	7	1	1	1
9	7	7	7	7	7	7	1	1	1
10	7	7	7	4	7	2	1	1	1
11	7	6	7	7	7	7	2	3	3
12	7	7	7	3	3	3	1	1	1
13	7	7	6	7	6	4	4	4	3
14	7	7	7	6	6	6	3	2	2
15	7	7	7	4	3	4	1	1	1
16	7	7	7	5	6	6	1	1	1
17	5	6	6	7	7	7	1	1	1

Figure 15.10.　Variables and correlation matrix for Example 15.2.

Sample Correlation Matrix (S)

1.000	0.846	0.850	0.157	0.151	0.116	-0.337	-0.352	-0.341
0.846	1.000	0.841	0.124	0.089	0.079	-0.318	-0.316	-0.315
0.850	0.841	1.000	0.148	0.121	0.100	-0.349	-0.342	-0.326
0.157	0.124	0.148	1.000	0.636	0.617	-0.103	-0.136	-0.140
0.151	0.089	0.121	0.636	1.000	0.612	-0.125	-0.134	-0.151
0.116	0.079	0.100	0.617	0.612	1.000	-0.144	-0.148	-0.175
-0.337	-0.318	-0.349	-0.103	-0.125	-0.144	1.000	0.825	0.836
-0.352	-0.316	-0.342	-0.136	-0.134	-0.148	0.825	1.000	0.821
-0.341	-0.315	-0.326	-0.140	-0.151	-0.175	0.836	0.821	1.000

Factor Loadings Matrix (Λ)

0.000	0.000	0.000	0.000	0.000	0.000	0.000	0.000	0.000
0.000	0.000	0.000	0.000	0.000	0.000	0.000	0.000	0.000
0.000	0.000	0.000	0.000	0.000	0.000	0.000	0.000	0.000

Factor Correlation Matrix (Φ)

1.000	0.000	0.000
0.000	1.000	0.000
0.000	0.000	1.000

Estimated Correlation Matrix (Σ)

1.000	0.000	0.000	0.000	0.000	0.000	0.000	0.000	0.000
0.000	1.000	0.000	0.000	0.000	0.000	0.000	0.000	0.000
0.000	0.000	1.000	0.000	0.000	0.000	0.000	0.000	0.000
0.000	0.000	0.000	1.000	0.000	0.000	0.000	0.000	0.000
0.000	0.000	0.000	0.000	1.000	0.000	0.000	0.000	0.000
0.000	0.000	0.000	0.000	0.000	1.000	0.000	0.000	0.000
0.000	0.000	0.000	0.000	0.000	0.000	1.000	0.000	0.000
0.000	0.000	0.000	0.000	0.000	0.000	0.000	1.000	0.000
0.000	0.000	0.000	0.000	0.000	0.000	0.000	0.000	1.000

Unique Variances/Covariance Matrix (Θ)

1.000	0.000	0.000	0.000	0.000	0.000	0.000	0.000	0.000
0.000	1.000	0.000	0.000	0.000	0.000	0.000	0.000	0.000
0.000	0.000	1.000	0.000	0.000	0.000	0.000	0.000	0.000
0.000	0.000	0.000	1.000	0.000	0.000	0.000	0.000	0.000
0.000	0.000	0.000	0.000	1.000	0.000	0.000	0.000	0.000
0.000	0.000	0.000	0.000	0.000	1.000	0.000	0.000	0.000
0.000	0.000	0.000	0.000	0.000	0.000	1.000	0.000	0.000
0.000	0.000	0.000	0.000	0.000	0.000	0.000	1.000	0.000
0.000	0.000	0.000	0.000	0.000	0.000	0.000	0.000	1.000

$S\Sigma^{-1}$

1.000	0.846	0.850	0.157	0.151	0.116	-0.337	-0.352	-0.341
0.846	1.000	0.841	0.124	0.089	0.079	-0.318	-0.316	-0.315
0.850	0.841	1.000	0.148	0.121	0.100	-0.349	-0.342	-0.326
0.157	0.124	0.148	1.000	0.636	0.617	-0.103	-0.136	-0.140
0.151	0.089	0.121	0.636	1.000	0.612	-0.125	-0.134	-0.151
0.116	0.079	0.100	0.617	0.612	1.000	-0.144	-0.148	-0.175
-0.337	-0.318	-0.349	-0.103	-0.125	-0.144	1.000	0.825	0.836
-0.352	-0.316	-0.342	-0.136	-0.134	-0.148	0.825	1.000	0.821
-0.341	-0.315	-0.326	-0.140	-0.151	-0.175	0.836	0.821	1.000

Discrepancy	6.656
Chi-Square	6656.327
Degrees of Freedom	36
Significance	0.000
RMSEA	0.429
Chi-Square (Null)	6656.327
Df (Null)	36
CFI	0.000
GFI	0.403

	Composite Reliability	Average Variance Extracted
Factor 1	0.000	0.000
Factor 2	0.000	0.000
Factor 3	0.000	0.000

Figure 15.11.　The null CFA model for Example 15.2.

	A	B	C	D	E	F	G	H	I	J	K	L	M	N	O	P	Q	R	S
1	Sample Correlation Matrix (S)										Factor Loadings Matrix (Λ)								
2	1.000	0.846	0.850	0.157	0.151	0.116	-0.337	-0.352	-0.341		0.924	0.908	0.917	0.173	0.152	0.131	-0.408	-0.411	-0.402
3	0.846	1.000	0.841	0.124	0.089	0.079	-0.318	-0.316	-0.315		0.000	0.000	0.000	0.000	0.000	0.000	0.000	0.000	0.000
4	0.850	0.841	1.000	0.148	0.121	0.100	-0.349	-0.342	-0.326		0.000	0.000	0.000	0.000	0.000	0.000	0.000	0.000	0.000
5	0.157	0.124	0.148	1.000	0.636	0.617	-0.103	-0.136	-0.140										
6	0.151	0.089	0.121	0.636	1.000	0.612	-0.125	-0.134	-0.151		Factor Correlation Matrix (Φ)								
7	0.116	0.079	0.100	0.617	0.612	1.000	-0.144	-0.148	-0.175		1.000	0.000	0.000						
8	-0.337	-0.318	-0.349	-0.103	-0.125	-0.144	1.000	0.825	0.836		0.000	1.000	0.000						
9	-0.352	-0.316	-0.342	-0.136	-0.134	-0.148	0.825	1.000	0.821		0.000	0.000	1.000						
10	-0.341	-0.315	-0.326	-0.140	-0.151	-0.175	0.836	0.821	1.000										
11																			
12	Estimated Correlation Matrix (Σ)										Unique Variances/Covariance Matrix (Θ)								
13	1.000	0.839	0.848	0.160	0.141	0.121	-0.377	-0.380	-0.372		0.146	0.000	0.000	0.000	0.000	0.000	0.000	0.000	0.000
14	0.839	1.000	0.832	0.157	0.138	0.119	-0.370	-0.373	-0.365		0.000	0.176	0.000	0.000	0.000	0.000	0.000	0.000	0.000
15	0.848	0.832	1.000	0.159	0.139	0.120	-0.374	-0.377	-0.369		0.000	0.000	0.159	0.000	0.000	0.000	0.000	0.000	0.000
16	0.160	0.157	0.159	1.000	0.026	0.023	-0.071	-0.071	-0.070		0.000	0.000	0.000	0.970	0.000	0.000	0.000	0.000	0.000
17	0.141	0.138	0.139	0.026	1.000	0.020	-0.062	-0.062	-0.061		0.000	0.000	0.000	0.000	0.977	0.000	0.000	0.000	0.000
18	0.121	0.119	0.120	0.023	0.020	1.000	-0.053	-0.054	-0.053		0.000	0.000	0.000	0.000	0.000	0.983	0.000	0.000	0.000
19	-0.377	-0.370	-0.374	-0.071	-0.062	-0.053	1.000	0.168	0.164		0.000	0.000	0.000	0.000	0.000	0.000	0.833	0.000	0.000
20	-0.380	-0.373	-0.377	-0.071	-0.062	-0.054	0.168	1.000	0.165		0.000	0.000	0.000	0.000	0.000	0.000	0.000	0.831	0.000
21	-0.372	-0.365	-0.369	-0.070	-0.061	-0.053	0.164	0.165	1.000		0.000	0.000	0.000	0.000	0.000	0.000	0.000	0.000	0.838
22																			
23	$S\Sigma^{-1}$																		
24	1.000	0.040	0.012	-0.003	0.010	-0.005	0.049	0.034	0.037		Discrepancy		3.402				Average		
25	0.049	1.000	0.056	-0.034	-0.050	-0.040	0.063	0.068	0.059		Chi-Square		3401.920				Composite Variance		
26	0.014	0.050	1.000	-0.012	-0.019	-0.020	0.031	0.042	0.051		Degrees of Freedom		27				Reliability Extracted		
27	-0.022	-0.188	-0.071	1.000	0.624	0.604	-0.039	-0.078	-0.084		Significance		0.000				Factor 1	0.400	0.343
28	0.070	-0.279	-0.118	0.628	1.000	0.603	-0.076	-0.086	-0.107		RMSEA		0.354				Factor 2	0.000	0.000
29	-0.032	-0.225	-0.124	0.613	0.606	1.000	-0.109	-0.114	-0.146		Chi-Square (Null)		6656.327				Factor 3	0.000	0.000
30	0.278	0.296	0.163	-0.034	-0.065	-0.092	1.000	0.791	0.801		Df (Null)		36						
31	0.192	0.322	0.221	-0.067	-0.074	-0.096	0.789	1.000	0.783		CFI		0.490						
32	0.212	0.280	0.268	-0.073	-0.092	-0.124	0.806	0.789	1.000		GFI		0.581						

Figure 15.12. Worksheet for unidimensional model for Example 15.2.

matrix (Λ) in cells K2:S4 is unused in this worksheet because the null model assumes no factors. The factor correlation matrix (Φ) in cells K7:M9 is also irrelevant to the null model. Only the main diagonal of the error terms matrix Θ in cells K13:S21 is used to fit the variances (covariances are ignored), which consumes only $p = 9$ degrees of freedom. Accordingly, the degrees of freedom for the null model are $df_{null} p(p+1)/2 - p = 36$. The discrepancy measure in cell N24 is $F_{MLE} = 6.656$ and $\chi^2_{null} = nF_{MLE} = 1000(6.656) = 6656.527$ in cell N25. Cell N28 shows RMSEA $= 0.429$ and cell N31 contains CFI $= 0$.

15.2.2.2 *Unidimensional model*

The worksheet for the unidimensional model is displayed in Figure 15.12. The locations of the **S**, Λ', Φ, and Θ matrices in Figure 15.12 are the same as in Figure 15.11. However, in this model, the nine cells in the first row of the transpose of the loadings matrix (cells K2:S2) are enabled as changing cells for GRG Nonlinear engine of the Excel Solver, in addition to the nine main diagonal elements of Θ in cells K13:S21. Accordingly, the degrees of freedom for this model are df $= p(p+1)/2 - p - p = 27$.

Equation (15.22) is used to compute Σ in cells A13:I21 with the aid of the MMULT and TRANSPOSE functions. The MMULT and MINVERSE function are used to compute $\mathbf{S}\Sigma^{-1}$ in cells A24:I32 because this matrix is necessary for computation of the discrepancy measure in cell N24 using Equation (15.23).

When using the GRG Nonlinear engine for this example, it is necessary to populate the changing cells prior to running the Solver. If zeros are the starting values for the loadings and the diagonal elements of Θ, then the error message 'Solver encountered an error value in the Objective Cell or a Constraint cell' appears. We populated cell values K2:S2 with the values of 0.1 and did likewise with the main diagonal elements in the matrix associated with cells K13:S21. We then ran the Solver and obtained values nearly identical to those in Figure 15.12 but with some differences in the third decimal place. We then ran the Solver once more, yielding the values reported in Figure 15.12.

The inadequacy of the unidimensional model for Example 15.2 is readily apparent from the results in Figure 15.12. There are several pieces of evidence that can be noted. First, only the loadings for the food quality variables are substantial, as the loadings for the other six variables in cells K2:S2 are below 0.5 in absolute value. Second, the value of RMSEA = 0.354 in cell N28 is still quite large. Typically values of 0.08 or smaller are desirable for this measure. Third, the value of CFI = 0.490 in cell N31 is well below the standard threshold of 0.95. Fourth, the composite reliability of $CR_1 = 0.400$ in cell R27 is well below the common threshold of 0.7. Fifth, the average variance extracted value of $AVE_1 = 0.343$ in cell S27 is below the common threshold of 0.5.

15.2.2.3   *Uncorrelated three-factor model*

The worksheet for the uncorrelated three-factor model is displayed in Figure 15.13. The layout of the worksheet for the three-factor model in Figure 15.13 is virtually identical to that of the unidimensional model in Figure 15.12. The major difference is that all the elements of the transpose of the loadings matrix in cells K2:S4 have now been enabled as decision variables. However, only variables x_1, x_2, and x_3 can load on the first factor in row 2 (cells K2:M2) and all other variables in that row (N2:S2) are constrained to zero. In a similar

	A	B	C	D	E	F	G	H	I	J	K	L	M	N	O	P	Q	R	S
1	Sample Correlation Matrix (S)										Factor Loadings Matrix (Λ)								
2	1.000	0.846	0.850	0.157	0.151	0.116	-0.337	-0.352	-0.341		0.924	0.915	0.919	0.000	0.000	0.000	0.000	0.000	0.000
3	0.846	1.000	0.841	0.124	0.089	0.079	-0.318	-0.316	-0.315		0.000	0.000	0.000	0.800	0.794	0.771	0.000	0.000	0.000
4	0.850	0.841	1.000	0.148	0.121	0.100	-0.349	-0.342	-0.326		0.000	0.000	0.000	0.000	0.000	0.000	0.917	0.900	0.912
5	0.157	0.124	0.148	1.000	0.636	0.617	-0.103	-0.136	-0.140										
6	0.151	0.089	0.121	0.636	1.000	0.612	-0.125	-0.134	-0.151		Factor Correlation Matrix (Φ)								
7	0.116	0.079	0.100	0.617	0.612	1.000	-0.144	-0.148	-0.175		1.000	0.000	0.000						
8	-0.337	-0.318	-0.349	-0.103	-0.125	-0.144	1.000	0.825	0.836		0.000	1.000	0.000						
9	-0.352	-0.316	-0.342	-0.136	-0.134	-0.148	0.825	1.000	0.821		0.000	0.000	1.000						
10	-0.341	-0.315	-0.326	-0.140	-0.151	-0.175	0.836	0.821	1.000										
11																			
12	Estimated Correlation Matrix (Σ)										Unique Variances/Covariance Matrix (Θ)								
13	1.000	0.846	0.850	0.000	0.000	0.000	0.000	0.000	0.000		0.145	0.000	0.000	0.000	0.000	0.000	0.000	0.000	0.000
14	0.846	1.000	0.841	0.000	0.000	0.000	0.000	0.000	0.000		0.000	0.163	0.000	0.000	0.000	0.000	0.000	0.000	0.000
15	0.850	0.841	1.000	0.000	0.000	0.000	0.000	0.000	0.000		0.000	0.000	0.155	0.000	0.000	0.000	0.000	0.000	0.000
16	0.000	0.000	0.000	1.000	0.636	0.617	0.000	0.000	0.000		0.000	0.000	0.000	0.360	0.000	0.000	0.000	0.000	0.000
17	0.000	0.000	0.000	0.636	1.000	0.612	0.000	0.000	0.000		0.000	0.000	0.000	0.000	0.369	0.000	0.000	0.000	0.000
18	0.000	0.000	0.000	0.617	0.612	1.000	0.000	0.000	0.000		0.000	0.000	0.000	0.000	0.000	0.406	0.000	0.000	0.000
19	0.000	0.000	0.000	0.000	0.000	0.000	1.000	0.825	0.836		0.000	0.000	0.000	0.000	0.000	0.000	0.160	0.000	0.000
20	0.000	0.000	0.000	0.000	0.000	0.000	0.825	1.000	0.821		0.000	0.000	0.000	0.000	0.000	0.000	0.000	0.189	0.000
21	0.000	0.000	0.000	0.000	0.000	0.000	0.836	0.821	1.000		0.000	0.000	0.000	0.000	0.000	0.000	0.000	0.000	0.168
22																			
23	$S\Sigma^{-1}$																		
24	1.000	0.000	0.000	0.102	0.085	0.001	-0.089	-0.183	-0.116		Discrepancy		0.229				Average		
25	0.000	1.000	0.000	0.115	0.017	-0.002	-0.126	-0.121	-0.111		Chi-Square		228.916				Composite	Variance	
26	0.000	0.000	1.000	0.120	0.045	-0.001	-0.184	-0.148	-0.050		Degrees of Freedom		27				Reliability	Extracted	
27	0.144	-0.065	0.079	1.000	0.000	0.000	0.094	-0.104	-0.134		Significance		0.000				Factor 1	0.943	0.846
28	0.246	-0.154	0.041	0.000	1.000	0.000	0.021	-0.041	-0.134		RMSEA		0.086				Factor 2	0.831	0.622
29	0.153	-0.087	0.043	0.000	0.000	1.000	0.017	-0.022	-0.170		Chi-Square (Null)		6656.327				Factor 3	0.935	0.828
30	-0.134	-0.026	-0.213	0.002	-0.060	-0.108	1.000	0.000	0.000		Df (Null)		36						
31	-0.222	0.002	-0.155	-0.052	-0.048	-0.087	0.000	1.000	0.000		CFI		0.970						
32	-0.206	-0.048	-0.110	-0.028	-0.058	-0.122	0.000	0.000	1.000		GFI		0.952						

Figure 15.13. Worksheet for uncorrelated three-factor model for Example 15.2.

fashion, only x_4, x_5, and x_6 can have non-zero loadings on the second factor in row 3, and only cells x_7, x_8, and x_9 can have non-zero loadings on the third factor in row 4. So, the degrees of freedom for the uncorrelated three-factor model are the same as they are for the unidimensional model, that is, df $= p(p+1)/2 - p - 3 - 3 - 3 = 27$. The difference is that, instead of nine variables loading on one factor in the unidimensional model, three variables are allowed to load on each of three factors in the three-factor model.

Like the unidimensional model, the Excel Solver should be implemented using non-zero values in the cells for the relevant loadings and error terms. Therefore, we used values of 0.1 in cells K2:M2, N3:P3, Q4:S4, and in the main diagonal of the matrix in cells K13:S21 when making an initial run of the Solver. This was followed by a second run to fine-tune the results.

The substantial improvement in fit for Example 15.2 that is realized from the three-factor model for Example 15.2 is evident from the results in Figure 15.13. First, the loadings for the variables associated with the first (food quality, cells K2:M2) and third (service efficiency, cells Q4:S4) factors all equal or exceed 0.9. Although the loadings for the second factor (aesthetics, cells N3:P3) are not quite as strong, they still range from 0.77 to 0.8. Second, the value of

RMSEA $= 0.086$ in cell N28 is only slightly larger than the standard of 0.08 or smaller for this measure. The value of CFI $= 0.970$ in cell N31 exceeds the standard threshold of 0.95. Fourth, the composite reliabilities of $CR_1 = 0.943$, $CR_2 = 0.831$, and $CR_3 = 0.935$ in cells R27:R29 are all well above the common threshold of 0.7. Fifth, the average variance extracted values of $AVE_1 = 0.846$, $AVE_2 = 0.622$, and $AVE_3 = 0.828$ in cells S27:S29 are all well above the common threshold of 0.5. The fourth and fifth observations are evidence of the *convergent* validity of the factors (Malhotra, 2016).

15.2.2.4   Correlated three-factor model

The worksheet for the correlated three-factor model is displayed in Figure 15.14. The layout of the worksheet for the correlated three-factor model in Figure 15.14 is virtually identical to that of the uncorrelated three-factor model in Figure 15.13. The only difference is that now the three-factor correlations (φ_{12}, φ_{13}, and φ_{23}) in the lower triangle of the Φ matrix in cells K7:M9 have been enabled as decision variables (note: the upper triangle elements of this matrix are constrained to equal their corresponding element in the lower triangle). So, the degrees of freedom for the correlated

	A	B	C	D	E	F	G	H	I	J	K	L	M	N	O	P	Q	R	S
1	Sample Correlation Matrix (S)										Factor Loadings Matrix (Λ)								
2	1.000	0.846	0.850	0.157	0.151	0.116	-0.337	-0.352	-0.341		0.926	0.013	0.920	0.000	0.000	0.000	0.000	0.000	0.000
3	0.846	1.000	0.841	0.124	0.089	0.079	-0.318	-0.316	-0.315		0.000	0.000	0.000	0.800	0.794	0.771	0.000	0.000	0.000
4	0.850	0.841	1.000	0.148	0.121	0.100	-0.349	-0.342	-0.326		0.000	0.000	0.000	0.000	0.000	0.000	0.916	0.901	0.912
5	0.157	0.124	0.148	1.000	0.636	0.617	-0.103	-0.136	-0.140										
6	0.151	0.089	0.121	0.636	1.000	0.612	-0.125	-0.134	-0.151		Factor Correlation Matrix (Φ)								
7	0.116	0.079	0.100	0.617	0.612	1.000	-0.144	-0.148	-0.175		1.000	0.169	-0.398						
8	-0.337	-0.318	-0.349	-0.103	-0.125	-0.144	1.000	0.825	0.836		0.169	1.000	-0.193						
9	-0.352	-0.316	-0.342	-0.136	-0.134	-0.148	0.825	1.000	0.821		-0.398	-0.193	1.000						
10	-0.341	-0.315	-0.326	-0.140	-0.151	-0.175	0.836	0.821	1.000										
11																			
12	Estimated Correlation Matrix (Σ)										Unique Variances/Covariance Matrix (Θ)								
13	1.000	0.846	0.851	0.125	0.124	0.121	-0.337	-0.332	-0.336		0.143	0.000	0.000	0.000	0.000	0.000	0.000	0.000	0.000
14	0.846	1.000	0.840	0.124	0.123	0.119	-0.333	-0.328	-0.332		0.000	0.165	0.000	0.000	0.000	0.000	0.000	0.000	0.000
15	0.851	0.840	1.000	0.124	0.124	0.120	-0.335	-0.330	-0.334		0.000	0.000	0.154	0.000	0.000	0.000	0.000	0.000	0.000
16	0.125	0.124	0.124	1.000	0.636	0.617	-0.141	-0.139	-0.141		0.000	0.000	0.000	0.360	0.000	0.000	0.000	0.000	0.000
17	0.124	0.123	0.124	0.636	1.000	0.612	-0.140	-0.138	-0.140		0.000	0.000	0.000	0.000	0.370	0.000	0.000	0.000	0.000
18	0.121	0.119	0.120	0.617	0.612	1.000	-0.136	-0.134	-0.136		0.000	0.000	0.000	0.000	0.000	0.405	0.000	0.000	0.000
19	-0.337	-0.333	-0.335	-0.141	-0.140	-0.136	1.000	0.825	0.835		0.000	0.000	0.000	0.000	0.000	0.000	0.161	0.000	0.000
20	-0.332	-0.328	-0.330	-0.139	-0.138	-0.134	0.825	1.000	0.822		0.000	0.000	0.000	0.000	0.000	0.000	0.000	0.188	0.000
21	-0.336	-0.332	-0.334	-0.141	-0.140	-0.136	0.835	0.822	1.000		0.000	0.000	0.000	0.000	0.000	0.000	0.000	0.000	0.168
22																			
23	$S\Sigma^{-1}$																		
24	1.000	0.003	-0.009	0.044	0.029	-0.049	0.047	-0.068	0.014		Discrepancy		0.040					Average	
25	0.003	1.000	0.007	0.057	-0.038	-0.052	0.007	-0.008	0.016		Chi-Square		40.145					Composite	Variance
26	-0.010	0.007	1.000	0.061	-0.011	-0.052	-0.050	-0.035	0.078		Degrees of Freedom		24					Reliability	Extracted
27	0.098	-0.103	0.036	1.000	0.001	-0.001	0.155	-0.052	-0.075		Significance		0.021			Factor 1		0.943	0.846
28	0.197	-0.194	-0.007	0.001	1.000	0.000	0.076	0.007	-0.080		RMSEA		0.026			Factor 2		0.831	0.622
29	0.100	-0.131	-0.007	-0.001	0.000	1.000	0.066	0.020	-0.123		Chi-Square (Null)		6656.327			Factor 3		0.935	0.828
30	0.001	0.087	-0.089	0.068	0.004	-0.052	1.000	0.000	0.003		Df (Null)		36						
31	-0.089	0.113	-0.032	0.014	0.015	-0.031	0.000	1.000	-0.004		CFI		0.998						
32	-0.070	0.066	0.015	0.037	0.005	-0.066	0.004	-0.003	1.000		GFI		0.991						

Figure 15.14.   Worksheet for the correlated three-factor model for Example 15.2.

three-factor model are three less than what they are for the uncorrelated three-factor model, that is, df $= p(p+1)/2 - p - 3 - 3 - 3 - 3 = 24$.

Once again, the Excel Solver was implemented using non-zero values in the cells for the relevant loadings and error terms. We used initial values of 0.1 in cells K2:M2, N3:P3, and Q4:S4, and in the main diagonal of the matrix in cells K13:S21. We also used initial values of 0.1 in the cells corresponding to the factor correlations in cells K8, K9, and L9. After making an initial run using these starting values, we made a second run to fine-tune the results.

The factor loadings for the correlated three-factor model in Figure 15.14 are almost identical to those for the uncorrelated three-factor model in Figure 15.13. However, a substantial improvement in RMSEA is realized from modeling the factor correlations. The value of RMSEA $= 0.026$ in cell N28 is substantially better than the value of 0.086 for the uncorrelated three-factor model and also well below the standard of 0.08 or smaller for this measure. The value of CFI $= 0.998$ in cell N31 is also a sizable improvement over the value of 0.970 for the uncorrelated three-factor model. The composite reliabilities of $CR_1 = 0.943$, $CR_2 = 0.831$, and $CR_3 = 0.935$ in cells R27:R29 are identical to those for the uncorrelated three-factor model and well above the common threshold of 0.7. The average variance extracted values of $AVE_1 = 0.846$, $AVE_2 = 0.622$, and $AVE_3 = 0.828$ in cells S27:S29 are also identical to those of the uncorrelated three-factor model, all well above the common threshold of 0.5. As noted previously, the CR and AVE values are evidence of the *convergent* validity of the factors (Malhotra, 2016). We also see evidence of *discriminant* validity, in that the square roots of the AVE values exceed the absolute values of the factor correlations (Malhotra, 2016).

15.2.2.5 *Confirmation of results using lavaan R package*

To confirm our results, we ran the unidimensional, uncorrelated three-factor, and correlated three-factor CFA models using the lavaan R package (Rosseel *et al.*, 2024). The null model is run automatically for each of these analyses. The commands and outputs for these analyses are displayed in the following. Loadings, variances, and relevant measures have been highlighted in bold font for easier comparison to the spreadsheet results.

RESULTS FOR THE UNIDIMENSIONAL MODEL

```
> library(lavaan)
lavaan is FREE software! Please report any bugs.
> dat <- read.table("c:/cfa/satisf.prn")
> X <- as.matrix(dat)
> m1a = 'f = V1 + V2 + V3 + V4 + V5 + V6 + V7
         + V8 + V9'
> m1a = 'f =~ V1 + V2 + V3 + V4 + V5 + V6 + V7
          + V8 + V9'
> res1 = cfa(m1a, data=X, std.lv=TRUE)
> summary(res1, fit.measures=TRUE,standardized=TRUE)
lavaan 0.6-9 ended normally after 23 iterations
```

Estimator	ML
Optimization method	NLMINB
Number of model parameters	18
Number of observations	1000
Model Test User Model:	
Test statistic	3401.920
Degrees of freedom	**27**
P-value (Chi-square)	0.000
Model Test Baseline Model:	
Test statistic	6656.327
Degrees of freedom	36
P-value	0.000
User Model versus Baseline Model:	
Comparative Fit Index (CFI)	**0.490**
Tucker-Lewis Index (TLI)	0.320
Loglikelihood and Information Criteria:	
Loglikelihood user model (H0)	-13648.276
Loglikelihood unrestricted model (H1)	-11947.316
Akaike (AIC)	27332.552
Bayesian (BIC)	27420.892
Sample-size adjusted Bayesian (BIC)	27363.723
Root Mean Square Error of Approximation:	
RMSEA	**0.354**
90 Percent confidence interval - lower	0.344
90 Percent confidence interval - upper	0.364
P-value RMSEA <= 0.05	0.000
Standardized Root Mean Square Residual:	
SRMR	0.234

```
Parameter Estimates:
  Standard errors                                  Standard
  Information                                       Expected
  Information saturated (h1) model                  Structured
Latent Variables:
              Estimate  Std.Err  z-value  P(>|z|)  Std.lv  Std.all
  f =~
    V1           1.234    0.033   37.946    0.000    1.234    0.924
    V2           1.235    0.034   36.787    0.000    1.235    0.908
    V3           1.220    0.033   37.446    0.000    1.220    0.917
    V4           0.228    0.042    5.366    0.000    0.228    0.173
    V5           0.200    0.043    4.697    0.000    0.200    0.152
    V6           0.171    0.042    4.034    0.000    0.171    0.131
    V7          -0.536    0.041  -13.117    0.000   -0.536   -0.408
    V8          -0.534    0.040  -13.202    0.000   -0.534   -0.411
    V9          -0.529    0.041  -12.897    0.000   -0.529   -0.402

Variances:
              Estimate  Std.Err  z-value  P(>|z|)  Std.lv  Std.all
   .V1          0.259    0.019   13.683    0.000    0.259    0.146
   .V2          0.326    0.021   15.502    0.000    0.326    0.176
   .V3          0.281    0.019   14.517    0.000    0.281    0.159
   .V4          1.671    0.075   22.319    0.000    1.671    0.970
   .V5          1.685    0.075   22.329    0.000    1.685    0.977
   .V6          1.680    0.075   22.337    0.000    1.680    0.983
   .V7          1.437    0.065   22.089    0.000    1.437    0.833
   .V8          1.406    0.064   22.085    0.000    1.406    0.831
   .V9          1.454    0.066   22.099    0.000    1.454    0.838
    f           1.000                               1.000    1.000
```

RESULTS FOR THE UNCORRELATED THREE-FACTOR MODEL

```
> m1c = 'f1 =~ V1 + V2 + V3
+        f2 =~ V4 + V5 + V6
+        f3 =~ V7 + V8 + V9
+        f1 ~~ 0*f2
+        f1 ~~ 0*f3
+        f2 ~~ 0*f3'
> res2 <- cfa(m1c, data=X, std.lv=TRUE)
> summary(res2, fit.measures=TRUE,standardized=TRUE)
lavaan 0.6-9 ended normally after 23 iterations

  Estimator                                         ML
  Optimization method                           NLMINB
  Number of model parameters                        18
  Number of observations                          1000

Model Test User Model:
  Test statistic                               228.916
```

Degrees of freedom   **27**
P-value (Chi-square)   0.000

Model Test Baseline Model:
  Test statistic   6656.327
  Degrees of freedom   36
  P-value   0.000

User Model versus Baseline Model:
  Comparative Fit Index (CFI)   **0.970**
  Tucker-Lewis Index (TLI)   0.959

Loglikelihood and Information Criteria:
  Loglikelihood user model (H0)   -12061.774
  Loglikelihood unrestricted model (H1)   -11947.316
  Akaike (AIC)   24159.548
  Bayesian (BIC)   24247.887
  Sample-size adjusted Bayesian (BIC)   24190.718

Root Mean Square Error of Approximation:
  RMSEA   **0.086**
  90 Percent confidence interval - lower   0.076
  90 Percent confidence interval - upper   0.097
  P-value RMSEA <= 0.05   0.000
Standardized Root Mean Square Residual:
  SRMR   0.171

Parameter Estimates:
  Standard errors   Standard
  Information   Expected
  Information saturated (h1) model   Structured

Latent Variables:

	Estimate	Std.Err	z-value	P(>|z|)	Std.lv	Std.all
f1 =~						
V1	1.234	0.033	37.933	0.000	1.234	**0.924**
V2	1.245	0.033	37.292	0.000	1.245	**0.915**
V3	1.223	0.033	37.571	0.000	1.223	**0.919**
f2 =~						
V4	1.050	0.038	27.327	0.000	1.050	**0.800**
V5	1.044	0.039	27.094	0.000	1.044	**0.794**
V6	1.007	0.039	26.159	0.000	1.007	**0.771**
f3 =~						
V7	1.204	0.032	37.191	0.000	1.204	**0.917**
V8	1.171	0.032	36.135	0.000	1.171	**0.900**
V9	1.201	0.033	36.904	0.000	1.201	**0.912**

Covariances:

	Estimate	Std.Err	z-value	P(>|z|)	Std.lv	Std.all
f1 ~~						

```
   f2          0.000                                      0.000    0.000
   f3          0.000                                      0.000    0.000
   f2 ~~
   f3          0.000                                      0.000    0.000

Variances:
             Estimate  Std.Err   z-value   P(>|z|)   Std.lv   Std.all
     .V1       0.259    0.019    13.542     0.000     0.259    0.145
     .V2       0.301    0.021    14.666     0.000     0.301    0.163
     .V3       0.274    0.019    14.192     0.000     0.274    0.155
     .V4       0.619    0.046    13.451     0.000     0.619    0.360
     .V5       0.636    0.046    13.796     0.000     0.636    0.369
     .V6       0.694    0.046    15.113     0.000     0.694    0.406
     .V7       0.276    0.021    13.385     0.000     0.276    0.160
     .V8       0.320    0.021    15.099     0.000     0.320    0.189
     .V9       0.291    0.021    13.877     0.000     0.291    0.168
      f1       1.000                                   1.000    1.000
      f2       1.000                                   1.000    1.000
      f3       1.000                                   1.000    1.000
```

RESULTS FOR THE THREE–FACTOR CORRELATED MODEL

```
>  m3a <- 'f1 =~ V1 + V2 + V3
+          f2 =~ V4 + V5 + V6
+          f3 =~ V7 + V8 + V9'
> res3 <- cfa(m3a, data=X, std.lv = TRUE)
> summary(res3, fit.measures=TRUE, standardized=TRUE)
lavaan 0.6-9 ended normally after 24 iterations

   Estimator                                      ML
   Optimization method                        NLMINB
   Number of model parameters                     21
   Number of observations                       1000

Model Test User Model:
   Test statistic                             40.145
   Degrees of freedom                             24
   P-value (Chi-square)                        0.021
Model Test Baseline Model:
   Test statistic                           6656.327
   Degrees of freedom                             36
   P-value                                     0.000

User Model versus Baseline Model:

   Comparative Fit Index (CFI)                 0.998
   Tucker-Lewis Index (TLI)                    0.996

Loglikelihood and Information Criteria:
   Loglikelihood user model (H0)          -11967.388
```

```
Loglikelihood unrestricted model (H1)      -11947.316
Akaike (AIC)                                23976.777
Bayesian (BIC)                              24079.840
Sample-size adjusted Bayesian (BIC)         24013.142
```

Root Mean Square Error of Approximation:

```
RMSEA                                            0.026
90 Percent confidence interval - lower           0.010
90 Percent confidence interval - upper           0.040
P-value RMSEA <= 0.05                            0.999
```
Standardized Root Mean Square Residual:
```
SRMR                                             0.015
```

Parameter Estimates:
```
Standard errors                             Standard
Information                                 Expected
Information saturated (h1) model            Structured
```

Latent Variables:

	Estimate	Std.Err	z-value	P(>|z|)	Std.lv	Std.all
f1 =~						
V1	1.236	0.032	38.032	0.000	1.236	**0.926**
V2	1.243	0.033	37.200	0.000	1.243	**0.914**
V3	1.224	0.033	37.616	0.000	1.224	**0.920**
f2 =~						
V4	1.050	0.038	27.405	0.000	1.050	**0.800**
V5	1.043	0.038	27.151	0.000	1.043	**0.794**
V6	1.008	0.038	26.231	0.000	1.008	**0.771**
f3 =~						
V7	1.203	0.032	37.180	0.000	1.203	**0.916**
V8	1.172	0.032	36.203	0.000	1.172	**0.901**
V9	1.201	0.033	36.935	0.000	1.201	**0.912**

Covariances:

	Estimate	Std.Err	z-value	P(>|z|)	Std.lv	Std.all
f1 ~~						
f2	0.169	0.035	4.859	0.000	0.169	**0.169**
f3	-0.398	0.029	-13.880	0.000	-0.398	**-0.398**
f2 ~~						
f3	-0.193	0.035	-5.565	0.000	-0.193	**-0.193**

Variances:

	Estimate	Std.Err	z-value	P(>|z|)	Std.lv	Std.all
.V1	0.255	0.019	13.534	0.000	0.255	**0.143**
.V2	0.306	0.020	14.956	0.000	0.306	**0.165**
.V3	0.273	0.019	14.271	0.000	0.273	**0.154**
.V4	0.619	0.046	13.569	0.000	0.619	**0.360**
.V5	0.637	0.046	13.934	0.000	0.637	**0.369**
.V6	0.693	0.046	15.184	0.000	0.693	**0.405**
.V7	0.278	0.020	13.625	0.000	0.278	**0.161**

.V8	0.318	0.021	15.164	0.000	0.318	**0.188**
.V9	0.291	0.021	14.035	0.000	0.291	**0.168**
f1	1.000				1.000	**1.000**
f2	1.000				1.000	**1.000**
f3	1.000				1.000	**1.000**

Chapter 16

Cluster Analysis

Like the preceding chapter on factor analysis, the input data for this chapter include an $n \times p$ matrix ($\mathbf{X}$) that contains measurements for n observations or cases (e.g., individuals, firms, or plants) on each of p variables. In factor analysis, the focus was on reducing the dimensionality of the columns of $\mathbf{X}$, that is, identifying a few ($q << p$) linear combinations of the p variables that explain a large proportion of the variation in the full dataset. In cluster analysis, the goal is to use the measurements on the p variables to establish a small number ($K <<< n$) of clusters or groups of homogeneous observations. Two popular approaches to cluster analysis are considered in this chapter.

In Section 16.1, we focus on the problem of partitioning the n observations into K clusters, so as to minimize the sum of squared Euclidean distances of observations from the centroid (i.e., the means of the p variables for the cluster) of the cluster to which they are assigned. This is a formidable non-convex optimization problem. Although branch-and-bound methods can be applied to small instances (Brusco & Stahl, 2005, Chapter 5; Diehr, 1985; Koontz *et al.*, 1975) and mathematical programming can be successfully used for somewhat larger problems (Aloise *et al.*, 2012, du Merle *et al.*, 2000), heuristic variants of the K-means clustering algorithm (Forgy, 1965; Jancey, 1966; MacQueen, 1967; Steinhaus, 1956) are, by far, the most popular approach. We implement what is arguably the most popular variant of K-means clustering using a VBA macro.

Mixture model clustering (also known as model-based clustering) is a popular alternative to K-means clustering. Rather than focusing on the minimization of sum of squares, the data are treated as arising from K subpopulations or classes and the goal is to estimate the mixing proportions for each class and other relevant model parameters so as to maximize likelihood. Class membership probabilities are obtained for each observation for each group, and a partition can be established by assigning the observation to the group having the largest probability. In Section 16.2, we describe a specific type of mixture model known as latent class analysis (Lazarsfeld, 1950). Here, the measurements in the data matrix $\mathbf{X}$ are binary and the underlying statistical model is the Bernoulli distribution. We will illustrate the estimation of model parameters using both the GRG engine of the Excel Solver and the popular expectation-maximization algorithm (E–M algorithm, Dempster *et al.*, 1977).

16.1 *K*-means Clustering

16.1.1 *Optimization problem and heuristic algorithm*

Our description of the problem of minimizing the within-cluster sum-of-squared distances of observations from their cluster centroids uses the following notation:

$n =$ the number of observations,

$p =$ the number of variables,

$K =$ the number of clusters,

$\mathbf{X} = [x_{ij}]$ the $n \times p$ data matrix of measurements for observation i on variable j (for $1 \leq i \leq n$ and $1 \leq j \leq p$),

$\Pi = \{\pi \in \Pi\}$ is the set of all partitions of the n observations into K clusters,

$\pi = \pi = \{S_1, \ldots, S_K\}$ is a partition of the observations into K clusters, where S_k contains the indices of the observations assigned to cluster k for all $1 \leq k \leq K$,

$\bar{x}_{jk} =$ the mean for variable j on cluster k (for $1 \leq j \leq p$ and $1 \leq k \leq K$),

$d_{ik} =$ squared Euclidean distance of observation i from the centroid of cluster k; $d_{ik} = \sum_{j=1}^{p}(x_{ij} - \bar{x}_{jk})^2$ for $1 \leq i \leq n$ and $1 \leq k \leq K$).

The relevant optimization problem is as follows:

$$\min_{\pi \in \Pi} f(\pi) = \sum_{k=1}^{K} \sum_{i \in S_k} d_{ik}. \tag{16.1}$$

One approach to solving this problem is to explicitly generate all partitions in Π and select the one that minimizes the total sum of squared distances of the observations from their cluster centroid in accordance with Equation (16.1). However, the number of possible partitions is a Stirling number of the second kind, which is computed as follows (see Brusco & Stahl, 2005; Hand, 1981):

$$\frac{1}{K!} \sum_{k=0}^{K} (-1)^k \binom{K}{k} (K - k)^n. \tag{16.2}$$

Even for a small dataset with $n = 50$ observations, the number of ways to partition 50 observations into $K = 5$ clusters is 7.7×10^{21}, which renders the generation of all partitions impossible. Branch-and-bound methods can sometimes enable the optimal partition to be identified for problems with up to $n = 150$ observations and $K = 6$ clusters (see Brusco, 2006), but these methods are very sensitive to the separation of the clusters and are generally impractical. Mathematical programming approaches, such as the one designed by Aloise *et al.* (2012), are state of the art for exact sum-of-squares clustering and they are sometimes successful for very large problem instances. Nevertheless, heuristic approaches are much more common.

There are a variety of K-means clustering heuristics that have been proposed over the decades (Forgy, 1965; Jancey, 1966; MacQueen, 1967; Steinhaus, 1956). A variety of metaheuristics have also been developed for sum-of-squares clustering, including simulated annealing (Klein & Dubes, 1989), genetic algorithms (Maulik & Bandyopadhyay, 2000), tabu search (Pacheco & Valencia, 2003), and variable neighborhood search (Hansen & Mladenovic, 2001). Brusco and Steinley (2007) provide a review and comparison of K-means heuristics, as well as metaheuristics, for the problem posed by Equation (16.1).

Despite all the research effort devoted to the development of exact and heuristic solution procedures to the problem posed by Equation (16.1), multiple restart (multistart) implementations of standard K-means clustering heuristics remain the method of choice for most applications. We use a VBA macro to implement K-means clustering in this chapter. A sketch of the algorithm is as follows:

Step 0. Choose K, the number of restarts, R, set $f^* = \infty$.

Step 1. Set $r = r + 1$.

Step 2. Randomly choose K observations and use their variable measurements to serve as initial centroids, ensuring that no two observations have exactly the same variable measurements. Using these centroids, compute d_{ik} for all $1 \leq i \leq n$ and $1 \leq k \leq K$. Establish a partition, $\pi' : (i \in S_l | d_{il} = \min_{(1 \leq k \leq K)} \{d_{ik}\})$, and compute f' using Equation (16.1).

Step 3. Recompute the centroids $\bar{x}_{jk} = \frac{1}{|S_k|} \sum_{i \in S_k} x_{ij}$ for $1 \leq j \leq p$ and $1 \leq k \leq K$.

Step 4. Reassign the observations by computing compute d_{ik} for all $1 \leq i \leq n$ and $1 \leq k \leq K$. Establish a partition, $\pi : (i \in S_l | d_{il} = \min_{(1 \leq k \leq K)} \{d_{ik}\})$, and compute f using Equation (16.1).

Step 5. If $f = f'$, then go to Step 6; otherwise, set $f' = f$ and $\pi' = \pi$ and return to Step 3.

Step 6. If $f' < f^*$, then set $f^* < f'$ and $\pi^* = \pi$.

Step 7. If $r = R$, then STOP; otherwise, return to Step 1.

The algorithm begins in Step 0 with the choice of K and the desired number of restarts, R. An upper bound on the best-found objective function value across all restarts is established as $f^* = \infty$. Steps 1 through 6 all pertain to an individual restart of the algorithm.

Step 1 increments the restart counter. In Step 2, K observations are randomly selected to serve as initial centroids. A check is conducted to ensure that no two selected observations have exactly the same measurements on all variables. An initial partition, π', is established by assigning each observation to the cluster for which its distance from the cluster centroid is the smallest and the objective function value (f') is computed for π'. The centroids are recomputed in Step 3 based on the assignments of observations to clusters made in Step 2. Reassignment of observations to clusters is then accomplished in Step 4 and a new objective function value is computed. If there is improvement in the objective function in Step 5, then the new partition and objective value are stored and processing returns to Step 3; otherwise, processing returns to Step 6. Essentially, the algorithm iterates between Steps 3 and 4 until convergence, that is, no improvement in the objective function value for Equation (16.1). At Step 6, the objective function value for the partition realized for the current restart (π') is evaluated against the objective function value associated with the best partition across all restarts thus far (π^*) and will replace that best partition if its objective function value is better. The algorithm terminates in Step 7 when all restarts have been completed.

We have also incorporated a couple of other desirable features in our VBA implementation of the K-means heuristic. First, there is an option that allows users to normalize the data (z-scores using population standard deviation) prior to implementation of the algorithm, which can be useful if some of the variables are measured on very different scales. Second, we have implemented some code to correct the problem of *degeneracy* that might occur in the reassignment of observations to clusters. Specifically, it is possible that fewer than K clusters might be used in the reassignment process. If this is the case, then, as necessary, we create a new cluster by assigning the case that is farthest from its current cluster centroid to its own new cluster.

16.1.2   *Example 16.1 — Fisher's iris data*

We apply our VBA macro for K-means clustering to Fisher's iris data. As we noted in our coverage of discriminant analysis, these data were originally collected by Anderson (1935) and subsequently analyzed by Fisher (1936, 1938) in his pioneering development of linear discriminant analysis. Specifically, the data were obtained from 50 samples of each of $g = 3$ species of iris flowers (Setosa, Versicolor, and Virginica), resulting in a total of $n = 150$ plants. Each plant was measured on $p = 4$ variables: (a) sepal length; (b) sepal width; (c) petal length; and (d) petal width. The variable measurements are located in cells B7:E156 of the 'Batch' worksheet displayed in Figure 16.1.

Prior to running the VBA macro, the user must specify the proper input in cells B1:B5: (i) Cell B1 — there are $n = 150$ observations in

Number of Observations (n)	150					SSE =	78.8514			
Number of Clustering Variables (p)	4									
Number of Clusters (k)	3	Run K-means								
Standardize	0									
Number of Restarts	50				Cluster Assigned			Cluster		

X1	X2	X3	X4 X5 X6 X7 X8 X9 10 11 12 13 14 15 16 17 18 19 20	Cluster Assigned		Size	Var 1	Var 2	Var 3	Var 4
5.1	3.5	1.4	0.2	1	Cluster 1	50	5.006	3.428	1.462	.246
4.9	3	1.4	0.2	1	Cluster 2	62	5.902	2.748	4.394	1.434
4.7	3.2	1.3	0.2	1	Cluster 3	38	6.850	3.074	5.742	2.071
4.6	3.1	1.5	0.2	1	Cluster 4					
5	3.6	1.4	0.2	1	Cluster 5					
5.4	3.9	1.7	0.4	1	Cluster 6					
4.6	3.4	1.4	0.3	1	Cluster 7					
5	3.4	1.5	0.2	1	Cluster 8					
4.4	2.9	1.4	0.2	1	Cluster 9					
4.9	3.1	1.5	0.1	1	Cluster 10					
5.4	3.7	1.5	0.2	1						
4.8	3.4	1.6	0.2	1						
4.8	3	1.4	0.1	1						
4.3	3	1.1	0.1	1						
5.8	4	1.2	0.2	1						
5.7	4.4	1.5	0.4	1						
5.4	3.9	1.3	0.4	1						
5.1	3.5	1.4	0.3	1						
5.7	3.8	1.7	0.3	1						
5.1	3.8	1.5	0.3	1						

Figure 16.1.   Worksheet for Example 16.1.

the iris dataset; (ii) Cell B2 — there are $p = 4$ variable measurements; (iii) Cell B3 — we have selected $K = 3$ clusters; (iv) Cell B4 — we set this cell to zero for no standardization of the variable measurements; and (v) Cell B5 — we have selected $R = 50$ restarts of the algorithm.

Clicking on the 'Run K-means' button will run the macro. The key pieces of output are the best-found value of the sum-of-squared distance objective function in cell Y1 (which is 78.8514), the vector of cluster labels for the observations in column W, the number of observations in each cluster in column Z, and the cluster centroids in Figure 16.1 spanning cells AA7 to AT16.

At present, the VBA macro is configured to run for up to $n = 5000$ observations, $p = 20$ variables, and $K = 20$ clusters. However, the computation time would be sizable for a problem near these limits. Whereas several hundred or even several thousand restarts are possible for a dataset the size of the iris data, the number of restarts must be chosen judiciously for larger problems. It is important to emphasize here that our goals are illustrative in nature. Like other statistical worksheets described in this book, in no way are we advocating our VBA macro as a practical platform for large-scale K-means clustering.

One of the reasons why we selected the iris data for illustrative purposes is that globally optimal partitions for the unstandardized data are available for $K = 2$ to $K = 6$ clusters (see Brusco, 2006).

This enabled us to benchmark the performance of our VBA macro. In Figure 16.2, we report the results for five separate runs of the macro for $2 \leq K \leq 6$ clusters, for both $R = 50$ and $R = 500$ restarts, and for both the standardized and unstandardized data. We note that a user running their own five trials might obtain results that differ from those in Figure 16.2 because the generation of initial centroids is random.

The results in Figure 16.2 indicate that the performance of the VBA macro for the unstandardized data is very stable, regardless of the number of clusters or the number of restarts. Even at 50 restarts, the VBA macro obtained the globally optimal partition on all five runs for all numbers of clusters. There was more variability for the standardized data at 50 restarts, and, particularly at $K = 5$ and $K = 6$ clusters, we see the potential for a fairly poor result with only 50 restarts. At 500 restarts, the results for the standardized data are much more stable. Nevertheless, based on recommendations originally offered by Steinley (2003), we recommend at least 5000 restarts whenever possible. Of course, for a larger dataset, 5000 restarts could be rather impractical using our VBA macro.

Although cluster analysis is typically used when group memberships are unknown, the group memberships are known for the plants in the iris data. Figure 16.3. provides a summary of the three-cluster solutions for both the unstandardized and standardized data. We see that classification performance is actually better for the unstandardized data.

K	50 Restarts - Unstandardized data					50 Restarts - Standardized data				
	run 1	run 2	run 3	run 4	run 5	run 1	run 2	run 3	run 4	run 5
2	152.3480	152.3480	152.3480	152.3480	152.3480	222.3617	222.3617	222.3617	222.3617	222.3617
3	78.8514	78.8514	78.8514	78.8514	78.8514	139.8205	139.8205	139.8205	139.8205	139.8205
4	57.2285	57.2285	57.2285	57.2285	57.2285	114.0925	114.0922	114.0925	114.0922	114.0925
5	46.4462	46.4462	46.4462	46.4462	46.4462	90.8076	90.8073	90.8073	90.8073	90.8331
6	39.0400	39.0400	39.0400	39.0400	39.0400	81.4674	79.9986	80.0225	80.0369	79.9986

K	500 Restarts - Unstandardized data					500 Restarts - Standardized data				
2	152.3480	152.3480	152.3480	152.3480	152.3480	222.3617	222.3617	222.3617	222.3617	222.3617
3	78.8514	78.8514	78.8514	78.8514	78.8514	139.8205	139.8205	139.8205	139.8205	139.8205
4	57.2285	57.2285	57.2285	57.2285	57.2285	114.0922	114.0922	114.0922	114.0922	114.0922
5	46.4462	46.4462	46.4462	46.4462	46.4462	90.8073	90.8073	90.8073	90.8073	90.8073
6	39.0400	39.0400	39.0400	39.0400	39.0400	79.9989	79.9986	79.9986	79.9986	79.9986

Figure 16.2.   Summary of some computational results for Example 16.1.

K = 3, Unstandardized		Z =		78.85144		K = 3, Standardized		Z =		139.8205
	Cluster Size	Var 1	Var 2	Var 3	Var 4		Cluster Size	Var 1	Var 2	Var 3	Var 4
Cluster 1	50	5.006	3.428	1.462	0.246	Cluster 1	50	-1.015	0.853	-1.305	-1.255
Cluster 2	62	5.902	2.748	4.394	1.434	Cluster 2	53	-0.050	-0.883	0.348	0.282
Cluster 3	38	6.850	3.074	5.742	2.071	Cluster 3	47	1.136	0.088	0.996	1.018
Cluster 1		50	100.00%			Cluster 1		50	100.00%		
Cluster 2		48	96.00%			Cluster 2		39	78.00%		
Cluster 3		36	72.00%			Cluster 3		36	72.00%		
Overall		134	89.33%			Overall		125	83.33%		

Figure 16.3. Summary of three-cluster solutions for Example 16.1.

Although both the unstandardized and standardized solutions correctly classify the same percentage of plants in groups 1 and 3, far more plants in the Versicolor group are correctly classified in the unstandardized solution. Overall, 89.33% of the plants are clustered in their correct group in the unstandardized solution, whereas only 83.33% are correctly classified in the standardized solution. However, as reported by Steinley and Brusco (2008), other standardization approaches perform much better.

16.2 Latent Class Analysis

16.2.1 *Optimization problem and estimation methods*

Our description of the optimization problem associated with latent class analysis uses the following notation:

$n =$ the number of observations.

$p =$ the number of variables.

$K =$ the number of classes.

$\mathbf{X} = [x_{ij}]$ the $n \times p$ data matrix of binary $\{0, 1\}$ measurements for observation i on variable j (for $1 \leq i \leq n$ and $1 \leq j \leq p$); we will also denote a row of $\mathbf{X}$ as $\mathbf{x}_i$, that is, the measurements for observation i on the p variables.

$\boldsymbol{\Theta} =$ the collection of model parameters, $\boldsymbol{\Theta} = \{\boldsymbol{\lambda}, \Pi\}$.

$\boldsymbol{\lambda}$ = a $K \times 1$ vector of mixing proportions with elements λ_k indicating the probability that an observation belongs to class k (for $1 \leq k \leq K$).

$\boldsymbol{\Pi}$ = a $p \times K$ matrix of probabilities with elements π_{jk} indicating the probability that the measurement for variable j will be equal to one for class k (for $1 \leq j \leq p$ and $1 \leq k \leq K$).

Using these definitions, the latent class model is expressed as follows:

$$f(\mathbf{x}_i|\boldsymbol{\Theta}) = \sum_{k=1}^{K} \lambda_k \prod_{j=1}^{p} \pi_{jk}^{x_{ij}} (1 - \pi_{jk})^{1-x_{ij}}. \tag{16.3}$$

The probability that observation i with measurement vector $\mathbf{x}_i$ has membership in class k is

$$f(k|\mathbf{x}_i, \boldsymbol{\Theta}) = \frac{\lambda_k \prod_{j=1}^{p} \pi_{jk}^{x_{ij}} (1 - \pi_{jk})^{1-x_{ij}}}{f(\mathbf{x}_i|\boldsymbol{\Theta})}. \tag{16.4}$$

The likelihood function for latent class analysis, which is computed across all n observations, is as follows:

$$L = \prod_{i=1}^{n} \left[\sum_{k=1}^{K} \lambda_k \prod_{j=1}^{p} \pi_{jk}^{x_{ij}} (1 - \pi_{jk})^{1-x_{ij}} \right]. \tag{16.5}$$

The corresponding log-likelihood function, which is easier to use for maximum likelihood estimation purposes is

$$LL = \sum_{i=1}^{n} log \left[\sum_{k=1}^{K} \lambda_k \prod_{j=1}^{p} \pi_{jk}^{x_{ij}} (1 - \pi_{jk})^{1-x_{ij}} \right]. \tag{16.6}$$

We consider two approaches for estimating the parameters, $\boldsymbol{\Theta}$, so as to maximize the value of the log-likelihood in Equation (16.6). The first approach is direct estimation using the nonlinear GRG engine of the Excel Solver. The second approach is to use the E–M algorithm (Dempster *et al.*, 1977), which works as follows:

Step 0. Begin with initial values for model parameters $\boldsymbol{\lambda}$ and $\boldsymbol{\Pi}$.
Step 1. E-Step: Use Equations (16.3) and (16.4) to establish class membership probabilities for all n observations.

Step 2. M-Step: Obtain new estimates for the model parameters $\boldsymbol{\lambda}$ and $\boldsymbol{\Pi}$ using the following equations:

$$\pi_{jk} = \frac{\sum_{i:x_{ij}=1} f(k|\mathbf{x}_i, \boldsymbol{\Theta})}{\sum_{i=1}^{n} f(k|\mathbf{x}_i, \boldsymbol{\Theta})} \quad \text{for} \quad 1 \le j \le p \quad \text{and}$$

$$1 \le k \le K, \tag{16.7}$$

$$\lambda_k = \frac{\sum_{i=1}^{n} f(k|\mathbf{x}_i, \boldsymbol{\Theta})}{\sum_{k=1}^{K}\sum_{i=1}^{n} f(k|\mathbf{x}_i, \boldsymbol{\Theta})} \quad \text{for} \quad 1 \le k \le K. \tag{16.8}$$

Step 3. Repeat steps 1 and 2 until the increase in LL is less than some user-specified threshold.

16.2.2 *Example 16.2 — Four-item math test*

Example 16.2 uses data from a math test with $p = 4$ items that was taken by $n = 142$ students (Macready & Dayton, 1977). We will assume $K = 2$ classes, a mastery and a non-mastery class. Following Templin (2015), we provide a succinct representation of the data by identifying the number of students associated with each of the $2^4 = 16$ possible binary strings for four items in cells B11:B26 of the 'Solver' worksheet in Figure 16.4. For example, there were 15 students who got all four items correct, that is, $x_1 = x_2 = x_3 = x_4 = 1$. There were seven students who got the first three items correct ($x_1 = x_2 = x_3 = 1$) and the fourth item wrong ($x_4 = 0$), and so on. Forty-one students got all four questions wrong, $x_1 = x_2 = x_3 = x_4 = 0$.

	X1	X2	X3	X4		Log Likelihood Value	−331.7637		83.29133	58.70867
Class 1	.58656	.75343	.78026	.43159	.70754	Posterior probabilities for groups 1 and 2				
Class 2	.41344	.20859	.06832	.01792	.05226					

Number of Cases	X1	X2	X3	X4	Group 1				prob of x	Group 2				prob of x	Total prob of x	LOG(p)	posterior prob group 1	posterior prob group 2
15	1	1	1	1	75343	78026	43159	70754	.10530	20859	06832	01792	05225	.00001	.10530	−2.25091	.99995	.00005
7	1	1	1	0	75343	78026	43159	29246	.04352	20859	06832	01792	94775	.00010	.04362	−3.13215	.99771	.00229
23	1	1	0	1	75343	78026	56841	70754	.13868	20859	06832	98208	05225	.00030	.13898	−1.97343	.99782	.00218
7	1	1	0	0	75343	78026	56841	29246	.05732	20859	06832	98208	94775	.00548	.06280	−2.76773	.91268	.08732
1	1	0	1	1	75343	21974	43159	70754	.02965	20859	93168	01792	05225	.00008	.02973	−3.51562	.99747	.00253
3	1	0	1	0	75343	21974	43159	29246	.01226	20859	93168	01792	94775	.00136	.01362	−4.29609	.89983	.10017
6	1	0	0	1	75343	21974	56841	70754	.03905	20859	93168	98208	05225	.00412	.04318	−3.14244	.90451	.09549
13	1	0	0	0	75343	21974	56841	29246	.01614	20859	93168	98208	94775	.07479	.09093	−2.39788	.17753	.82247
4	0	1	1	1	24657	78026	43159	70754	.03446	79141	06832	01792	05225	.00002	.03448	−3.36732	.99939	.00061
2	0	1	1	0	24657	78026	43159	29246	.01424	79141	06832	01792	94775	.00038	.01462	−4.22511	.97404	.02596
5	0	1	0	1	24657	78026	56841	70754	.04539	79141	06832	98208	05225	.00115	.04653	−3.06761	.97535	.02465
6	0	1	0	0	24657	78026	56841	29246	.01876	79141	06832	98208	94775	.02081	.03957	−3.22979	.47413	.52587
4	0	0	1	1	24657	21974	43159	70754	.00970	79141	93168	01792	05225	.00029	.00999	−4.60614	.97143	.02857
1	0	0	1	0	24657	21974	43159	29246	.00401	79141	93168	01792	94775	.00518	.00919	−4.68982	.43658	.56342
4	0	0	0	1	24657	21974	56841	70754	.01278	79141	93168	98208	05225	.01564	.02843	−3.56049	.44985	.55036
41	0	0	0	0	24657	21974	56841	29246	.00528	79141	93168	98208	94775	.28374	.28902	−1.24124	.01828	.98172

Figure 16.4. Worksheet for Example 16.2 — template for the Solver approach.

16.2.2.1 *Excel Solver approach*

There are nine decision variables, which are entered in the 'By Changing Cell Values' of the Solver dialog box as shown in Figure 16.5. They are the estimated model parameters for λ_1 in cell B2 and the eight π_{jk} values in cells G2:J3. The estimate for λ_2 in cell B3 is not an unknown variable because it is determined by the relationship $\lambda_2 = 1 - \lambda_1$. As observed in Figure 16.5, we place upper bounds of $1 - \varepsilon$ and lower bounds of ε on the variables in the 'Subject to the Constraints' portion of the Solver dialog box because they are probabilities.

For each possible binary string, the parameter estimates in cell B2 and cells G2:J2 are used, in conjunction with the variable

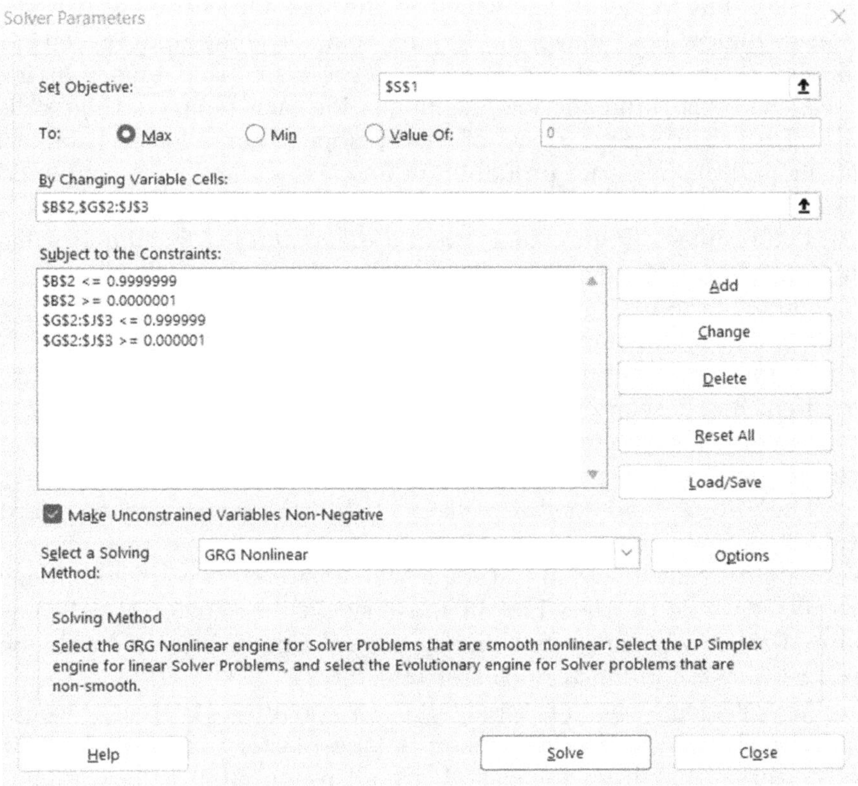

Figure 16.5. Solver dialog box for Example 16.2 — template for the Solver approach.

measurements, to establish a numerator for $f(k = 1|\mathbf{x}_i, \boldsymbol{\Theta})$ in equation (16.4) via cells G11:K26. Likewise, cell B3 and cells G3:J3 are used, in conjunction with the variable measurements, to establish a numerator for $f(k = 2|\mathbf{x}_i, \boldsymbol{\Theta})$ in equation (16.4) via cells M11:Q26. Adding the numerator terms for class 1 in cells K11:K26 to the numerator terms for class 2 in cells Q11:Q26 yields the denominator for Equation (16.4) in cells S11:S26. Equation (16.4) then establishes the posterior probabilities for class 1 in cells V11:V26 by dividing the values in K11:K26 by the values in S11:S26, and posterior probabilities for class 2 in cells W11:W26 by dividing the values in Q11:Q26 by the values in S11:S26.

The log-likelihood for each binary string in cells T11:T26 is obtained by taking the natural logarithm of the values in cells S11:S26. The value of *LL*, which is the 'Set Objective' cell in the Solver dialog box shown in Figure 16.5, is computed in cell S1 as the sumproduct of the number of cases for each binary string (cells A11:A26) and the log-likelihoods for each binary string (cells T11:T26). The sumproducts of the number of cases for each binary string and the posterior probabilities for classes 1 and 2 are contained in cells V1 and W1, respectively.

The results in Figure 16.4 reveal a maximum log-likelihood value of -331.7637. The overall group membership probabilities are $\lambda_1 = 0.58656$ for the mastery class and $\lambda_2 = 0.41344$ for the non-mastery class. For the mastery class, the probabilities of getting items 1, 2, and 4 correct are all in the 0.7 to 0.8 range; however, the probability for item 3 is only about 0.43. For the non-mastery class, the probabilities of getting items 2, 3, and 4 correct are all less than 0.07 and the probability for item 1 is roughly 0.21.

The use of bold font in cells V11:W26 indicates the standard rule of assigning cases to *clusters* based on largest class membership probability. Succinctly, any student who answered two or more items correctly would be placed in the mastery cluster (cluster 1), whereas any student who answered one or fewer items correctly would be placed in the non-mastery cluster (cluster 2).

Any student who answered at least two items correctly would definitively be assigned to cluster 1 (the mastery cluster) as their posterior probabilities for class 1 are all roughly 0.9 or larger (and often larger than 0.95). At the other end of the spectrum, the 41 students who did not answer any item correctly are definitively placed in cluster 2, as their posterior probability for class 2 exceeds 0.98.

The 13 students who answered *only* item 1 correctly could confidently be placed in cluster 2 because their posterior probability for class 2 is approximately 0.82 and their posterior probability for class 1 is only about 0.18. We also see that students who answered only item 2, only item 3, or only item 4 correctly would be placed in cluster 2; however, those assignments would be made with less confidence. For example, for the six students who answered only item 2 correctly, the posterior probabilities for classes 1 (0.47) and 2 (0.53) are very close.

16.2.2.2 *E–M algorithm approach*

The design of the worksheet 'expmax' in Figure 16.6 for implementing the E–M algorithm is similar to the one for the Solver approach. However, there are some important differences. Like the Solver worksheet, cells B2:B3 and G2:J3 contain the model parameters that are used to compute the posterior probabilities. However, because of the iterative nature of the E–M algorithm, once the posterior probabilities are computed in the E-step (Step 1) of the algorithm, they must be used in the M-step (Step 2) to re-estimate the model parameters. The updated values of λ are computed from the posterior probabilities in cells B5:B6 and the updated values of Π are computed in cells G5:J6. Cells X11:Y26 have been added to the worksheet and

Parameter block

	B	X1	X2	X3	X4			
Class 1	.60000	.50000	.40000	.30000	.20000	Log Likelihood Value	-406.928	93.19290 48.80710 53.78663 15.21337
Class 2	.40000	.40000	.30000	.20000	.10000	Posterior probabilities for groups 1 and 2		
Class 1	.65629	.60611	.57715	.32025	.53574			
Class 2	.34371	.37934	.31170	.14659	.24736			

Cells B2:B3 are input lambdas and G2:J3 are input pis. They are used for computation in the E-step.
Cells B5:B6 are computed lambdas and G5:J6 are computed pis. They are computed from the posterior probabilities in the M-step.
The macro pastes the computed values to the input values

Number of Number of Cases / Cases times times

Data block

Cases	X1	X2	X3	X4	G1				prob of x	G2				prob of x	Total prob of x	LOG(p)	post prob group 1	post prob group 2	num group 1	num group 2
15	1	1	1	1	50000	40000	30000	20000	00720	40000	30000	20000	10000	00098	00816	-4.80851	.88235	.11765	13.23529	1.76471
7	1	1	1	0	50000	40000	30000	80000	02880	40000	30000	20000	90000	00864	03744	-3.28502	.76923	.23077	5.38462	1.61538
23	1	1	0	1	50000	40000	70000	20000	01680	40000	30000	80000	10000	00384	02064	-3.88052	.81395	.18605	18.72093	4.27907
7	1	1	0	0	50000	40000	70000	80000	06720	40000	30000	80000	90000	03456	10176	-2.28514	.66038	.33962	4.62264	2.37736
1	1	0	1	1	50000	60000	30000	20000	01080	40000	70000	20000	10000	00224	01304	-4.33973	.82822	.17178	0.82822	0.17178
3	1	0	1	0	50000	60000	30000	80000	04320	40000	70000	20000	90000	02016	06336	-2.75892	.68182	.31818	2.04545	0.95455
6	1	0	0	1	50000	60000	70000	20000	02520	40000	70000	80000	10000	00896	03416	-3.37670	.73770	.26230	4.42623	1.57377
13	1	0	0	0	50000	60000	70000	80000	10080	40000	70000	80000	90000	08064	18144	-1.70683	.55556	.44444	7.22222	5.77778
4	0	1	1	1	50000	40000	30000	20000	00720	60000	30000	20000	10000	00144	00864	-4.75135	.83333	.16667	3.33333	0.66667
2	0	1	1	0	50000	40000	30000	80000	02880	60000	30000	20000	90000	01296	04176	-3.17582	.68966	.31034	1.37931	0.62069
5	0	1	0	1	50000	40000	70000	20000	01680	60000	30000	80000	10000	00576	02256	-3.79158	.74468	.25532	3.72340	1.27660
6	0	1	0	0	50000	40000	70000	80000	06720	60000	30000	80000	90000	05184	11904	-2.12830	.56452	.43548	3.38710	2.61290
4	0	0	1	1	50000	60000	30000	20000	01080	60000	70000	20000	10000	00336	01416	-4.25733	.76271	.23729	3.05085	0.94915
1	0	0	1	0	50000	60000	30000	80000	04320	60000	70000	20000	90000	03024	07344	-2.61129	.58824	.41176	0.58824	0.41176
4	0	0	0	1	50000	60000	70000	20000	02520	60000	70000	80000	10000	01344	03864	-3.25347	.65217	.34783	2.60870	1.39130
41	0	0	0	0	50000	60000	70000	80000	10080	60000	70000	80000	90000	12096	22176	-1.50616	.45455	.54545	18.63636	22.36364

M-Step

Figure 16.6. Worksheet for Example 16.2 — E–M algorithm approach (starting).

328   *Linear and Nonlinear Optimization Using Spreadsheets*

they contain the products of the number of cases for each binary string and the posterior probabilities. They assist with the computation of the numerators of Equation (16.7) when updating the **Π** matrix. Clicking on the 'M-Step' button runs a macro that copies the values from B5 and G5:J6 and pastes them, respectively, to cells B2 and G2:J3, which completes another iteration of the E–M algorithm. (Cell B6 does not need to be copied to cell B3 because cell B3 is determined by the relationship B3 = 1−B2.)

The worksheet in Figure 16.6 is clearly suboptimal with an $LL = -406.928$. Arbitrary starting values have been entered in cells B2 and G2:J6. However, the worksheet has completed the E-step and automatically updated the parameter values in cells B5 and G5:J6. Clicking the M-step button repeatedly will run iterations of the E–M algorithm. When the numbers in the parameter cells stop changing, we conclude that the algorithm has converged. After clicking this button 30–50 times, we obtained the results displayed in Figure 16.7. Faster or slower convergence might be realized with different starting values. The algorithm is generally robust for most starting solutions. However, an absurd starting solution, such as using 0.5 for all parameter values, can lead to an absurd result (i.e., two classes that are exactly the same).

Figure 16.7 reveals the same solution as Figure 16.4. The E–M algorithm is a flexible and efficient tool for model estimation; however, like the K-means algorithm, it can be sensitive to the initial solution and is not guaranteed to provide an optimal solution.

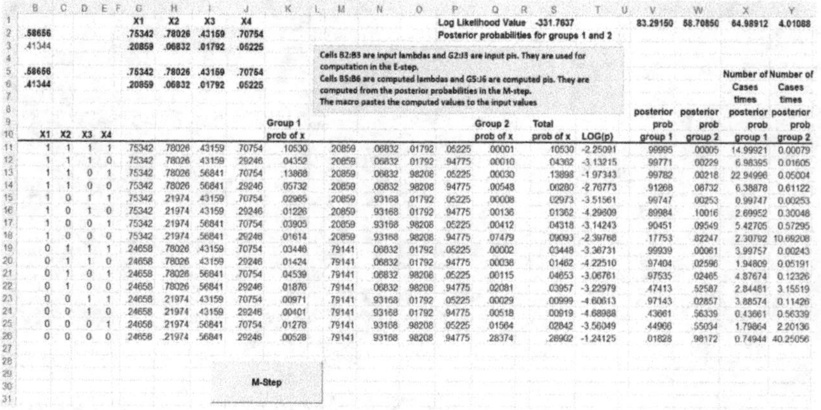

X1	X2	X3	X4	x1	x2	x3	x4	prob of x	x1	x2	x3	x4	prob of x	Total prob of x	LOG(p)	post prob grp1	post prob grp2	cases×post grp1	cases×post grp2
1	1	1	1	.75342	.78026	.43159	.70754	.10530	.20859	.06832	.01792	.05225	.00001	.10530	-2.25091	.99995	.00005	14.99921	0.00079
1	1	1	0	.75342	.78026	.43159	.29246	.04352	.20859	.06832	.01792	.94775	.00010	.04362	-3.13215	.99771	.00229	6.98365	0.01605
1	1	0	1	.75342	.78026	.56841	.70754	.13868	.20859	.06832	.98208	.05225	.00030	.13898	-1.97343	.99782	.00218	22.94998	0.05004
1	1	0	0	.75342	.78026	.56841	.29246	.05732	.20859	.06832	.98208	.94775	.00548	.06280	-2.76773	.91268	.08732	6.38878	0.61122
1	0	1	1	.75342	.21974	.43159	.70754	.02905	.20859	.93168	.01792	.05225	.00008	.02973	-3.51561	.99747	.00253	0.99747	0.00253
1	0	1	0	.75342	.21974	.43159	.29246	.01226	.20859	.93168	.01792	.94775	.00136	.01362	-4.29809	.89984	.10016	2.69952	0.30048
1	0	0	1	.75342	.21974	.56841	.70754	.03905	.20859	.93168	.98208	.05225	.00412	.04318	-3.14243	.90451	.09549	5.42705	0.57295
1	0	0	0	.75342	.21974	.56841	.29246	.01614	.20859	.93168	.98208	.94775	.07479	.09093	-2.39768	.17753	.82247	2.30702	10.66208
0	1	1	1	.24658	.78026	.43159	.70754	.03446	.79141	.06832	.01792	.05225	.00002	.03448	-3.36731	.99939	.00061	3.99757	0.00243
0	1	1	0	.24658	.78026	.43159	.29246	.01424	.79141	.06832	.01792	.94775	.00038	.01462	-4.22510	.97404	.02596	1.94809	0.05191
0	1	0	1	.24658	.78026	.56841	.70754	.04539	.79141	.06832	.98208	.05225	.00115	.04653	-3.06761	.97535	.02465	4.87674	0.12326
0	1	0	0	.24658	.78026	.56841	.29246	.01876	.79141	.06832	.98208	.94775	.02081	.03957	-3.22979	.47413	.52587	2.84481	3.15519
0	0	1	1	.24658	.21974	.43159	.70754	.00971	.79141	.93168	.01792	.05225	.00029	.00999	-4.60613	.97143	.02857	3.88574	0.11426
0	0	1	0	.24658	.21974	.43159	.29246	.00401	.79141	.93168	.01792	.94775	.00518	.00919	-4.68988	.43661	.56339	0.43661	0.56339
0	0	0	1	.24658	.21974	.56841	.70754	.01278	.79141	.93168	.98208	.05225	.01564	.02842	-3.56049	.44966	.55034	1.79864	2.20136
0	0	0	0	.24658	.21974	.56841	.29246	.00528	.79141	.93168	.98208	.94775	.28374	.28902	-1.24125	.01828	.98172	0.74944	40.25056

Log Likelihood Value -331.7637   83.29150 88.70850 64.98912 4.01088

M-Step

Figure 16.7.   Worksheet for Example 16.2 — E–M algorithm approach (ending).

References

Addinsoft (2021) XLSTAT (version 2021.2). Accessed April 30, 2021, https://www.xlstat.com/en/.

Adler, I., Erera, A.L., Hochbaum, D.S., & Olinick, E.V. (2002). Baseball, optimization, and the world wide web. *INFORMS Journal on Applied Analytics*, 32(2), 12–22.

Agbemava, E., Nyarko, I.K., Adade, T.C., & Bediako, A.K. (2016). Logistic regression analysis of predictors of loan defaults by customers of non-traditional banks in Ghana. *European Science Journal*, 12(1), 175–189.

Akaike, H. (1973). Maximum likelihood identification of Gaussian autoregressive moving average models. *Biometrika*, 60, 255–265.

Akaike, H. (1974). A new look at the statistical identification model. *IEEE Transactions on Automatic Control*, 19, 716–723.

Aktas, E., Ozaydin, O., Bozkaya, B., Ulengin, F., & Onsel, S. (2013). Optimizing fire station locations for the Istanbul Metropolitan Municipality. *Interfaces*, 43(3), 240–255.

Alarcón, F., Durán, G., Guajardo, M., Miranda, J., Muñoz, H., Ramírez, L., Ramírez, M., Sauré, D., Siebert, M., Souyris, S., Weintraub, A., Wolf-Yadlin, R., & Zamorano, G. (2017). Operations research transforms the scheduling of Chilean soccer leagues and South American World Cup qualifiers. *INFORMS Journal on Applied Analytics*, 47(1), 52–69.

Alboqami, H., Al-Karaghouli, W., Baeshen, Y., Erkan, I., Evans, C., & Ghoneim, A. (2015). Electronic word of mouth in social media: The common characteristics of retweeted and favourited marketer-generated content posted on Twitter. *International Journal of Internet Marketing Advertising*, 9(4), 338–358.

Aloise, D., Hansen, P., & Liberti, L. (2012). An improved column generation algorithm for minimum sum-of-squares clustering. *Mathematical Programming*, 131(1–2), 195–220.

Anderson, D.R., Sweeney, D.J., Williams, T.A., Camm, J.D., Cochran, J.J., & Fry, M.J., Ohlmann, J.W. (2019). *An Introduction to Management Science*, 15th Edition. Boston: Cengage Learning.

Anderson, E. (1935). The irises of the Gaspé peninsula. *Bulletin of the American Iris Society*, 59, 2–5.

Armour, G.C. & Buffa, E.S. (1963). A heuristic algorithm and simulation approach to relative location of facilities. *Management Science*, 9(2), 294–309.

Arnold, M.J. & Reynolds, K.E. (2003). Hedonic shopping motivations. *Journal of Retailing*, 79, 77–95.

Azen, R. & Traxel, N. (2009). Using dominance analysis to determine predictor importance in logistic regression. *Journal of Educational and Behavioral Statistics*, 34(3), 319–347.

Baker, K.R. (1976). Workforce allocation in cyclical scheduling problems: A survey. *Operational Research Quarterly*, 27(1), 155–167.

Baker, K.R. (2015). *Optimization Modeling with Spreadsheets*, 3rd Edition. Wiley: Hoboken, New Jersey.

Balinski, M.L. (1965). Integer programming: methods, uses, computations. *Management Science*, 12(3), 253–313.

Bass, F.M. (1969). A new product growth for model consumer durables. *Management Science*, 15(5), 215–227.

Bejaei, M., Wiseman, K., & Cheng, K.M.T. (2015). Developing logistic regression models using purchase attributes and demographics to predict the probability of purchases of regular and specialty eggs. *British Poultry Science*, 56(4), 1–11.

Bertsimas, D. & King, A. (2016). OR Forum — An algorithmic approach to linear regression. *Operations Research*, 64(1), 2–16. https://doi.org/10.1287/opre.2015.1436.

Bertsimas, D. & King, A. (2017). Logistic regression: from art to science. *Statistical Science*, 32(3), 367–384. doi: 10.1214/16-STS602.

Bertsimas, D. & Van Parys, B. (2020). Sparse high-dimensional regression: Exact scalable algorithms and phase transitions. *The Annals of Statistics*, 4(1), 300–323. https://doi.org/10.1214/18-AOS1804.

Bertsimas, D., King, A., & Mazumder, R. (2016). Best subset selection via a modern optimization lens. *The Annals of Statistics*, 44(2), 813–852. doi: 10.1214/15-AOS1388.

Bertsimas, D., Pauphilet, J., & Van Parys, B. (2020a). Sparse regression: scalable algorithms and empirical performance. *Statistical Science*, 35(4), 555–578. https://doi.org/10.1214/19-STS701.

Bertsimas, D., Pauphilet, J., & Van Parys, B. (2020b). Rejoinder: Sparse regression: Scalable algorithms and empirical performance. *Statistical Science*, 35(4), 623–624. https://doi.org/10.1214/20-STS701REJ.

Breiman, L. (1995). Better subset regression using the nonnegative garrote. *Technometrics*, 37, 373–384.

Brimberg, J. (1995). The Fermat-Weber location problem revisited. *Mathematical Programming*, 71, 71–76.

Brimberg, J., Hansen, P., Mladenović, N., & Taillard, E.D. (2000). Improvements and comparison of heuristics for solving the uncapacitated multisource Weber problem. *Operations Research*, 48(3), 444–460.

Brusco, M. (2018). Demonstrating the mechanics of principal component analysis via spreadsheets. *Spreadsheets in Education*, 11(1). Retrieved from: https://sie.scholasticahq.com/article/6895-demonstrating-the-mechanics-of-principalcomponent-analysis-via-spreadsheets Licensed under CC-BY-NC-ND 4.0.

Brusco, M. (2019). An Excel spreadsheet and VBA macro for model selection and predictor importance using all-possible-subsets regression. *Spreadsheets in Education*, 12(1). Retrieved from: https://sie.scholasticahq.com/article/8064-an-excel-spreadsheet-andvba-macro-for-model-selection-and-predictor-importance-using-all-possible-subsetsregression Licensed under CC-BY-NC-ND 4.0.

Brusco, M.J. (2006). A repetitive branch-and-bound algorithm for minimum within-cluster sums of squares partitioning. *Psychometrika*, 71(2), 347–363.

Brusco, M.J. (2022a). Spreadsheet-based analysis of multisource continuous facility location problems. *Decision Sciences Journal of Innovative Education*, 20(2), 102–111.

Brusco, M.J. (2022b). Solving classic discrete facility location problems using Excel spreadsheets. *INFORMS Transactions on Education*, 22(3), 160–171. https://doi.org/10.1287/ited.2021.0245, used under a Creative Commons Attribution License: https://creativecommons.org/licenses/by/4.0/.

Brusco, M.J. (2022c). Logistic regression via Excel spreadsheets: Mechanics, model selection, and relative predictor importance. *INFORMS Transactions on Education*, 23(1), 1–11. https://doi.org/10.1287/ited.2021.0263, used under a Creative Commons Attribution License: https://creativecommons.org/licenses/by/4.0/.

Brusco, M.J. & Jacobs, L.W. (2000). Optimal models for meal-break and start-time flexibility in continuous tour scheduling. *Management Science*, 46(12), 1630–1641.

Brusco, M.J. & Showalter, M.J. (1993). Constrained nurse staffing analysis. *Omega, 21* (2), 175–186.

Brusco, M.J. & Stahl, S. (2005). *Branch-and-Bound Applications in Combinatorial Data Analysis*. New York: Springer-Verlag.

Brusco, M.J. & Steinley, D. (2007). A comparison of heuristic procedures for minimum within-cluster sums of squares partitioning. *Psychometrika*, 72(4), 583–600.

Brusco, M.J. & Steinley, D. (2011). Exact and approximate algorithms for variable selection in linear discriminant analysis. *Computational Statistics and Data Analysis*, 55, 123–131.

Brusco, M.J., Jacobs, L.W., Bongiorno, R. J., Lyons, D., & Tang, B. (1995). Improving personnel scheduling at airline stations. *Operations Research*, 43(5), 741–751.

Brusco, M.J., Steinley, D., & Cradit, J. D. (2009). An exact algorithm for finding hierarchically well-formulated subsets in second-order polynomial regression. *Technometrics*, 51(3), 306–315.

Budescu, D.V. (1993). Dominance analysis: A new approach to the problem of relative importance of predictors in multiple regression. *Psychological Bulletin*, 114, 542–551.

Camm, J.D., Chorman, T.E., Dill, F.A., Evans, J.R., Sweeney, D.J., & Wegryn, G.W. (1997). Blending OR/MS, judgment, and GIS: Restructuring P&G's supply chain. *Interfaces*, 27(1), 128–142.

Cánovas, L., Cañavate, R., & Marin, A. (2002). On the convergence of the Weiszfeld algorithm. *Mathematical Programming*, 93, 327–330.

Charnes, A., Cooper, W.W., & Rhodes, E. (1978). Measuring the efficiency of decision making units. *European Journal of Operational Research*, 2(6), 429–444.

Chickering, D.M. & Heckerman, D. (2003). Targeted advertising on the Web with inventory management. *Interfaces*, 33(5), 71–77.

Church, R.L. & ReVelle C. (1974). The maximal covering location problem. *Papers of the Regional Science Association*, 32(1), 101–118.

Cocchi, G., Galligari, A., Nicolino, F.P., Piccialli, V., Schoen, F., & Sciandrone, M. (2018). Scheduling the Italian national volleyball tournament. *INFORMS Journal on Applied Analytics*, 48(3), 271–284.

Cooper, L. (1964). Heuristic methods for location-allocation problems. *SIAM Review*, 6(1), 37–53.

Cooper, L. & Katz, I.N. (1981). The Weber problem revisited. *Computers and Mathematics with Applications*, 7(3), 225–234.

Dantzig, G.B., Orden, A., & Wolfe, P. (1954). *Notes on Linear Programming: Part I: The Generalized Simplex Method for Minimizing a Linear Form Under Linear Equality Constraints*. Santa Monica, CA: The Rand Corporation.

Daskin M.S. (2013). *Network and Discrete Location: Models, Algorithms and Applications*, 2nd Edition. Hoboken, NJ: John Wiley & Sons.

De, P., Dunne, E.J., Ghosh, J.B., Wells, C.E. (1995). The discrete time-cost tradeoff problem revisited. *European Journal of Operational Research*, 81(2), 225–238.

DeCani, J.S. (1969). Maximum likelihood paired comparison ranking by linear programming. *Biometrika*, 56, 537–545.

Dekle, J., Lavieri, M.S., Martin, E., Emir-Farinas, H., & Francis, R.L. (2005). A Florida county locates disaster recovery centers. *Interfaces*, 35(2), 133–139.

Dempster, A.P., Laird, N.M., & Rubin, D.B. (1977). Maximum likelihood from incomplete data via the EM algorithm (with discussion). *Journal of the Royal Statistical Society Series B. Methodological*, 39, 1–38.

Diehr, G. (1985). Evaluation of a branch and bound algorithm for clustering. *SIAM Journal for Scientific and Statistical Computing*, 6, 268–284.

Duarte Silva, A.P. (2001). Efficient variable screening for multivariate analysis. *Journal of Multivariate Analysis*, 76, 35–62.

du Merle, O., Hansen, P., Jaumard, B., & Mladenović, N. (2000). An interior point algorithm for minimum sum-of-squares clustering. *SIAM Journal on Scientific Computing*, 21, 1485–1505.

Eckart, C. & Young, G. (1936). The approximation of one matrix by another of lower rank. *Psychometrika*, 1, 211–218.

Efroymson, M.A. (1960). Multiple regression analysis. In A. Ralston & H.S. Wilf (Eds.), *Mathematical Methods for Digital Computers* (pp. 191–203). New York: Wiley.

Fisher, R.A. (1936). The use of multiple measurements in taxonomic problems. *Annals of Eugenics*, 7, 179–188.

Fisher, R.A. (1938). The statistical utilization of multiple measurements. *Annals of Eugenics*, 8, 376–386.

Forgy, E.W. (1965). Cluster analyses of multivariate data: Efficiency versus interpretability of classifications. *Biometrics*, 21, 768.

Fulkerson, D.R. (1961). A network flow computation for project cost curve. *Management Science*, 7(2), 167–178.

Furnival, G.M. & Wilson, R.W. (1974). Regression by leaps and bounds. *Technometrics*, 16, 499–512.

Garside, M.J. (1971). Some computational procedures for the best subset problem. *Applied Statistics*, 20, 8–15.

Goodnight, J.H. (1979). A tutorial on the SWEEP operator. *The American Statistician*, 33(3), 149–158.

Gomory, R.E. (1958). Outline of an algorithm for integer solutions to linear programs. *Bulletin of the American Mathematical Society*, 64, 275–278.

Gower, J.C. & Hand, D.J. (1996). *Biplots*. London: Chapman and Hall.

Greenacre, M.J. (1984). *Theory and Applications of Correspondence Analysis*. London: Academic Press.

Grömping, U. (2015). Variable importance in regression models. *WIREs Computational Statistics*, 7, 137–152.

Hakimi, S.L. (1964). Optimum locations of switching centers and the absolute centers and medians of a graph. *Operations Research*, 12(3), 450–459.

Hakimi, S.L. (1965). Optimum distribution of switching centers in a communication network and some graph theoretic problems. *Operations Research*, 13(3), 462–475.

Hand, D.J. (1981). *Discrimination and Classification.* New York: Wiley.

Hansen, P. & Mladenović, N. (2001). J-Means: A new local search heuristic for minimum sum of squares clustering. *Pattern Recognition*, 34, 405–413.

Heizer, J., Render, B., & Munson, C. (2017). *Operations Management: Sustainability and Supply Chain Management*, 12th Edition. Boston: Pearson.

Hitchcock, F.L. (1941). The distribution of a product from several sources to numerous localities. *Journal of Mathematics and Physics*, 20(1–4), 224–230.

Hoerl, A.E. & Kennard, R.W. (1970). Ridge regression: Biased estimation for nonorthogonal problems. *Technometrics*, 12(1), 55–67.

Hoffman, P.J. (1960). The paramorphic representation of clinical judgment. *Psychological Bulletin*, 57, 116–131.

Holland, J.H. (1975). *Adaptation in Natural and Artificial Systems: An Introductory Analysis with Applications to Biology, Control, and Artificial Intelligence.* Ann Arbor, MI: University of Michigan Press.

Hosmer, D.W., Lemeshow, S., & Sturdivant, R.X. (2013). *Applied Logistic Regression*, 3rd Edition. New York: Wiley.

Huggins, E., Bailey, M., & Guardiola, I. (2020). Converting point spreads into probabilities: A case study for teaching business analytics. *INFORMS Transactions on Education*, 21(1), 61–63.

Huse, C. & Brusco, M.J. (2021). A tale of two linear programming formulations for crashing project networks. *INFORMS Transactions on Education*, 21(2), 82–95. https://doi.org/10.1287/ited.2019.0236, used under a Creative Commons Attribution License: https://creativecommons.org/licenses/by/4.0/.

Jacobs, L.W. & Brusco, M.J. (1996). Overlapping start-time bands in implicit tour scheduling. *Management Science*, 42 (9), 1247–1259.

Jancey, R.C. (1966). Multidimensional group analysis. *Australian Journal of Botany*, 14, 127–130.

Johnson, R.A. & Wichern, D.W. (2007). *Applied Multivariate Statistical Analysis*, 6th Edition. Upper Saddle River, NJ: Pearson Prentice Hall.

Jolliffe, I.T. (2002). *Principal Component Analysis*, 2nd Edition. New York: Springer.

Kaiser, H.F. (1958). The varimax criterion for analytic rotation in factor analysis. *Psychometrika*, 23, 187–200.

Kao, E.P.C. & Queyranne, M. (1985). Budgeting costs of nursing in a hospital. *Management Science*, 31(5), 608–621.

Kelley, J.E., Jr. (1961). Critical-path planning and scheduling: Mathematical basis. *Operations Research*, 9(3), 296–320.

Klein, R.W. & Dubes, R.C. (1989). Experiments in projection and clustering by simulated annealing. *Pattern Recognition*, 22, 213–220.

Kuhn, H.W. & Kuenne, R.E. (1962). An efficient algorithm for the numerical solution of the generalized weber problem in spatial economics. *Journal of Regional Science*, 4(2), 21–34.

Koontz, W.L.G., Narendra, P.M., & Fukunaga, K. (1975). A branch and bound clustering algorithm. *IEEE Transactions on Computers*, C-24, 908–915.

Korte, B. & Oberhofer, W. (1971). Triangularizing input-output matrices and the structure of production. *European Economic Review*, 2, 493–522.

Krajewski, L.J., Malhotra, M.K., & Ritzman, L.P. (2016). *Operations Management: Processes and Supply Chains*, 11th Edition. Boston: Pearson.

Kuehn, A.A. & Hamburger, M.J. (1963). A heuristic program for locating warehouses. *Management Science*, 9, 643–666.

Kuo, C.-C. & White, R.E. (2004). A note on the treatment of the center-of-gravity method in operations management textbooks. *Decision Sciences Journal of Innovative Education*, 2(2), 219–227.

Kvam, P.H., Sokol, J. (2004). Teaching statistics with sports examples. *INFORMS Transactions on Education*, 5(1), 75–87.

LaMotte, L.R. & Hocking, R.R. (1970). Computational efficiency in the selection of regression variables. *Technometrics*, 12, 83–93.

Land, A.H. & Doig, A.G. (1960). An automatic method of solving discrete programming problems. *Econometrica*, 28, 497–520.

Lawler, E.L. (1964). A comment on minimum feedback arc sets. *IEEE Transactions on Circuit Theory*, 11, 296–297.

Lattin, J., Carroll, J.D., & Green, P.E. (2003). *Analyzing multivariate data*. Pacific Grove, CA: Brooks/Cole – Thomson Learning.

Lay, D.C., Lay, S.R., & McDonald, J.J. (2015). *Linear Algebra and Its Applications*, 5th Edition. New York: Pearson.

Lazarsfeld, P.F. (1950). The logical and mathematical foundations of latent structure analysis. In S. A. Stouffer (Ed.), *Measurement and Prediction* (pp. 362–412). Princeton, NJ: Princeton University Press.

Lindeman, R.H., Mirenda, P.F., & Gold, R.Z. (1980). *Introduction to Bivariate and Multivariate Analysis*. Glenview, IL: Scott, Foresman.

MacQueen, J.B. (1967). Some methods for classification and analysis of multivariate observations. In L.M. Le Cam & J. Neyman (Eds.), *Proceedings of the Fifth Berkeley Symposium on Mathematical Statistics and Probability* (Vol. 1, pp. 281–297), Berkeley, CA: University of California Press.

Macready, G.B. & Dayton, C.M. (1977). The use of probabilistic models in the assessment of mastery. *Journal of Educational Statistics*, 33, 379–416.

Mallows, C.L. (1973). Some comments on *Cp*. *Technometrics*, 15, 661–675.

Markowitz, H. (1952). Portfolio selection. *The Journal of Finance*, 7, 77–91.

Markowitz, H. (1956). The optimization of a quadratic function subject to linear constraints, *Naval Research Logistics Quarterly*, 3, 111–133.

Mason A.J. (2012) OpenSolver — An open source add-in to solve linear and integer programmes in Excel. D. Klatte, H.-J. Lüthi, & K. Schmedders (Eds.), *Operations Research Proceedings 2011*, Berlin: Springer, 401–406.

Maulik, U. & Bandyopadhyay, S. (2000). Genetic algorithm-based clustering technique. *Pattern Recognition*, 33, 1455–1465.

Menard, S. (2010). *Logistic Regression: From Introductory to Advanced Concepts and Applications*, Thousand Oaks, CA: Sage.

Metters, R., Queenan, C., Ferguson, M., Harrison, L., Higbie, J., Ward, S., Barfield, B., Farley, T., Kuyumcu, H.A., & Duggasani, A. (2008). The "Killer Application" of revenue management: Harrah's Cherokee Casino & Hotel. *Interfaces*, 38(3), 161–175.

Microsoft Corporation. (2018). *Microsoft Excel*. Retrieved from https://office.microsoft.com/excel.

Miles, J.N.V. (2005). Confirmatory factor analysis using Microsoft Excel. *Behavior Research Methods*, 37(4), 672–676.

Miller, A.J. (2002). *Subset Selection in Regression*, 2nd Edition. London: Chapman and Hall.

Morrison, K.E. (2010). The FedEx problem. *The College Mathematics Journal*, 41 (3), 222–232.

Nimon, K.F. & Oswald, F.L. (2013). Understanding the results of multiple linear regression: Beyond standardized regression coefficients. *Organizational Research Methods*, 16, 650–674.

Olejnik, S., Mills, J., & Keselman, H. (2000). Using Wherry's adjusted R2 and Mallows' Cp for model selection from all possible regressions. *Journal of Experimental Education*, 68, 365–380.

Pacheco, J. & Valencia, O. (2003). Design of hybrids for the minimum sum-of-squares clustering problem. *Computational Statistics and Data Analysis*, 43, 235–248.

Pratt, J.W. (1987). Dividing the indivisible: Using simple symmetry to partition variance explained. In T. Pukkila & S. Putanen (Eds.), *Proceedings of the Second International Conference in Statistics* (pp. 245–260). Tampere, Finland: University of Tampere.

Ragsdale, C.T. (2003). A new approach to implementing project networks in spreadsheets. *INFORMS Transactions on Education*, 3(3), 76–85.

Ragsdale, C.T. (2018). *Spreadsheet Modeling and Decision Analysis: A Practical Introduction to Business Analytics*, 8th Edition. Boston: Cengage.

Ragsdale C.T. & Stam, A. (1992). Introducing discriminant analysis to the business statistics curriculum. *Decision Sciences*, 23(3), 724–745.

Reinig, B.A. & Horowitz. I. (2018). Using mathematical programming to select and seed teams for the NCAA tournament. *INFORMS Journal on Applied Analytics*, 48(3), 181–188.

Robinson, E.P., Gao, L.-L., & Muggenborg, S.D. (1993). Designing an integrated distribution system at DowBrands, Inc.. *Interfaces* 23(3), 107–117.

Rosseel, Y., Jorgensen, T.D., & De Wilde, L. (2024). Package lavaan: Latent variable analysis. Retrieved from: https://cran.r-project.org/web/packages/profileR/index.html.

Salamah, U. & Ramayanti, D. (2018). Implementation of logistic regression algorithm for complaint text classification in Indonesian ministry of marine and fisheries. *International Journal of Computing Techniques*, 5(5), 74–78.

Schärlig, A. (1973). About the confusion between the center of gravity and Weber's optimum. *Regional and Urban Economics*, 3(4), 371–382.

Schield, M. (2017). Teaching logistic regression using ordinary least squares in Excel. 2017 *JSM Proc. Papers Presented Joint Statistical Meetings* (American Statistical Association, Baltimore), 2963–2987. Accessed December 29, 2020, http://www.statlit.org/pdf/2017-Schield-ASA.pdf.

Schwartz, G. (1978). Estimating the dimension of a model. *Annals of Statistics*, 6, 461–464.

Smith, D.M. (1991). Algorithm AS 268, all possible subset regressions using the QR decomposition. *Applied Statistics*, 40, 502–513.

Späth, H. (1980). *Cluster Analysis Algorithms for Data Reduction and Classification of Objects*. New York: Wiley.

Spearman, C. (1904). General intelligence, objectively determined and measured. *American Journal of Psychology*, 15, 201–292.

Steinhaus, H. (1956). Sur la division des corps matériels en parties (On the separation of objects into groups). *Bulletin de l'Académie Polonaise des Sciences*, Classe III, IV(12), 801–804.

Steinley, D. (2003). Local optima in K-means clustering: What you don't know may hurt you. *Psychological Methods*, 8, 294–304.

Steinley, D. & Brusco, M. J. (2008). A new variable weighting and selection procedure for K-means cluster analysis. *Multivariate Behavioral Research*, 43(1), 77–108.

Stevenson, W.J. (2018). *Operations Management*, 13th Edition. New York: McGraw-Hill.

Templin, J. (2015). Latent class analysis. Retrieved from: https://www. yumpu.com/en/document/view/37463059/latent-class-analysis-jonat han-templins-website-university-of-.

Thomas D.R., Hughes, E., & Zumbo, B.D. (1998). On variable importance in linear regression. *Social Indicators Research*, 45, 253–275.

Thomas D.R., Kwan, E., & Zumbo, B.D. (2018). In defense of Pratt's variable importance axioms: A response to Grömping. *WIREs Computational Statistics* e1433. Available at https://doi.org/10.1002 /wics.1433 Accessed June 5, 2018.

Tibshirani, R. (1996). Regression shrinkage and selection via the lasso. *Journal of the Royal Statistical Society B*, 58, 268–288.

Toregas, C. & ReVelle, C.S. (1972). Optimal location under time or distance constraints. *Papers of the Regional Science Association*, 28(1), 133–144.

Toregas, C., Swain R., ReVelle C. S., & Bergman, L. (1971). The location of emergency service facilities. *Operations Research*, 19(6), 1363–1373.

Vardi, Y. & Zhang, C.-H. (2000). The multivariate L1-median and associated data depth. *Proceedings of the National Academy of Sciences*, 97(4), 1423–1426.

Weber, A. (1909). *Über den Standort der Industrien*, Tübingen, J.C.B. Mohr — English translation: *The Theory of the Location of Industries*, Chicago: Chicago University Press, 1929.

Weiszfeld, E. (1937). Sur le point pour lequel la somme des distances de n points donnes est minimum. *Tohoku Mathematical Journal*, 43, 355–386.

Welling, M. (2006). Fisher linear discriminant analysis. Retrieved 12/25/23 from: https://pdfcoffee.com/fisher-linear-discriminant-analysis-pdf-free.html.

Westphal, S. (2014). Scheduling the German basketball league. *INFORMS Journal on Applied Analytics*, 44(5), 498–508.

Younger, D.H. (1963). Minimum feedback arc sets for a directed graph. *IEEE Transactions on Circuit Theory*, 10, 238–245.

Zaghdoudi, T. (2013). Bank failure prediction with logistic regression. *International Journal of Economic and Financial Issues*, 3(2), 537–543.

Zou, H. (2006). The adaptive lasso and its oracle properties. *Journal of the American Statistical Association*, 101(476), 1418–1429.

Index

Printed in the USA
CPSIA information can be obtained
at www.ICGtesting.com
LVHW022348191024
794127LV00002B/122

9 789811 294044